COMPUTATIONAL MATHEMATICS AND ANALYSIS

AN ESSENTIAL GUIDE TO FUZZY SYSTEMS

COMPUTATIONAL MATHEMATICS AND ANALYSIS

Additional books and e-books in this series can be found on Nova's website under the Series tab.

COMPUTATIONAL MATHEMATICS AND ANALYSIS

AN ESSENTIAL GUIDE TO FUZZY SYSTEMS

MICHAEL GR. VOSKOGLOU
EDITOR

nova
science publishers
New York

NOTICE TO THE READER

Library of Congress Cataloging-in-Publication Data

ISBN: 978-1-53616-128-1

Published by Nova Science Publishers, Inc. † New York

CONTENTS

PREFACE

Situations appear often in the everyday life where certain definitions have no clear boundaries, like “the young people of a city”, “the good players of a team”, “the excellent students of a class”, etc. The need to model mathematically such kind of situations was one of the main reasons that laid to the development of the notion of fuzzy set. Since Zadeh introduced the fuzzy set theory in 1965, many efforts have been made by the specialists for improving its effectiveness to deal with uncertain, ambiguous and vague situations. As a result a series of extensions and generalizations of the ordinary fuzzy set followed and several theories have been proposed as alternatives to the fuzzy set theory.

Those new mathematical tools gave to the scientists the opportunity to model under conditions which are vague or not precisely defined, thus succeeding to solve mathematically problems whose statements are expressed in our natural language. As a result the spectre of their applications has been rapidly extended covering the Physical Sciences, Economics and Management, Expert Systems like financial planners, diagnostic, meteorological, information-retrieval, control systems, etc., Industry, Robotics, Decision Making, Programming, Medicine, Biology, Humanities, Education and almost all the other sectors of the human activity, including human reasoning itself.

The target of the present book is to serve as an essential guide to fuzzy sets and systems. Chapter One, after an introduction to fuzzy sets and the generated by them infinite-valued fuzzy logic, discusses the headlines of the grey system theory and presents assessment methods of computational thinking skills based on the use of triangular fuzzy numbers and grey numbers respectively. Chapter Two presents a general method for evaluating fuzzy data that frequently appear nowadays in science and technology and uses fuzzy relation equations as a tool for assessing student mathematical model building abilities. Chapter Three is devoted to an introduction to fuzzy graph theory illustrated by a series of interesting examples. Chapter Four studies the controllability of nonlinear fuzzy fractional differential systems by using the Banach's fixed point theorem. Chapter Five reviews the hybrids of neural networks and fuzzy logic and discusses how their combination can improve the quality of learning, decision making, inference, and computation. In Chapter Six the fuzzy probability and fuzzy possibility theory is used for an integration of voltage sag type detection and its impact is examined on equipment behavior. Chapter Seven presents an algorithm based on the Bhattacharyya distance method to highlight the importance of using the statistical methods in the field of the pattern recognition. Chapter Eight studies the Goursat problem for a fuzzy hyperbolic equation under the fractional Caputo g-derivative for fuzzy-valued multivariable functions. Chapter Nine introduces the space of single valued neutrosophic numbers and the granular calculus related to single valued neutrosophic functions. In Chapter Ten a hybrid fuzzy potential field method is applied for the navigation of Sumo robots. The following Commentary reviews the main extensions and generalizations of fuzzy sets and the related to them theories.

Chapter 1 - Computational thinking is a hybrid of several modes of thinking (abstract, logical, constructive, modeling thinking, etc.) that synthesizes critical thinking and existing knowledge and applies them for solving complex real life and technological problems, designing systems and understanding human behavior, by drawing on fundamental principles of computer science. In this Chapter grey numbers are used for evaluating computational thinking skills under fuzzy conditions and applications are presented to mathematical modeling and to case-based reasoning systems. The outcomes of this new assessment method are compared to the corresponding outcomes of the classical method of calculating the GPA index and of a similar method developed earlier by the author that uses as tools triangular fuzzy numbers instead of grey numbers. The required background from grey system theory and fuzzy logic and the management of the uncertainty in a fuzzy environment are also discussed.

Chapter 2 - In large and complex systems that frequently appear nowadays in science, technology and real life situations many different and constantly changing factors are usually involved, the relationships among which are indeterminate. As a result the data obtained from their operation mechanisms cannot be easily determined precisely and in practice estimates of them are used. A model is proposed in this Chapter for evaluating such kind of data with the help of the corresponding membership degrees and possibilities. The general character of that model enables its application to a variety of human and machine activities. The examples presented here concern the process of learning a subject-matter in the classroom and a market's research. On the other hand Fuzzy Relation Equations, which are associated with the composition of Fuzzy Binary Relations, have currently become a valuable tool for applications of fuzzy mathematics to many topics. After presenting the headlines of the corresponding theory, an application is developed here on the assessment of student mathematical modelling skills.

Chapter 3 - The graphs are very useful tools for modelling many problems of practical importance. Applications of graphs concern the database management, ecosystems, organic molecules, physical networks, signal-flow diagrams, map colorings, tracing mazes, flow of a computer programs and several other topics. However, our nowadays world is full of uncertainly. Consequently, the classical graphs, otherwise called crisp graphs, cannot model all real problems completely and properly. For tackling such kind of problems membership values can be assigned to the vertices and/or to the edges of the corresponding graphs. Such kind of graphs are known as fuzzy graphs. This Chapter is devoted to an introduction to the fuzzy graph theory. Several widely used types of fuzzy graphs are defined like the fuzzy planar graphs, the fuzzy tolerance graphs, the fuzzy competition graphs, etc. Many interesting properties of those types of fuzzy graphs are presented accompanied by suitable examples and some other related topics are highlighted.

Chapter 4 - Controllability plays a major role in the development of modern mathematical control theory; it enables many applications in modern engineering. The purpose of this Chapter is to prove the controllability of fuzzy fractional order stochastic systems. The existence and uniqueness are acquired via Banach's fixed point theorem, where the controllability of fuzzymapping is defined by using the Mittag-Leffler matrix function. In particular, the α-level controllability results are investigated for the nonlinear fuzzy fractional stochastic differential systems.

Chapter 5 - Nowadays, we are exposed to a vast stream of information that can be used in many areas, including image/speech processing, system identification, estimation, and control. Fuzzy Logic and Artificial Neural Networks are two well-known intelligent methodologies which have been employed to fulfill these tasks with various architectures and hybridization configurations. This Chapter reviews the hybrids of neural networks and fuzzy logic and discusses how their combination can improve the quality of learning, decision making, inference, and computation. In particular the deep neural network is studied having more than three layers (including input and output layers) of neurons that can simulate and model a system with certain level of complexity of dynamics.

Chapter 6 - Maintenance of power quality is important for the smooth operation of equipment connected to an electrical system. When the system voltage drops by 10-90% of its nominal value for a few cycles to one minute, the situation is known as voltage sag. In this Chapter the overall impact of voltage sag on equipment as well as on industrial production processes is determined. Four basic types of voltage sag occurrence due to balanced and unbalanced fault types are considered to observe the characteristics of several equipment in the system. This categorization is done by using fuzzy probability theory. For individual sag type and for different operating criterion in voltage sag condition the equipments' behavior is determined by fuzzy possibility theory. Voltage tolerance limits and voltage sag type are the inputs of fuzzy possibility theory. The determination of the equipment trip due to different types of voltage sag is the goal of this approach.

Chapter 7 - Pattern recognition refers to the classification or description of objects or patterns. Several mathematical methods of statistical decision theory can be used for this purpose and this approach is known as statistical pattern recognition. The target of this Chapter is to highlight the importance of using statistical methods in pattern recognition underpinning the development of corresponding neural methods. It is well-known that some of the classification algorithms are more efficient in spaces with low dimension and fail to be practical when they are applied in spaces of higher dimension. Therefore, the transformation of the original n-dimensional feature space to a space with lower dimension is strongly recommended. To this aim a classification algorithm is presented here based on the Bhattacharyya distance method, where the criterion of the feature selection is directly related to the probability of the Bayes error.

Chapter 8 - Fractional calculus has recently appeared as an important tool for the study of dynamic systems where classical methods reveal strong limitations and in the modeling of phenomena in physics, mechanics, elasticity, dynamics, economics, biology, chemistry, etc. In this Chapter the well-known from Physics Goursat problem for a fuzzy hyperbolic equation under the fractional Caputo g-derivative is studied. The well posed of this problem is proven and some applications to real world situations are investigated. The fractional integral and the fractional Caputo g-derivatives are also defined for fuzzy-valued multivariable functions and some properties between fractional integral and Caputo g-derivatives are established illustrated by suitable examples.

Chapter 9 - In this Chapter the single valued triangular neutrosophic numbers are introduced and a foundation of neutrosophic calculus is constructed. Basic arithmetic operations on the space of single valued neutrosophic numbers are established allowing the direct introduction of neutrosophic functions in usual mathematical formulas. In particular, in terms of a metric defined on the space of single valued neutrosophic numbers, which is induced from the Hamming distance, a new type of derivative of neutrosophic functions is introduced, namely the granular derivative, and the corresponding granular integral is studied. Several numerical examples are presented to illustrate the obtained results.

Chapter 10 - The potential field method has been widely used in robotics for generating paths for mobile robots navigation. This method creates a field that directs the robot to the goal position. When the robots have to deal with unknown environments and imprecise information, the navigational problem becomes extremely complex. In this Chapter a hybrid fuzzy potential field method is applied for the navigation of Sumo robots. In this hybrid approach the Mamdani model is used to obtain the necessary variables to determine the velocity and direction to be followed by the robot. Two fuzzy systems are used, both with three inputs and one output. The difference between the two fuzzy systems is in the number of the linguistics terms. Simulations have been performed by the MATLAB software to verify the enhancement of the hybrid method with respect to the conventional potential field method. The simulations represent the arena used for the Sumo robot competitions and it is possible to determine the trajectory of the robot until finding the target. Accordingly to the simulations performed, the hybrid method using potential field and fuzzy systems was 14% more effective than the simulation using just the conventional potential field method.

Commentary - A review of the main extensions and generalizations of fuzzy sets and the related to them theories is presented here from the time that Zadeh introduced the concept of fuzzy set in 1965 until nowadays. This material could be used as an initial framework for those wanting to study deeper those topics.

From this position I feel the need to thank once more all the authors of the above Chapters for their excellent collaboration during the preparation of the book. Thanks are also due to the staff of the Nova Publishers for their help and support on the technical matters of this Edition. I close by expressing the hope that the present book will become useful to the readers as an essential guide to fuzzy systems and to the related to them theories and useful to them willing to study deeper those topics.

In: An Essential Guide to Fuzzy Systems ISBN: 978-1-53616-128-1
Editor: Michael Gr. Voskoglou

Chapter 1

FUZZY SETS, GREY SYSTEM THEORY AND COMPUTATIONAL THINKING

Michael Gr. Voskoglou*
Graduate Technological Educational Institute of Western Greece, Koukouli, Patras, Greece

ABSTRACT

Computational thinking is a hybrid of several modes of thinking (abstract, logical, constructive, modeling, etc.) that synthesizes critical thinking and existing knowledge and applies them for solving complex real life and technological problems, designing systems, and understanding human behavior, by drawing on fundamental principles of computer science. Modelling thinking could be considered as the fundamental component of computational thinking, because it utilizes all the other modes of thinking involved in it and applies them for constructing the model of the real system that leads to the solution of the corresponding problem. The case-based reasoning systems are characteristic and very useful examples of systems in which computational thinking plays an important role On the other hand, a grey system is characterized by lack of adequate information about its

* Corresponding Author's Email: mvoskoglou@gmail.com

components and (or) its function and the corresponding theory has found nowadays many applications to real life, science and engineering. In grey system theory the main tool for handling approximate data is the use of the grey numbers, which are indeterminate numbers defined with the help of the closed real intervals. In the present work grey numbers are used for evaluating computational thinking skills under fuzzy conditions and applications are presented on mathematical modeling and on case-based reasoning systems illustrating our results. The outcomes of this new assessment method are compared to the corresponding outcomes of the classical method of calculating the GPA index and of a similar method developed earlier by the author that uses as tools triangular fuzzy numbers instead of grey numbers. The background about grey system theory and about fuzzy logic, required for the better understanding of this work, is also presented. In addition, the calculation of a fuzzy system's uncertainty is also considered as an assessment method and its disadvantages are described with respect to the other methods.

Keywords: Problem Solving (PS), Computational Thinking (CT), Case-Based Reasoning (CBR), Mathematical Modelling (MM), Fuzzy Sets (FS), Uncertainty of Fuzzy Systems, Centre of Gravity (COG) Defuzzification Technique, Triangular Fuzzy Numbers (TFNs), Grey System (GS), Grey Numbers (GNs)

INTRODUCTION

An organized *system* is defined to be a set of interacting components forming an integrated whole and working together for achieving a common target. A factory, a hospital, a bank, etc. are common examples of systems, whereas in general one can distinguish between physical, biological, social, economic, engineering, abstract knowledge systems, etc. As a multi-perspective domain systems' theory serves as a bridge for an interdisciplinary dialogue between autonomous areas of study [1]. In the present work we deal with *Computational Thinking (CT)* systems.

The assessment of a system's performance is a very important part of the systems' theory, because it enables the system's designer to correct its weaknesses and therefore to increase its effectiveness. When the performance of a system's components is evaluated with numerical scores, then the traditional way for assessing the system's *mean performance* is the calculation of the mean value of those scores. However, in order to comfort the user's existing uncertainty about the exact value of the numerical scores corresponding to each of the system's components, frequently in practice the assessment is made not by numerical scores but by qualitative linguistic expressions, like excellent, very good, good, etc, which makes the calculation of their mean value impossible.

A popular in such cases method for evaluating the overall system's performance is the calculation of the *Grade Point Average (GPA) index* ([47], Chapter 6, p.125). However, GPA is a weighted average in which greater coefficients (weights) are assigned to the higher grades, which means that it reflects not the mean, as we wish, but the *quality* system's *performance*.

In order to overcome this difficulty we have utilized in earlier works the system's *total uncertainty* under fuzzy conditions (created by the qualitative assessment of the system's components) as a measure of its effectiveness ([47], Chapter 5). This manipulation is based on a fundamental principle of the Information Theory according to which the reduction of a system's uncertainty is connected to the increase of information obtained by a system's activity. In other words, lower uncertainty indicates a greater amount of information and therefore a better system's performance with respect to the corresponding activity. However, this method needs laborious calculations, cannot give a precise qualitative characterization of a system's performance and, most importantly, it is applicable for comparing the performance of two different systems with respect to a common activity only under the assumption that the existing uncertainty is the same in the two systems before the activity. This was the reason for turning our attention in later works to the use of *Fuzzy Numbers (FNs)* for the assessment of a system's mean performance under fuzzy conditions ([47], Chapter 7).

In the present work *Grey Numbers (GNs)* are also used for assessing the performance of *Computational Thinking (CT)* systems, in which vagueness and uncertainty are frequently appearing. This method is proved to be equivalent to the use of a special form of FNs, the *Triangular FNs*

(TFNs), but it has the advantage of reducing significantly the required computational burden. Moreover, the GNs are defined easily with the help of the closed intervals of real numbers, in contrast to the TFNs that need the knowledge of basic elements of the theory of *Fuzzy Sets (FS).*

The rest of the Chapter is formulated as follows: The next Section describes the CT process, the process of *modelling* which is a fundamental component of CT and the *Case-Based Reasoning (CBR)* systems, probably the most important type of CT systems. The third Section is devoted to Fuzzy Logic and the assessment methods of a system's performance using the system's uncertainty and the TFNs as their main tools respectively. The fourth Section deals with the basic principles of the *Grey System (GS)* theory and the assessment method using the GNs, whereas in the fifth Section examples are provided on the assessment of CT systems. The Chapter closes with our general conclusions and a short discussion on the perspectives of future research on the subject.

COMPUTATIONAL THINKING

Computational Thinking in Problem Solving

The importance of *Problem Solving (PS)* has been realized for such a long time that in a direct or indirect way affects our daily lives for ages [39, 46]. However, even graduates have nowadays difficulty in solving real life problems. Somehow, they cannot apply theory into practice, or theorize/reflect on practice [6]. In fact, it is the human mind at the end that has to be applied in a problematic situation and its capacity to solve the problem is directly related to the *knowledge* stored in the mind. But knowledge is the product of thinking that varies from a very simple and mundane thought to a very sophisticated and complex one. The nature of a problem dictates the level of thinking required to solve it.

A simple problem can be solved by applying a lower-order thinking that requires minimum cognitive effort. On the contrary, the solution of composite, non - routine problems require a higher - order thinking that is referred as *Critical Thinking (CrT)*; e.g., see [21]. CrT has its roots in the ancient Greek philosopher Socrates, who perfected the art of questioning, whereby by asking pertinent questions he would show that people could not rationally justify their confident claims of knowledge. CrT can be conceptualized as a complex mode of thinking involving abstraction, uncertainty, application of multiple criteria, reflection, self-regulation, making judgments, analysis and synthesis and so on. Such kind of thinking, which often generates multiple solutions, facilitates the *transfer of knowledge*, i.e., the use and transformation of already existing knowledge for creating new knowledge. In concluding, creativity and innovation driven by tacit knowledge and CrT driven by logic become the necessary tools for solving non-routine problems.

However, living in a knowledge era and an ever increasing progress in technology, combining knowledge and technology to solve problems becomes the mode rather than the exception. In an attempt by humans to increase the power of the mind (Artificial Intelligence), the holder of a Nobel prize Herb Simon saw thinking as information-processing and computers started taking over as a kind of "thinking machines" [20]. If technology is added as a tool, then *Computational Thinking (CT)* is another essential component of thinking required for solving complex real life or technological problems. The term CT was first introduced by S. Papert [23], who is widely known as the "father" of the Logo software. However, it was brought to the forefront of the computer society by Wing [54] in 2006 to describe how to think like a computer scientist. Wing has described CT as "solving problems, designing systems and understanding human behavior by drawing on concepts fundamental to computer science."

The main characteristics of CT include [53]:

1. Analyzing and logically organizing data
2. Data modelling, data abstractions, and simulations
3. Formulating problems in a way such that computers may assist
4. Identifying, testing, and implementing possible solutions
5. Automating solutions via algorithmic thinking

6. Generalizing and applying this process to other problems

It can be concluded that CrT is a prerequisite to knowledge acquisition and application to solve problems, but not a sufficient condition when we are faced with complex real life or technological problems, the solution of which requires also a pragmatic way of thinking such as CT. The way in which CrT and CT are simultaneously acting for the solution of such complex problems has not been clearly established yet. In an earlier work [41] we have attempted at shedding some light into this action. According to our approach, which is graphically represented in Figure 1, the given problem is considered as an obstacle to be passed through.

The three cognitive components acting during the PS process are the *already existing knowledge, CrT and CT.* Our hypothesis is that, once awareness of the problem is made, the existing information is extracted and critically analyzed. Then, if there exists sufficient background knowledge, the necessary new knowledge is retrieved through CrT and the solver, thinking in a computer like scientific way (CT), applies that knowledge to solve the problem. In other words, CT synthesizes CrT and existing knowledge and applies them for the solution of the problem. In concluding, CT does not propose that problems need to be solved in the same way a computer tackles them, but rather it encourages the critical thinking using computer science concepts and techniques. This approach could be used as a model for the process of solving complex real life or technological problems.

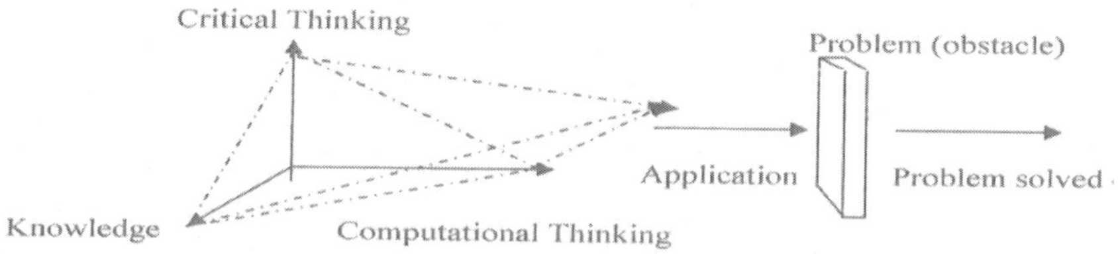

Figure 1. Voskoglou's and Buckley's problem solving model.

Computer Science is not just about programming, it is about an entire way of thinking, which has become nowadays an intrinsic part of our lives.

Computation is an increasingly essential tool for doing scientific research. It is expected that future generations of scientists and engineers will need to engage and understand computing in order to work effectively with scientific objects, technologies and methodologies. CT builds on the power and limits of computing processes whether they are executed by a human or by a machine. Therefore, it is a fundamental skill to everyone, not just for scientists and engineers. To reading, writing and arithmetic, we should add CT to every child's analytical ability [54].

It is widely accepted that motivation and involvement are imperative in retaining students in computer science and in order to do so we need to suitably engage students in the process of learning programming by building more effective mechanisms and tools for the development of programming skills. However, this is not an easy task and one of the core aims of learning programming should be to constantly highlight that programming is not only coding, but also thinking computationally and acquiring skills to develop solid solutions through understanding of concrete problems. Recent studies in this field address the necessity to become trained in thinking computationally *before* learning programming, and conclude that the education of programming along with the theory of computing needs to be represented in a way that would make sense to students within the computer science discipline [10].

People today using the convenient small calculators can make quickly and accurately all kinds of numerical operations. Also, the existence of a variety of mathematical packages gives the possibility to solve automatically with the help of computers all kinds of equations, to make any kind of algebraic operations, to calculate limits, derivatives, integrals, etc, and even more to obtain automatically the proofs of all the known mathematical theorems. As a matter of fact a number of experts in Computer Science have already concluded that teachers will not be needed in future for the development of students' knowledge base and learning skills, since everything could be done by the computers (possibly at home). "The use of horses is not necessary, from the time that cars were invented," they argue parallelizing those two things.

Nevertheless, this is actually an illusion! In fact, although the acquisition of information is valuable for the learner, the most important thing is to learn how to think rationally and creatively. The latter is impossible to be succeeded with the help of computers only. In fact, computers (the hardware) have been created by humans, and come into 'life' through programming, which is also performed by humans. And through programming it is possible to input information and get an output almost at the speed of light. However, the old credo "garbage in, garbage out" is still valid!

Concerning mathematics, the student practice with numerical, algebraic and analytic calculations, with the solution of problems and the rediscovery of proofs of the mathematical theorems must be continued for ever. Otherwise people will gradually loose the sense of numbers and symbols, the sense of space and time, and they will become unable to create new knowledge and technology!

Computational Thinking and Modelling

According to Liu & Wang [19], CT is a hybrid of abstract, algorithmic, logical, constructive and modelling thinking.

Abstract thinking is essential in order to understand the main body of a complex problem. In solving such a problem abstraction of thinking is one very general purpose heuristic that can help to attack the problem. Informally, abstract thinking can be thought as the mapping from a ground representation to a new but simpler representation.

Logical thinking is the process in which one uses reasoning consistency to come to a conclusion. The problems or situations involving logical thinking always call for mathematics structure, for relationships between some hypotheses and given statements, and for a sequence of reasoning that makes the conclusion more reasonable.

The *algorithmic thinking* is necessary for executing the problem's solution, whereas *constructive thinking* is any well-defined procedure that takes some value, or set of values as input and produces some value, or set of values as output.

One could claim that *modelling thinking* constitutes the essence of CT, since it synthesizes all the other modes of thinking for constructing a *model* for the solution of the corresponding problem. The study of a system's behaviour through the model saves time and reduces the relevant cost.

It is recalled that the model is a simplified representation of the real system including only the system's entities and features related to the corresponding problem. The construction of it usually involves a preliminary deep abstracting process on identifying the system's dominant variables and the relationships governing them. The resulting structure of this activity is known as the *assumed real system*. The model, being a further abstraction of the assumed real system, identifies and simplifies the relationships among these variables in a form amenable to analysis. The modelling process is graphically illustrated in Figure 2.

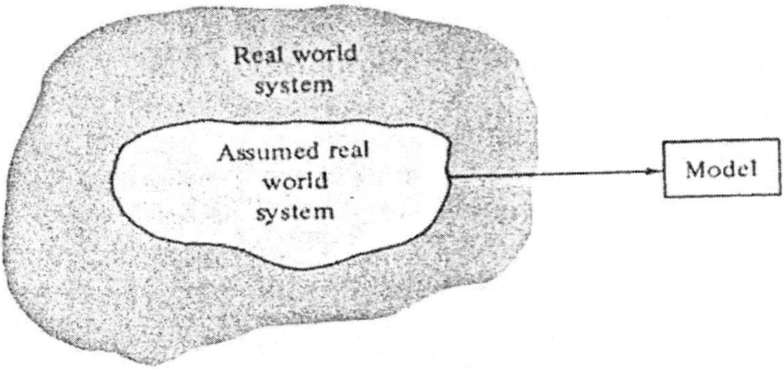

Figure 2. A graphical representation of the modelling process.

There are several types of models in use according to the form of the corresponding problem ([33], Section 1.3.1). The representation of a system's operation through the use of a *mathematical model* is achieved by a set of mathematical expressions (equalities, inequalities, etc.) and functions properly related to each other. The solutions provided by a mathematical model are more general and accurate than those provided by the other types of models. However, in cases where a system's operation is too complicated to be described in mathematical terms (e.g., biological systems, socio-economic systems, etc.), or the corresponding mathematical relations are too difficult to deal with in providing the problem's solution, a *simulation* model can be used that "initiates" the system's behaviour over a period of time. This is usually achieved with the help of computers equipped with a set of suitable orders in order to mimic the real system's operation. Other kind of models include the *figurative models* (maps, bas-relief representations, etc.) that are used in the case of relatively simple systems and the *analogical models* (graphs, logical diagrams, etc.) representing the quantitative relationship between two system's parameters (e.g., time and speed, speed and distance, height and weight, etc.). In case of a complex system a *heuristic model* can be also used, which is actually a search procedures that move from one solution point to another with the objective of improving the value of the model criterion. This eventually leads to the best (approximate) solution for the model.

Until the middle of 1970's *Mathematical Modelling (MM)* was mainly a tool in hands of scientists and engineers for solving real world problems related to their disciplines (physics, industry, constructions, economics, etc.). However, the failure of introduction of the "new mathematics" in school education turned the attention of researchers and educators to PS processes and in particular to the process of "MM and Applications" [44].One of the first who described the process of MM in such a way that it could be used for teaching mathematics was Pollak [24]. He represented the interaction between mathematics and the real world with the scheme shown in Figure 3, which is known as the "*circle of MM*."

The most important feature of Pollak's scheme is the direction of the arrows, representing a looping between the other (real) world (including all the other sciences and the human activities of everyday life) and the "universe" of mathematics: Starting from a real problem of the real world we transfer to the other part of the scheme, where we use or develop suitable mathematics for its solution. Then we return to the other world interpreting and testing on the real situation the mathematical results obtained. If these results are not giving a satisfactory solution to the real problem, then we repeat the same circle again one or more times.

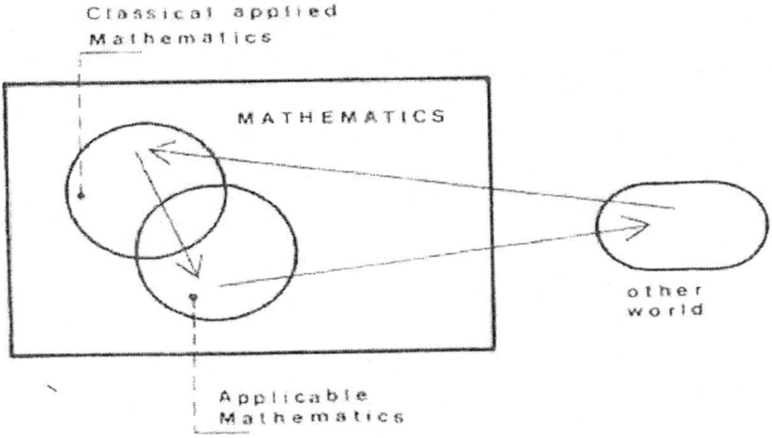

Figure 3. The Pollak's circle of MM.

From the time that Pollak presented this scheme in ICME-3 (Karlsruhe, 1976) until nowadays much effort has been placed to analyze in detail the process of MM. A brief but comprehensive account of the different models used for the description of the MM process can be found in Haines & Crouch [8] including the present author's model [36, 38], where the MM circle is treated as a Markov chain process dependent upon the transition between the successive discrete steps of the MM process. These steps involve:

- S_1: Analysis of the problem (understanding the statement and recognizing the restrictions and requirements of the real system.
- S_2: Mathematization (formulation of the problem and construction of the model).
- S_3: Solution of the model.
- S_4: Validation (control) of the model, which is usually achieved by reproducing, through the model, the behaviour of the real system under the conditions existing before the solution of the model and by comparing it to the existing, from the previous "history" of the corresponding real system, real data. In cases of systems having no past history, an extra simulation model can be used for the validation of the initial mathematical model.

- S_5: Interpretation of the final mathematical results and implementation of them to the real system, in order to give the "answer" to the real world problem.

The flow-diagram of the Voskoglou's model is depicted in Figure 4, where the arrows represent the possible transitions between steps. When the solution of a problem is completed (step S_5), it is assumed that a new problem is given for solution and therefore the MM process starts again from the beginning (step S_1).

Figure 4. Voskoglou's model for the MM process.

MM appears today as a dynamic tool for teaching and learning mathematics, because it connects mathematics with our everyday life problems giving the possibility to students to understand its usefulness in practice and therefore increasing their interest about it [44]. But we must be careful. The process of MM could not be considered as a general and therefore applicable in all cases method for teaching mathematics. In fact, such a consideration could lead to far-fetched situations, where more emphasis is given to the search of the proper applications rather, than to the consolidation of the new mathematical knowledge [45]!

Case-Based Reasoning Systems

Case-Based Reasoning (CBR) is the process of solving new problems by adapting the solution of similar (analogous) problems solved in the past, which are referred as *past cases*. A lawyer, who advocates a particular outcome in a trial based on legal precedents, or an auto mechanic, who fixes an engine by recalling another car that exhibited similar symptoms, or even a physician, who considers the diagnosis and treatment of a previous patient having similar symptoms to determine the disease and treatment for the patient in front of him, are using CBR; in other words CBR is a prominent kind of analogy making.

The driving force behind CBR methods has to a large extent come from the *machine learning* community, and CBR is regarded as a subfield of machine learning. In fact, the notion of CBR does not only denote a particular reasoning method, irrespective of how the cases are acquired, it also denotes a machine learning paradigm that enables sustained learning by updating the case base (library) after a problem has been solved. A case - library can be a powerful corporate resource allowing everyone in an organization to tap in it when handling a new problem. A CBR system, usually designed and functioning with the help of computers, allows the case-library to be developed incrementally, while its maintenance is relatively easy and can be carried out by domain experts. It becomes evident that a great part of the CBR process involves CT, which means that the CBR systems can be considered as a *special case of CT systems*.

The CBR approach has got a lot of attention over the last 30-40 years, because as an intelligent - systems method enables information managers to increase efficiency and reduce cost by substantially automating processes. There are two styles of CBR, the *PS style* and the *interpretive style*. The PS style can support a variety of tasks including planning, diagnosis and design in Medicine, Industry, Robotics, etc. The interpretive style is useful for situation classification, evaluation or justification of solution, argumentation, interpretation or plan and projection of the effects of a decision of plan. Lawyers and managers making strategic decisions use the interpretive style. In general the main domains of the CBR applications include:

- Diagnostic systems, whose symptom lists are similar in nature to that of the new case, (e.g., see Section 4 of [42] for such applications in Medicine).
- Help-desk applications in the customer service area.
- Decision-making systems, in which people, when faced with a complex problem, often look for analogous problems for possible solutions.
- Assessment processes, where CBR systems are used to determine values for variables on comparing them to the known value of something similar. Such tasks are quite common in finance and marketing domains.
- Systems to support human designers in architectural and industrial design. These systems assist the user only for retrieving past cases and would need to be combined with other forms of reasoning to support the full design process.

CBR first appeared in commercial tools in the early 1990's and since then has been sued to create numerous applications in a wide range of domains. Schank's model of *dynamic memory* [29] was the basis of the earliest CBR systems that might be called *case-based reasoners*, Kolodner's CYRUS [13] and Lebowitz's IPP [16]. The basic idea of Schank's model is to organize specific cases, which share similar properties, under a more general structure called a *generalized episode (GE)*. During storing of a new case, when a feature of it matches a feature of an existing past case, a new GE is created. Thus the organization and structure of memory is dynamic, i. e. changes over time. An alternative approach for the representation of cases in a CBR system is the *category and exemplar model,* produced by the work of Bruce Porter and his group at the University of Texas [25]. In this model the case memory is embedded in a network of categories, cases and index pointers. The above model applied first to the PROTOS system, where emphasis is given to the combination of the general with the specific knowledge obtained through the study of cases. Another case memory model was produced by the work of Edwina Rissland and her group at the University of Massachusetts, interested in the role of precedence reasoning in legal judgments [26]. This work resulted in the HYPO and CABARET systems, where cases are grouped under a set of domain-specific dimensions. Other early significant contributions to CBR include, the *Memory-Based Reasoning (MBR)* model of Stanfill and Waltz [31], designed for parallel computation rather than knowledge-based matching, the study of Phyllis Koton [15] at MIT on the use of CBR to optimize performance in an existing knowledge based system resulted in the CASEY system, etc. Organizations as diverse as IBM, Visa International, Volkswagen, British Airways, Lokheed, General Dynamics and NASA have already made use of CBR in applications such as customer support, quality assurance, aircraft maintenance, process planning, decision support and many other applications. More details about the historical development and the applications of CBR can be found in [40, 42].

CBR has been formalized for purposes of computer and human reasoning as a four step process, often referred as the "four R's." These steps involve:

1. R_1: *Retrieve* the most similar to the new problem past case.
2. R_2: *Reuse* the information and knowledge of the retrieved case for the solution of the new problem.
3. R_3: *Revise* the proposed solution.
4. R_4: *Retain* the part of this experience likely to be useful for future problem solving.

The first three of the above steps are not linear, characterized by a backward - forward flow among them. A simplified flow - chart of the CBR process, which is adequate for the purposes of the present paper, is presented in Figure 5:

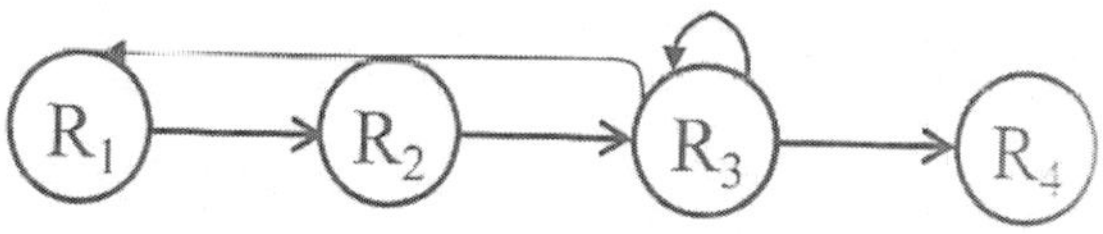

Figure 5. A simplified flow-chart of the CBR process.

A detailed functional diagram illustrating the four steps of the CBR process is presented in Voskoglou & Salem [42].

FUZZY SYSTEMS

Preliminaries

A few years ago *Probability* used to be the unique tool in hands of the experts for dealing with situations of uncertainty appearing in problems of science, technology and of the everyday life. Nevertheless, things have been changed today. In fact, the development of FS theory and in extension of *Fuzzy Logic (FL)*, which is based on it, gave to the experts the opportunity to model under conditions which are not precisely defined. As a result the spectre of their applications has been rapidly extended during the last 50 years covering, apart from the Physical Sciences, sectors like Economics and Management, Industry, Robotics, Decision Making, Programming, Fuzzy Control, Medicine, Biology, Humanities, Education and almost all the other areas of human activity including human reasoning itself ([11]; Chapter 6, [47], etc.). Expert systems, like financial planners, diagnostic, meteorological, information-retrieval, control systems, etc, have been the most obvious recipients of the benefits of FL [34] since their domain is often characterized by vagueness. The first major commercial application of FL was in cement kiln control [57], followed by a navigation system for automatic cars, a fuzzy controller for automatic operation of trains, laboratory level controllers, controllers for robot vision, graphics controllers for automated police sketchers and many others. It must be mentioned that fuzzy mathematics has been also significantly developed in theoretical level providing important contributions even in branches of the classical mathematics, like Algebra, Analysis, Geometry, etc.

FL constitutes one of the three portals of the *Computational Intelligence,* which is a topic of the wide field of *Artificial Intelligence.* The other two portals of Computational Intelligence are the *Neural Networks* and the *Evolutionary Computing.*

History

Human Reasoning was dominated for centuries by the fundamental *"Laws of Thought"* [14], introduced by Aristotle (384-322 BC, Figure 6) and the philosophers that preceded him, which include:

- The principle of identity
- The law of the excluded middle
- The law of contradiction

In particular, the second of the above laws, stating that every proposition has to be either "True" or "False," was the basis for the genesis of the Aristotle's bi-valued Logic. The precision of the traditional mathematics owes undoubtedly a large part of its success to this Logic.

However, even when Parmenides proposed, around 400 BC, the first version of the law of the excluded middle, there were strong and immediate objections. For example, Heraclitus opposed that things could be simultaneously true and not true, whereas the Buddha Sidhartha Gautama, who lived in India a century earlier, had already indicated that almost every notion contains elements from its opposite one. The ancient Greek philosopher Plato (427-377 BC, Figure 6) laid the foundation of what it was later called FL by claiming that there exists a third area beyond "True" and "False," where these two opposite notions can exist together. More modern philosophers like Hegel, Marx, Engels and others adopted and further cultivated the above Plato's belief.

The Polish philosopher Jan Lukasiewicz (1878-1956) was the first to propose a systematic alternative of the bi-valued logic introducing in the early 1900's a three valued logic by adding the term "Possible" between "True" and "False" [17]. Eventually he developed an entire notation and axiomatic system from which he hoped to derive modern mathematics. Later he also proposed four and five valued Logics and he finally arrived to the conclusion that axiomatically nothing could prevent the derivation of an infinite valued Logic. Similar ideas were also expressed by the famous logician and mathematician Alfred Tarski (1901-1983).

Figure 6. Plato (left) and Aristotle in a Raphael's fresco (1509).

But it was not until relatively recently that an infinite-valued Logic was introduced, called FL [56], because it is based on the notion of FS initiated in 1965 by Lotfi Aliasker Zadeh [55], Professor at the University of Berkeley, California. An important goal of FL is that through it algorithmic procedures can be devised which translate the "fuzzy" terminology into numerical values, perform reliable operations upon those values and then return natural language statements in a reliable manner.

Zadeh (1921–2017, Figure 7) was born in Baku, Azerbaijan of USSR, to a Russian Jewish mother (Fanya Koriman), who was a pediatrician, and an Iranian Azeri father (Rahim Aleskerzade), who was a journalist on assignment from Iran.

At the age of 10, when Stalin introduced collectivization of farms in USSR, the Zadeh family moved to Iran. In 1942 Zadeh graduated from the University of Tehran with a degree in electrical engineering and in 1944 he moved to the USA. He received a Master's degree from MIT in 1946 and a Ph.D. in electrical engineering from Columbia University in 1949. He taught for ten years in Columbia, where he promoted to a Full Professor in 1957, before moving to Berkeley in 1959.

Figure 7. L. A. Zadeh (1921–2017).

Apart from FL, Zadeh introduced jointly with J.R. Ragazzini in 1962 the pioneering *z-transform method*, which is used today in the digital analysis [2]. His more recent works include computing with words and perceptions [59] and an outline towards a generalized theory of uncertainty [60]. It has been estimated that Zadeh, who died in Berkeley on 6

September 2017, aged 96, counted in 2011 more than 180 000 citations by other researchers, 90000 of which concern his seminal paper on FS [55].

As it was expected, the far-reaching theory of fuzzy systems aroused some objections to the scientific community. While there have been generic complaints about the fuzziness of assigning values to linguistic terms, the most cogent criticisms come from Haack [7]. She argued that there are only two areas – the nature of Truth and Falsity and the fuzzy systems' utility – in which FL could be possibly needed, and then maintained that in both cases it can be shown that FL is unnecessary. Fox [5] responded to her objections indicating that FL is useful in three areas: To handle real-world relationships which are inherently fuzzy, to calculate the frequently existing in real world situations fuzzy data and to describe the operation of some inferential systems which are inherently fuzzy. His most powerful arguments were that traditional and FL need not be seen as competitive, but as complementary and that FL, despite the objections of classical logicians, has found its way into practical applications and has proved very successful there.

The Concept of Fuzzy Set

Cases appear frequently in real life where some definitions have not clear boundaries. Examples of such kind of cases are "the young people of a city," "the good players of a team," "the diligent students of a class," etc. The need to model mathematically such kind of situations was one of the main reasons that laid to the development of the FS theory.

Let U be the set of the discourse. Then, according to Zadeh [55], a *fuzzy subset A of U* (or for brevity a FS in U) can be defined with the help of its *membership function* $m_A: U \to [0,1]$, which assigns to each element x of U a real value $m_A(x)$ in [0, 1], called the *membership degree of x in* ***A***. The closer is $m_A(x)$ to 1, the more x satisfies the characteristic property of A. Then one defines A as a set of ordered pairs of the form:

$$A = \{(x, m_A(x)) : x \in U\}$$

Many authors, for reasons of simplicity, identify the FS *A* with its membership function m_A. A FS can be also denoted in the form of a symbolic sum, or a symbolic power series, or a symbolic integral, when *U* is a finite or numerable set or it has the power of the continuous respectively. For general facts on FS and the uncertainty connected to them we refer to the book of Klir and Folger [11].

Example 1: The young human ages

Let *U* be the set of the non-negative integers not exceeding 140 (considered as the upper bound of human life) representing the human ages. The set of all ages not exceeding a given integer in *U*, e.g., 20, is a crisp subset of *U*. On the contrary the set *A* of the young human ages, being not precisely defined, is a FS in *U*. The membership function of *A* can be defined by

$$m_A(x) = \begin{cases} \left[1+(0.04x)^2\right]^{-1}, & x \leq 40 \\ 0, & x > 40 \end{cases}$$

Therefore, the age of a recently born baby has membership degree $m_A(0) = 1$, the age of 25 years has membership degree $m_A(25) = (1+1^2)^{-1} = 0.5$, etc.

Remarks:

1) The membership function of a FS is not uniquely determined, its definition depending on the designer's personal criteria, which are based on statistical, empirical or logical observations. Nevertheless the creditability of a FS for modeling a vague real situation depends on the successful or not definition of its membership function.

2) A crisp subset *B* of *U* can be considered as a FS in *U* with membership function $m_B(x) = \begin{cases} 1, x \in B \\ 0, x \notin B \end{cases}$. On the basis of this consideration most of the notions and operations concerning the crisp sets, like subset,

complement, union, intersection, Cartesian product, binary and other relations, etc., can be extended to FS. Moreover, the Zadeh's *extension principle* [11] enables the transform of models with crisp variables to corresponding models with fuzzy variables. Therefore, the argument that FL is a natural extension of the bi-valued logic is strongly justified by those facts.

3) It must be emphasized that, despite to the fact that FS and Probability theories act on the same real interval [0, 1], they *essentially differ* to each other. For example, the expression "The probability that John is tall is 85%" has a completely different meaning from the FL statement "John's membership degree in the FS of the tall men is 0.85." In fact, the former expression, in terms of the law of the excluded middle, means that John, being an unknown to the observer person, is either tall or short, but the probability to be tall is 85%. On the contrary, the latter statement, since John's membership degree in the FS of the tall men approaches 1, means that John is a rather tall person. There are also other differences between the above two theories mainly arising from the way of defining the corresponding notions and operations. For instance, whereas the sum of the probabilities of all the single events (singleton subsets) of *U* is always 1 (probability of the certain event), this is not necessarily true for the membership degrees. Consequently, a probability distribution could be used to define membership degrees, but the converse is not true in general.

4) From the time that Zadeh introduced the theory of *Fuzzy Set (FS)* in 1965 many efforts have been made for improving its effectiveness to deal with uncertain, ambiguous and vague situations. As a result a series of extensions and generalizations of the ordinary FS followed and relevant theories have been proposed as alternatives to the FS theory. For more details see the author's Commentary in the present book.

Types of Uncertainty in Fuzzy Systems

Uncertainty is the shortage of precise knowledge and of complete information on data, which describe together the state of the corresponding system. One of the key problems of artificial intelligence is the modelling of the uncertainty for solving real life problems and several models have been proposed for this purpose.

The amount of information obtained by an action can be measured by the reduction of the uncertainty resulting from this action. Therefore, a system's uncertainty is connected to its capacity for obtaining relevant information. Accordingly a measure of uncertainty could be adopted as a measure of a system's effectiveness in solving related problems. The greater is the decrease of the uncertainty resulting from the action (i.e., the difference of the existing uncertainty before and after the action), the better the system's performance with respect to the action. Obviously, the measurement of the uncertainty focuses on the system's mean performance.

In classical probability theory a system's uncertainty and the information connected to it are measured by the Shannon's formula, better known as the Shannon's entropy [30]. This term comes from the mathematical definition of the information, say I, by $I = -\frac{\Delta(\log P)}{\log 2}$, where P is the probability of appearance of each of the equally probable cases of the evolution of the corresponding real situation. This expression appears to be analogous to the well-known from Physics formula $\Delta S = \frac{\Delta Q}{T}$, where ΔS is the increase of a physical system's entropy caused by an increase of the heat by ΔQ, when the absolute temperature T remains constant.

Let U denote the universal set of the discourse. For use in a fuzzy environment Shannon's formula has been adapted ([12], p. 20) to the form:

$$H = -\frac{1}{\ln n}\sum_{s=1}^{n} m_s \ln m_s \tag{1}$$

In formula (1) $m_s = m(s)$ denotes the membership degree of the element s of U in the corresponding fuzzy set and n denotes the total number of the elements of U. The sum is divided by the natural logarithm of n in order to be normalized. Thus H takes values within the real interval [0, 1]. Formula (1) measures a fuzzy system's probabilistic uncertainty.

It is recalled that the fuzzy probability of an element s of U is defined in a way analogous to the crisp probability, i.e., by

$$P_s = \frac{m_s}{\sum_{s \in U} m_s} \quad (2)$$

However, according to the British economist Shackle [28] and many other researchers after him, human reasoning can be formulated more adequately by the possibility rather, than by the probability theory. The possibility, say r_s, of an element s of U is defined by

$$r_s = \frac{m_s}{\max\{m_s\}} \quad (3)$$

where $\max\{m_s\}$ denotes the maximal value of m_s, for all s in U. In other words, the possibility of s expresses the relative membership degree of s with respect to $\max\{m_s\}$.

Within the domain of possibility theory uncertainty consists of strife (or discord), which expresses conflicts among the various sets of alternatives, and of non-specificity (or imprecision), which indicates that some alternatives are left unspecified, i.e., it expresses conflicts among the sizes (cardinalities) of the various sets of alternatives ([16], p. 28). For a better understanding of the above two types of uncertainty we give the following example:

Example 2: Let U be the set of integers from 0 to 120 representing human ages and let Y = young, A = adult and O = old be fuzzy subsets of U defined by the membership functions m_Y, m_A and m_O respectively. People are considered as young, adult or old according to their outer appearance. Then, given x in U, there usually exists a degree of uncertainty about the values that the membership degrees $m_Y(x)$, m_A (x) and $m_O(x)$ could take, resulting to a conflict among the fuzzy subsets Y, A and O of

U. For instance, if x = 18, values like $m_Y(x) = 0.8$ and $m_A(x) = 0.3$ are acceptable, but they are not the only ones. In fact, values like $m_Y(x) = 1$ and $m_A(x) = 0.5$ are also acceptable, etc. The existing conflict becomes even greater if x =50. In fact, it is not reasonable in this case to take $m_Y(x) = 0$, because sometimes people being 50 years old look much younger than others aged 40 or even 30 years. But, there exist also people aged 50 who look older from others aged 70, or even 80 years! All the above are examples of the type of uncertainty that we have termed as strife.

On the other hand, non - specificity is connected to the question: How many x in U should have non zero membership degrees in Y, A and O respectively? In other words, the existing in this case uncertainty creates a conflict among the cardinalities (sizes) of the fuzzy subsets of U. It is recalled that the cardinality of a fuzzy subset, say B, of U is defined to be the sum $\sum_{x \in U} m_B(x)$ of all membership degrees of the elements of U in B.-

Strife is measured ([12]; p.28) by the function ST(r) on the ordered possibility distribution r: $r_1=1 \geq r_2 \geq \ldots\ldots \geq r_n \geq r_{n+1}$ of a group of a system's entities defined by

$$ST(r) = \frac{1}{\log 2}[\sum_{i=2}^{n}(r_i - r_{i+1})\log\frac{i}{\sum_{j=1}^{i} r_j}] \tag{4}$$

Under the same conditions non-specificity is measured ([17]; p.28) by the function

$$N(r) = \frac{1}{\log 2}[\sum_{i=2}^{n}(r_i - r_{i+1})\log i] \tag{5}$$

The sum T(r) = ST(r) + N(r) measures the fuzzy system's total possibilistic uncertainty.

Example 3: Table 1 depicts the performance of two equivalent student groups in solving problems involving CT with respect to the linguistic grades A = excellent, B = very good, C = good, D = fair and F = unsatisfactory.

Table 1. Student results

Grade	G_1	G_2
A	1	10
B	13	6
C	4	3
D	3	0
F	0	1
Total	21	20

It is asked to evaluate the performance of the two student groups by calculating the existing in them total possibilistic and probabilistic uncertainty.

(i) *Total possibilistic uncertainty:* Defining the membership function in terms of the frequencies of the student grades one can represent the two student groups as fuzzy sets on the set U = {A, B, C, D. F} in the form $\{(x, \frac{n_x}{n}): x \in U\}$, where n_x and is the number of students who received the grade x and n is the total number of the students in each group. In other words we can write

$$G_1 = \{(A, \frac{1}{21}), (B, \frac{13}{21}), (C, \frac{4}{21}), (D, \frac{3}{21}), (F, 0)\} \text{ and}$$

$$G_2 = \{(A, \frac{10}{20}), (B, \frac{6}{20}), (C, \frac{3}{20}), (D, 0), (F, \frac{1}{20})\}.$$

The maximal membership degree in G_1 is equal to $\frac{13}{21}$ and therefore the possibilities of the elements of U in G_1 are: $r(A) = \frac{1}{13}$, $r(B) = 1$, $r(C)$

$=\frac{4}{13}$, r(D) $=\frac{3}{13}$, r(F) = 0. Therefore the ordered possibility distribution defined on G_1 is

$$\boldsymbol{r} : r_1 = 1 > r_2 = \frac{4}{13} > r_3 = \frac{3}{13} > r_4 = \frac{1}{13} > r_5 = 0 \qquad (6)$$

Working in the same way one finds that the ordered possibility distribution on G_2 is:

$$\boldsymbol{r} : r_1 = 1 > r_2 = \frac{6}{10} > r_3 = \frac{3}{10} > r_4 = \frac{1}{10} > r_5 = 0 \qquad (7)$$

Formula (4) gives in our case that

$$ST(r) = \frac{1}{\log 2}\left[(r_2 - r_3)\log\frac{2}{r_1 + r_2} + (r_3 - r_4)\log\frac{3}{r_1 + r_2 + r_3} + (r_4 - r_5)\log\frac{4}{r_1 + r_2 + r_3 + r_4}\right]$$

Replacing the values of the possibility distribution $\boldsymbol{r}$ from (6) one finds for G_1 that

$$ST(r) = \frac{1}{\log 2}\left[\frac{1}{13}\log\left(\frac{26}{17}\right) + \frac{2}{13}\log\left(\frac{39}{20}\right) + \frac{1}{13}\log\left(\frac{42}{21}\right)\approx 0.27.\right.$$

Also, formula (5) gives for G_1 that

$$N(r) = \frac{1}{\log 2}\left[\frac{1}{13}\log 2 + \frac{2}{13}\log 3 + \frac{1}{13}\log 4\right] \approx 0.48.$$

Therefore, the total possibilistic uncertainty for G_1 is $T\ (r)\ \approx 0.27+$ 0.48 =0.75.

In the same way, replacing the values of ***r*** from (7) one finds that the total possibilistic uncertainty for G_2 is $T(r) \approx 0.33 + 0.82 = 1.15$.

Since the two student groups were chosen to be equivalent, the existing uncertainty before the solution of the problems was the same for both of them. Thus, the reduction of the uncertainty was greater for the group G_1, which therefore demonstrates a better mean performance than the group G_2.

iii) Probabilistic uncertainty: Replacing the membership degrees of G_1 to formula (1) one finds that the probabilistic uncertainty for the experimental group is equal to

$$H=-\frac{1}{\ln 5}\left(\frac{1}{21}\ln\frac{1}{21}+\frac{13}{21}\ln\frac{13}{21}+\frac{4}{21}\ln\frac{4}{21}+\frac{3}{21}\ln\frac{3}{21}\right)\approx 0.64.$$

In the same way one finds that for G_2 the probabilistic uncertainty is approximately equal to 0.71. Therefore group G_1 demonstrates again a better performance.

Remark: From the previous example it becomes evident that the measurement of the uncertainty can be used for assessing the performance of the two groups only under the assumption that they are equivalent, which means that the existing uncertainty before the solution of the problems was the same for both of them.

The COG Defuzzification Technique

The solution of a problem using principles and methods of FL involves the following steps:

- *Choice* of the set of the discourse.
- *Fuzzification* of the problem's data by introducing the proper membership functions to define the necessary for this reason FSs.

- *Evaluation* of the fuzzy data by using principles and methods of FL in order to find the solution of the given problem in the form of a *unique* FS.
- *Defuzzification* of the problem's fuzzy solution, i.e., representation of it with a crisp numerical value, in order to be expressed in the natural language.

Among the at least 30 known defuzzification methods two are the most frequently used in practical applications. In the *Maximum method* one of the values at which the FS expressing the problem's solution has its maximum truth is chosen as the crisp value for the solution's output. For example, this technique is used when the *Bellman-Zadeh's criterion* is applied for solving decision-making problems under fuzzy conditions (e.g., [47]; Chapter 4, pp. 97-99).

On the other hand, *the Centre of Gravity (COG) technique* is perhaps the most popular defuzzification method of fuzzy mathematics. According to this method, the crisp value for the solution's output is obtained by calculating the coordinates of the COG of the level's area S contained between the graph of the membership function of the FS expressing the problem's solution and the semi-axis OX [35].

Voskoglou developed in 1999 a fuzzy model for describing mathematically the process of learning a subject matter in the classroom [37] and later, considering the class as a fuzzy system, he calculated the existing in the class uncertainty to obtain a measure of the student mean performance ([47]; Chapter 5). Meanwhile, Subbotin et al. [32], based on the Voskoglou's model, adapted properly the COG defuzzification technique to be used as an assessment method of student learning skills. Since then, Subbotin and Voskoglou have applied jointly or separately to each other the COG assessment method in many other types of assessment problems ([47]; Chapter 6).

In this Chapter we are going to apply the COG technique for the defuzzification of the TFNs.

Triangular Fuzzy Numbers

FNs, which were introduced also by Zadeh [58], play an important role in fuzzy mathematics, analogous to the role played by the ordinary numbers in classical mathematics. A FN, say A, is defined to be a FS on the set ***R*** of real numbers, such that:

- A is *normal*, i.e., there exists x in ***R*** such that $m_A(x) = 1$.
- A is *convex*, i.e., all its *a-cuts* $A^a = \{x \in U: m_A(x) \geq a\}$, with a in [0, 1], are closed real intervals
- Its membership function $y = m(x)$ is a *piecewise continuous* function.

One can define the basic arithmetic operations on FNs in two, equivalent to each other, ways:

i) With the help of their *a*-cuts and the representation-decomposition Theorem of Ralesscou-Negoita ([27]; Theorem 2.1, p.16) stating that a FS A can be completely and uniquely expressed by the family of its *a*-cuts in the form $A = \sum_{a \in [0,1]} aA^a$.

ii) By applying the Zadeh's extension principle ([11], Section 1.4, p.20), which provides the means for any function f mapping the crisp set X to the crisp set Y to be generalized so that to map fuzzy subsets of X to fuzzy subsets of Y.

However, the above two general methods of the fuzzy arithmetic, requiring laborious calculations, are rarely used in practical applications, where the utilization of simpler forms of FNs is preferred.

A TFN (a, b, c), with a, b, c real numbers such that $a < b < c$ is the simplest form of a FN representing mathematically the fuzzy statement that "the value of b lies in the interval $[a, c]$." The membership function $y = m(x)$ of (a, b, c) is zero outside the interval $[a, c]$, whereas its graph in $[a, c]$ forms a triangle with the OX axis (Figure 8). Therefore we have:

$$y = m(x) = \begin{cases} \dfrac{x-a}{b-a}, & x \in [a,b] \\ \dfrac{c-x}{c-b}, & x \in [b,c] \\ 0, & x < a \text{ or } x > c \end{cases}$$

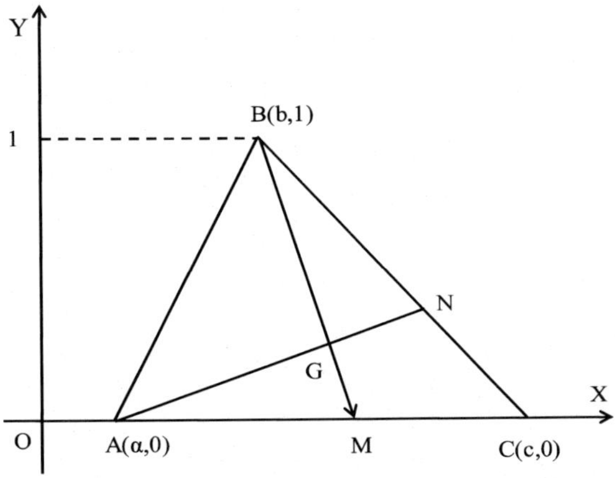

Figure 8. Graph and COG of the TFN (a, b, c).

The coordinates (*X*, *Y*) of the *Centre of Gravity (CoG)* of the graph of the TFN A = (*a*, *b*, *c*), being the intersection point G of the medians AN and BM of the triangle ABC (Figure 8), are calculated by the formulas

$$X\,(\mathrm{A}) = \frac{a+b+c}{3},\ Y(\mathrm{A}) = \frac{1}{3} \tag{8}$$

In fact, since M($\frac{a+c}{2}$. 0) and N($\frac{b+c}{2}$,$\frac{1}{2}$), equation (8) is easily obtained by calculating the equations of the medians AN and BM and by solving their linear system.

According to the COG technique the first of formulas (8) can be used to defuzzify the TFN A.

The two general methods for defining arithmetic operations on FNs lead to the following simple rules for the *addition* and *subtraction* of TFNs:

Let A = (a, b, c) and B = (a_1, b_1, c_1) be two TFNs. Then one defines:

- The sum A + B = ($a+a_1$, $b+b_1$, $c+c_1$).
- The difference A - B = A + (-B) = ($a-c_1$, $b-b_1$, $c-a_1$), where –B = ($-c_1$, $-b_1$, $-a_1$) is defined to be the opposite of B.

On the contrary, the *product* and the *quotient* of A and B are FNs, which are not TFNs in general, apart from some special cases.

The following two *scalar operations* can also be defined:

- k + A= (k+a, k+b, k+c), $k \in \boldsymbol{R}$
- kA = (ka, kb, kc), if k>0 and kA = (kc, kb, ka), if k<0, $k \in \boldsymbol{R}$.

Remark: Another simple form of FNs that are frequently used in practical applications are the Trapezoidal Fuzzy Numbers (TpFNs). The membership function of a TpFN (a, b, c, d), with a, b, c, d in **R** such that $a < b \leq c < d$, is zero outside the interval [a, d], whereas its graph in [a, d] forms a trapezoid with the OX axis. The TFN (a, b, d) is a special case of the TpFN (a, b, c, d) with b = c, i.e., the TpFNs are actually generalizations of the TFNs.

For general facts on FNs we refer to the book of Kaufmann and Gupta [9].

The Assessment Method Using TFNs

In [43] we have developed with the help of TFNs a method for assessing a system's mean performance, whose steps are the following:

- Let A_i, i = 1, 2,..., *n* be given TFNs, where *n* is a non-negative integer, $n \geq 2$. Then, we define the *mean value* of the A_i's to be the TFN

$$A = \frac{1}{n}(A_1 + A_2 + \dots + A_n).$$

- The qualitative grades A = excellent, B = very good, C = good, D = fair and F = unsatisfactory are considered for the assessment of the system's performance and a scale of numerical scores from 1 – 100 is assigned to them as follows: A (85 - 100), B (75 - 84), C (60 – 74), D (50 - 59) and F (0 - 49).

- Each of the above qualitative grades can be represented by a TFN, denoted for simplicity by the same letter, as follows: A = (85, 92.5, 100), B = (75, 79.5, 84), C (60, 67, 74), D (50, 54.5, 59) and F (0, 24.5, 49). The middle entry of each of the above TFNs is equal to the average of its other two entries. In other words, if A (a_1, b_1, c_1), B (a_2, b_2, c_2), C (a_3, b_3, c_3), D (a_4, b_4, c_4), F(a_5, b_5, c_5), then $b_i = \frac{a_i + c_i}{2}$, i = 1, 2, 3, 4, 5.

- In order to assess the total system's effectiveness, the performance of each of the system's components is evaluated by one of the above five qualitative grades, which means that one of the TFNs A, B, C, D, F can be assigned to each of those components.

- Let *n* be the total number of the system's components and let n_X be the number of the components corresponding to the TFN X, where X = A, B, C, D, F. Then the mean value M of all those TFNs is equal to the TFN

$$M(a, b, c) = \frac{1}{n}(n_A A + n_B B + n_C C + n_D D + n_F F) \tag{9}$$

- Since the calculation of the mean value of the qualitative grades is not possible, it looks logical to consider the TFN M as a fuzzy representative for evaluating the system's mean performance. Replacing the values of the TFNs A, B, C, D and F in equation (9) it is straightforward to check that for M (a, b, c) we have

$$a = \frac{85n_A + 75n_B + 60n_C + 50n_D + 0n_F}{n},$$

$$b = \frac{92.5n_A + 79.5n_B + 67n_C + 54.5n_D + 24.5n_F}{n},$$

$$c = \frac{100n_A + 84n_B + 74n_C + 59n_D + 49n_F}{n}.$$

- Then, equation (8) gives that

$$X\,(\mathrm{M}) = \frac{a+b+c}{3} = \frac{92.5n_A + 79.5n_B + 67n_C + 54.5n_D + 24.5n_F}{n} = \frac{a+c}{2} = b \tag{10}$$

Therefore, for the defuzzification of M (a, b, c) one needs only its middle component b. The value of X (M) provides a crisp representation of the TFN M evaluating the system's mean performance.

An analogous assessment method was also developed in [43] using TpFNs instead of TFNs.

GREY SYSTEM THEORY

Preliminaries

Frequently in the everyday life, as well as in many applications of science and engineering, a system's data cannot be easily determined precisely and in practice estimates of them are used. The reason for this is that in large and complex systems, like the socio – economic, the biological ones, etc., many different and constantly changing factors are usually involved, the relationships among which are indeterminate, making their operation mechanisms to be not clear. Two are the main tools today for handling such approximate data: FL and the theory of *Grey System (GS)* initiated by Deng [3] in 1982.

The GS theory was mainly developed in China and it has found many important applications in agriculture, economy, management, industry, ecology, environment, meteorology, geography, geology, forecast of earthquakes, history, military affairs, sports, traffic, material science, biological protection and in many other fields of the human activity; see [4] and its relevant references.

The systems which lack information, such as structure message, operation mechanism and behaviour document, are referred to as GSs. For example, the human body, the world economy, etc., are GSs. Usually, on the grounds of existing grey relations and elements one can identify where "grey" means poor, incomplete, uncertain, etc.

The aim of the GS theory is to provide techniques, notions and ideas for analyzing latent and intricate systems, including the establishment of non-function models, the development of a grey process replacing an existing stochastic process, the transformation of disorderly raw data into a more regular series by grey generating techniques instead of modelling with the original data; grey decision making, grey forecasting control, the study of feeling and emotion functions and of fields with whitening functions, etc. [4].

Grey Numbers

An effective tool for handling the approximate data of a GS is the use of GNs. A GN is an indeterminate number whose probable range is known, but which has unknown position within its boundaries. The GNs are defined with the help of the closed real intervals. More explicitly, if ***R*** denotes the set of real numbers, a GN, say A, can be expressed mathematically by

$A \in [a, b] = \{x \in \boldsymbol{R} : a \le x \le b\}$.

If $a = b$, then A is called a *white number* and if $A \in (-\infty, +\infty)$, then A is called a *black number*. A GN may enrich its uncertainty representation with respect to the interval [a, b] by a *whitening function* g: [a, b] → [0, 1] defining a *degree of greyness* g(x) for each x in [a, b]. The closer is g(x) to 1, the greater the probability for x to be the representative real value of the corresponding GN. For general facts on GNs we refer to [18].

The well-known arithmetic of the real intervals [22] has been used to define the basic arithmetic operations among the GNs. More explicitly, if $A \in [a_1, a_2]$ and $B \in [b_1, b_2]$ are given GNs and k is a real number, one defines:

- Addition by $A + B \in [a_1 + b_1, a_2 + b_2]$
- Subtraction by $A - B = A + (-B) \in [a_1 - b_2, a_2 - b_1]$, where $-B \in [-b_2, -b_1]$ is defined to be the opposite of B.
- Multiplication by

$A \times B \in [\min\{a_1b_1, a_1b_2, a_2b_1, a_2b_2\}, \max\{a_1b_1, a_1b_2, a_2b_1, a_2b_2\}]$

- Division by $A : B = A \times B^{-1} \in [\min\{\frac{a_1}{b_1}, \frac{a_1}{b_2}, \frac{a_2}{b_1}, \frac{a_2}{b_2}\}, \max\{\frac{a_1}{b_1}, \frac{a_1}{b_2}, \frac{a_2}{b_1}, \frac{a_2}{b_2}\}]$

with $b_1, b_2 \neq 0$ and $B^{-1} \in [\frac{1}{b_2}, \frac{1}{b_1}]$, which is defined to be the inverse of B.

- Scalar multiplication by $kA \in [ka_1, ka_2]$, if $k \geq 0$ and by $kA \in [ka_2, ka_1]$, if $k < 0$.

Observe that $B + (-B) \in [b_1 - b_2, b_2 - b_1] \neq [0, 0]$, $B + (-B) \neq (-B) + B \neq 0$ and $B \times B^{-1} = B^{-1} \times B \in [\frac{b_1}{b_2}, \frac{b_2}{b_1}] \neq [1, 1]$.

The white number with the greatest probability to be the representative real value of the GN A∈[*a*, *b*] is denoted by *W*(A). The technique of determining the value of *w*(A) is called *whitening* of A.

One usually defines $W(A) = (1- t)a + tb$, with *t* in [0, 1]. This is known as the *equal weight whitening*. In case that the distribution of A is unknown (i.e., no whitening function has been defined for A), one takes $t = \frac{1}{2}$, which gives that

$$W(A) = \frac{a+b}{2} \tag{11}$$

The Assessment Method with GNs

When using TFNs for the assessment of a system's mean performance we have found that $X(M) = b$, which means that only its middle component *b* is needed for the defuzzification of the mean value M(*a*, *b*, *c*). This remark gives the idea to search for a "formal" assessment method, analogous to that with TFNs, that possibly reduces the required computational burden. This idea led us to utilize GNs [48] instead of TFNs for the system's assessment. The steps of our new method are the following:

- Consider again the numerical scores A (100-85), B(84-75), C (74-60), D(59-50), F(49-0) attached to the corresponding qualitative grades.
- Correspond to each grade a GN, denoted for simplicity with the same letter, as follows: A∈ [85, 100], B∈[75, 84], C∈ [60, 74], D∈[50, 59] and F∈[0, 49].
- Assign to each of the system's components the GN assessing its individual performance.
- Using analogous notation and definitions as in the case of the TFNs calculate the *mean value*

$$M^* = \frac{1}{n}\,[n_A A + n_B B + n_C C + n_D D + n_F F] \tag{12}$$

of all those GNs and consider it as the grey representative of the system's mean performance.

- We have that $n_A A \in [85n_A,\ 100n_A]$, $n_B B \in [75n_B,\ 84n_B]$, $n_C C \in [60n_C,\ 74n_C]$, $n_D D \in [50n_D,\ 59n_D]$ and $n_F F \in [0n_F,\ 49n_F]$. Therefore it turns out that $M^* \in [m_1, m_2]$, with

$$m_1 = \frac{85n_A + 75n_B + 60n_C + 50n_D + 0n_F}{n},$$

$$m_2 = \frac{100n_A + 84n_B + 74n_C + 59n_D + 49n_F}{n}$$

- Since the distributions of the GNs A, B, C, D and F are unknown, the same happens with the distribution of M*. Therefore, one can take

$$W(M^*) = \frac{m_1 + m_2}{2} \tag{13}$$

The value of $W(M^*)$ provides a crisp representation of the GN M* evaluating the system's mean performance.

Remarks:

1) From equations (10) and (13) it turns out that $X(M) = W(M^*)$. This shows that the two assessment methods of a system's mean performance using as tools the TFNs and the GNs A, B, C, D and F respectively are *equivalent to each other*, providing the same assessment outcomes.

2) In the extreme case where the maximal possible numerical score corresponds to each system's component for each grade, i.e., the n_A scores corresponding to A are equal to 100, the n_B scores corresponding to B are equal to 84, etc., the mean value of all those scores is equal to c or m_2 respectively. Also, in the opposite extreme case, where the minimal

possible numerical score corresponds to each system's component for each linguistic grade, i.e., the n_A scores corresponding to A are equal to 85, the n_B scores corresponding to B are equal to 75, etc., the mean value of all those scores is equal to *a* or m_1 respectively. Consequently, the assessment methods with the TFNs and the GNs give a reliable approximation of the system's mean performance. Therefore they are useful when no numerical scores are used, but the system's performance is assessed by qualitative grades, because in this case the calculation of those grades is not possible in the traditional way.

3) Observe that for obtaining the mean value $M^* \in [m_1, m_2]$ one needs to calculate two components only, in contrast to the mean value M (a, b, c) where the calculation of three components is needed. Consequently the method with the GNs reduces the required computational burden.

APPLICATIONS TO THE ASSESSMENT OF COMPUTATIONAL THINKING SKILLS

As we have already seen, modelling thinking is the essence of CT combining all the other modes of thinking involved in CT for constructing a proper model for the solution of the given problem. As a matter of fact, our first example on the assessment of CT skills concerns the evaluation of student MM abilities.

Example 4*:* Table 2 depicts the performance of two student groups G_1 and G_2, in a common mathematical test involving the solution of two MM problems on the use of derivatives for finding the extreme values of a function in one variable.

Assigning to each student the corresponding GN we calculate the mean values M_1^* and M_2^* of all those GNs for the groups G_1 and G_2 respectively, which are approximately equal to

$$M_1^* = \frac{1}{60}(20A+15B+7C+10D+8F) \in [62.42, 79.33] \text{ and}$$

$$M_2^* = \frac{1}{85}(20A+30B+15C+10D+10F) \in [62.94, 78.94].$$

Therefore $W(M_1^*) \approx \frac{62.42+79.33}{2} \approx 70.88$ and

$$w(M_2^*) \approx \frac{62.94+78.94}{2} \approx 70.94.$$

Table 2. Student performance

Grade	G_1	G_2
A	20	20
B	15	30
C	7	15
D	10	10
F	8	10
Total	60	85

Consequently both groups demonstrated a good (C) mean performance, with the mean performance of the second group being slightly better.

As we had already the opportunity to realize, a special form of CT systems, with great importance for Artificial Intelligence, are the CBR systems and our next examples concern the evaluation of the effectiveness of such kind of systems.

Example 5: Consider two CBR systems designed for help desk applications with their libraries containing 105 and 90 past cases respectively. Assume that the two systems' designers have supplied them with the same mechanism (software) that enables the assessment of the degree of success of each one of their past cases at each step of the CBR process, when used for the solution of new similar problems. Table 3 depicts the degree of success of their past cases in each of the three first steps of the CBR process.

Here we shall compare the quality performance of the two systems by calculating the GPA index and their mean performance by applying our assessment methods with the TFNs and the GNs.

i) GPA Index: Denote by y_i, i = 1, 2, 3, 4, 5 the ratios of the CBR system's cases whose performance is characterized by F, D, C, B and A respectively to the total number of its cases (frequencies). Then the GPA index is calculated ([47], Chapter 6, p.125) by the formula

$$GPA = y_2 + 2y_3 + 3y_4 + 4y_5 \quad (14)$$

Table 3. Assessment of the past cases of the CBR systems

First System					
Steps	F	D	C	B	A
R_1	0	0	51	24	30
R_2	18	18	48	21	0
R_3	36	30	39	0	0

Second System					
Steps	F	D	C	B	A
R_1	0	18	45	27	0
R_2	18	24	48	0	0
R_3	36	27	27	0	0

In case of the ideal performance ($y_5 = 1$) we have GPA = 4, while in case of the worst performance ($y_1 = 1$) we have GPA = 0; therefore $0 \leq GPA \leq 4$. Consequently, values of GPA greater than 2 could be considered as corresponding to a more than satisfactory system's performance. In our case, the data of Table 3 give the following frequencies:

Table 4. Frequencies of the past cases of the CBR systems

FIRST SYSTEM					
Steps	Y_1	y_2	Y_3	Y_4	Y_5
R_1	0	0	$\frac{51}{105}$	$\frac{24}{105}$	$\frac{30}{105}$
R_2	$\frac{18}{105}$	$\frac{18}{105}$	$\frac{48}{105}$	$\frac{21}{105}$	0
R_3	$\frac{36}{105}$	$\frac{30}{105}$	$\frac{39}{105}$	0	0

SECOND SYSTEM					
Steps	Y_1	y_2	y_3	Y_4	Y_5
R_1	0	$\frac{18}{90}$	$\frac{45}{90}$	$\frac{27}{90}$	0
R_2	$\frac{18}{90}$	$\frac{24}{90}$	$\frac{48}{90}$	0	0
R_3	$\frac{36}{90}$	$\frac{27}{90}$	$\frac{27}{90}$	0	0

Replacing the values of frequencies from Table 3 in equation (14) one finds the following values for the GPA index:

First System: R_1: $\frac{294}{105} = 2.8$, R_2: $\frac{177}{105} \approx 1.69$, R_3: $\frac{108}{105} \approx 1.03$.

Second System: R_1: $\frac{189}{90} = 2.1$, R_2: $\frac{168}{90} \approx 1.87$, R_3: $\frac{81}{90} = 0.9$.

The above values of the GPA index show that the first system demonstrated a better quality performance at steps R_1 and R_3 (Retrieve, Revise), whereas the second one demonstrated a better performance at R_2 (Reuse). Further, the two systems' performance was proved to be more than satisfactory in R_1 and less than satisfactory in the other two steps, being worse at R_3. This was logically expected, since the success in each step depends on the success in the previous steps. Notice that the two systems' performance at the last step R_4 was not examined, since ALL the past cases, even the unsuccessful ones, are retained in a system's library for possible use in future with related new problems; the unsuccessful ones to help for exploring possible reasons of failure to find a solution for the new problem.

The mean values of the GPA index for the two systems at the three steps R_1, R_2 and R_3 are approximately equal to 1.84 and 1.62 respectively, showing that the first system demonstrated a better overall quality performance.

ii) Use of the TFNs: From the data of Table 2 one finds that for the first system and in step R_1 we have 51 TFNs equal to C(60, 67, 74), 24 TFNs equal to B(75, 79.5, 85) and 30 TFNs equal to A(85, 92.5, 100). The mean value of all those TFNs, denoted for simplicity by the same letter R_1, is equal to

$$R_1 = \frac{1}{105}(51C + 24B + 30A) = \frac{1}{105}[(3060, 3417, 3774) + (1800, 1908, 2016) + (2550, 2775. 3000) = \frac{1}{105}(7410, 8100, 8790) \approx (70.57, 77.14, 83.71).$$

Therefore, from equation (10) one gets that $X(R_1) = 77.14$, which shows that the first system demonstrated a very good (B) performance at step R_1.

In the same way one calculates for the first system the mean values

$$R_2 = \frac{1}{105}(18F + 18D + 48C + 21B) \approx (51, 60.07, 69.14)$$

and

$$R_3 = \frac{1}{105}(36F + 30D + 39C) \approx (36.57, 48.86, 61.14).$$

In this way one obtains the analogous conclusions for the system's performance at the steps R_2 and R_3 of the CBR process.

The overall system's performance can be assessed by the mean value

$$R = \frac{1}{3}(R_1 + R_2 + R_3) \approx (52.71, 62.02, 71.33),$$

Since $X(R) = 62.02$, the system demonstrated a good (C) mean performance.

A similar argument gives for the second system the values $R_1 = (62.5, 68.25, 74)$, $R_2 \approx (45.33, 55.17, 65)$, $R_3 = (33, .46.25, 59.5)$ and $R \approx (46.94, 56.56, 66.17)$. Therefore, analogous conclusions can be obtained for its mean performance at each step of the CBR process and for its overall mean performance.

iii) Use of the GNs: According to this approach in step R_1 we have 51 GNs equal to $C \in [60,74]$, 24 GNs equal to $B \in [75, 84]$ and 30 GNs equal to $A \in [85, 100]$. The mean value of all those GNs, denoted by R_1^*, is equal to

$$R_1^* = \frac{1}{105}(51C + 24B + 30A) \in [70.57, 83.71].$$

Therefore, $W(R_1^*) = \frac{70.57 + 83.71}{2} = 77.14$, etc.

As we have seen in the previous section, this approach provides in general the same assessment outcomes with the use of TFNs, but, as it becomes more evident from the present example, it reduces significantly the required computational burden.

Table 5 depicts the assessment outcomes obtained in this example.

The outcomes of Table 5 illustrate the logically accepted fact that the assessment of a system's quality performance could lead to different outcomes from the assessment of its mean performance (step R_2).

Table 5. Outcomes of the assessment methods used in Example 5

GPA index	The first system demonstrated a better overall quality performance and a better quality performance at steps R_1 and R_3, while the second system performed better at step R_2
TFNs or GNs	The first system demonstrated a better mean performance at all the steps of the CBR process

Example 6*:* Six different users of a CBR system ranked with scores from 0-100 the effectiveness of its following five past cases for solving new related problems:

C_1 (Case 1)**:** 43, 48, 49, 49, 50, 52, C_2**:** 81, 83. 85, 88, 91, 95, C_3: 76, 82, 89, 95, 95, 98, C_4: 86, 86, 87, 87, 87, 88, C_5: 35, 40, 44, 52, 59, 62.

The *mean value* of the given, 6X5 = 30 in total numerical scores is approximately equal to 72.07 demonstrating a good (C) mean performance of the system with respect to the above five past cases. For reasons of comparison of the two methods the system's mean performance will also be calculated by using GNs.

In fact, the given numerical scores correspond to 14 GNs equal to the GN A, 4 equal to B, 1 equal to C, 4 equal to D and 7 GNs equal to the GN F. The mean value of the above GNs is equal to

$$M^* = \frac{1}{30}(14A + 4B + C + 4D + 7F) \in [60.33, 79.63].$$

Therefore, from equation (13) one obtains that $W(M^*)$ = 69.98. Consequently the CBR system demonstrates a good mean performance with respect to the given five past cases. However the exact score corresponding to the system's mean performance is equal to the mean value 72.02 of the given numerical scores. In concluding, the assessment methods using GNs and TFNs, although they give a satisfactory approximation of the system's mean performance, they are practically useful only when the effectiveness of the system's past cases is evaluated by qualitative grades and not by numerical scores, because in this case the calculation of the mean value of those grades in the traditional way is not possible.

Our last example illustrates a method of solving algebraic systems with grey (or fuzzy) coefficients.

Example 7: The performance of a CBR system in solving two different problems is represented by the GNs A and B respectively, for which we know that A + B $\in$ [2, 4] and A.B $\in$ [1, 3]. It is asked to compare the system's performance in those two cases with its ideal performance, to which the numerical score 5 has being assigned.

Solution: The whitening of the given sum and product of the GNs A and B leads to the ordinary system $\left.\begin{matrix} x+y=3 \\ x.y=2 \end{matrix}\right\}$, which has the solutions x =2, y = 1 and x =1, y = 2. Therefore, one can take *W*(A) = 2 and *W*(B) =1 or *W*(A) = 1 and *W*(B) = 2. Consequently, the system's performance in the above two cases was equal to the 40% and the 20% respectively of its ideal performance.

Remark: The attempt to solve the above problem by applying the definitions of addition and multiplication of GNs presented before, leads to a system of equations with respect to the unknown numbers a_1, b_1, a_2, b_2 , which does not have a concrete solution. For more details on solving systems of equations with grey data see [49].

Concluding Remarks

A method using GNs as tools was developed for assessing a system's mean performance, which is useful when using qualitative grades and not numerical scores for this purpose. This new method was proved to be equivalent with an analogous method using TFNs instead of GNs developed in earlier works, but it reduces significantly the required computational burden, since it requires the calculation of two components only (instead of three in case of the TFNs) for obtaining the mean value of the GNs. Examples were also presented on the assessment of CT systems illustrating our results and showing that the system's quality performance, calculated by the traditional GPA index, may lead to different assessment conclusions.

Although the enormous development of technology during the last years makes easier and more comfortable the human life, it creates in parallel more and more complicated artificial systems, which are difficult to be managed by the traditional scientific methods. As a result, the applications of FL and of the GS theory have been rapidly expanded nowadays covering almost all sectors of the human activities. In particular the FNs and the GNs have been proved to be effective tools in handling approximate and /or uncertain data, playing an important role in fuzzy mathematics and in GS theory respectively, analogous to the role played by the ordinary numbers in crisp mathematics. Therefore, the attempt to use FNs or / and GNs in other applications, apart from the assessment purposes, to science and technology and to the everyday life, looks as a very interesting and promising direction for future research on the subject. Note that such efforts have been already started by the present author on solving equations, systems of equations and linear programming problems with fuzzy or grey data, connected to real life applications [50 - 52].

DISCUSSION AND CONCLUSION

Based on what was reported in this chapter, one can lead to the following conclusions:

- CT is a prerequisite for PS when tackling complex real life and technological problems. Modelling thinking is a fundamental component of CT, because it synthesizes all the other existing in it modes of thinking and applies them to construct the real system's model for solving the corresponding problem. The CBR systems are characteristic examples of situations where .CT plays an important role.

- The use of computers has become nowadays an increasingly essential tool for doing scientific research. Recent studies in this field address the necessity to become trained in thinking computationally before learning programming, However, the student practice with numerical, algebraic and analytic calculations, with the solution of problems and the rediscovery of proofs of the mathematical theorems must be continued for ever. Otherwise people will gradually loose the sense of numbers and symbols, the sense of space and time, and they will become unable to create new knowledge and technology.

- In the present work assessment methods are studied of student CT skills under fuzzy conditions. Those methods include utilization of principles of the Zadeh's FL (calculation of the uncertainty in a fuzzy environment, use of the TFNs) and of the Deng's GS theory (use of GNs).

- A few years ago Probability theory used to be the unique tool in hands of the scientists for dealing with situations characterized by uncertainty or / and vagueness. However, nowadays this is not the rule anymore. In fact, the introduction of FL, of the GS theory and of other relative theories (see the author's Commentary in the present book) provided new tools for dealing with such situations in real life, science and technology and enabled the solution of problems with fuzzy or approximate data that could not be solved with the traditional techniques of probability theory.

The target of our future research on the subject involves the effort to improve and extend our fuzzy and grey assessment methods and to apply the principles of FL and GS theory to other fields of human activity, like the solution of equations and algebraic systems with fuzzy or grey coefficients, the fuzzy and grey linear programming, the decision-making under fuzzy conditions, etc.

REFERENCES

[1] Balley, K. D., *Sociology and the New Systems Theory: Toward a Theoretical Synthesis*. New York: State of New York Press, 1994.

[2] Brule, J.F., *Fuzzy Systems – A Tutorial*, retrieved from http://austinlinks.com/Fuzzy/tutorial.html, 2016.

[3] Deng, J., Control Problems of Grey Systems, *Systems and Control Letters,* 288-294, 1982.

[4] Deng, J., Introduction to Grey System Theory, *The Journal of Grey System*, 1, 1-24, 1989.

[5] Fox, J., Towards a reconciliation of fuzzy logic and standard logic, *Int. J. of Man - Machine Studies,* 15, 213-220, 1981.

[6] Green, A. J. K. & Gillhooly, K., Problem solving, in Braisby, N. & Gelatly, A. (Eds.), *Cognitive Psychology*, Oxford University Press, Oxford, 2005.

[7] Haack, S., Do we need fuzzy logic? *Int. J. of Man-Machine Studies,* 11, 437-445, 1979.

[8] Haines, C. & Crouch, R., Remarks on a Modelling Cycle and Interpretation of Behaviors, in Lesh, R.A. et al. (Eds.); *Modelling Students' Mathematical Modelling Competencies (ICTMA 13)*, 145-154, Springer, USA, 2010.

[9] Kaufmann, A. & Gupta, M., *Introduction to Fuzzy Arithmetic*, Van Nostrand Reinhold Company, New York, 1991.

[10] Kazimoglu, C., Kiernan, M., Bacon, L. & MacKinnon, L., Understanding Computational Thinking Before Programming: Developing Guidelines for the Design of Games to Learn Introductory Programming through Game-Play, *International Journal of Game-Based Learning*, 1(3), 30-52, 2011.

[11] Klir, G. J. & Folger, T. A., *Fuzzy Sets, Uncertainty and Information*, Prentice-Hall, London, 1988.

[12] Klir, G. J., Principles of Uncertainty: What are they? Why do we need them? *Fuzzy Sets and Systems,* 74(1), 15-31, 1995.

[13] Kolodner, J., Reconstructive Memory: A Computer Model, *Cognitive Science*, 7, 281-328, 1983.

[14] Korner, S., Laws of Thought, *Encyclopaedia of Philosophy*, 4, 414-417, New York: Mac Millan, 1967.

[15] Koton, Ph., *Using experience in learning and problem solving*, MIT, Laboratory of Computer Science (Ph.D. Thesis.), MIT/LCS/TR-441, 1989.

[16] Lebowitz, M., Memory-Based Parsing, *Artificial Intelligence*, 21, 363-404, 1983.

[17] Lejewski, C., Jan Lukasiewicz, *Encyclopedia of Philosophy*, 5, 104-107, New York: Mac Millan, 1967.

[18] Liu, S. F. & Lin, Y. (Eds.), *Advances in Grey System Research*, Berlin – Heidelberg: Springer, 2010.

[19] Liu, J. & Wang, L., Computational Thinking in Discrete Mathematics, *Proceedings of the IEEE 2nd International Workshop on Education Technology and Computer Science,* 413-416, 2010.

[20] Mc Guinness, C., Teaching thinking: New signs for theories of cognition, *Educational Psychology,* 13(3-4), 305-316, 1993.

[21] Mc Peck, J. E., *Critical thinking and education*, Martin Robinson, Oxford, 1981.

[22] Moore, R. A., Kearfort, R. B. & Clood, M. J., *Introduction to Interval Analysis*, 2nd Printing, Philadelphia, SIAM, 1995.

[23] Papert, S., An exploration in the space of Mathematics Education, *International Journal of Computers for Mathematics*, 1(1), 95-123, 1996.

[24] Pollak, H. O., The interaction between Mathematics and other school subjects; *New Trends in Mathematics Teaching*, Volume IV, Paris, UNESKO, 1979.

[25] Porter, B. & Bareiss, B., PROTOS: An experiment in knowledge acquisition for heuristic classification tasks, *Proceedings of the 1st Intern. Meeting on Advances in Learning* (IMAL), 159-174, Les Arcs, France, 1986.

[26] Rissland, E., Examples in legal reasoning: Legal hypotheticals, in *Proceedings of the 8th International Joint Conference on Artificial Intelligence* (IJCAI), Karlsruhe, 1983.

[27] Sakawa, M., *Fuzzy Sets and Interactive Multiobjective Optimization*, Plenum press, NY and London, 1993.

[28] Shackle, G. L. S., *Decision, Order and Time in Human Affairs*, Cambridge University Press, Cambridge, UK, 1961.

[29] Schank, R., *Dynamic memory: A theory of reminding and learning in computers and people,* Cambridge Univ. Press, 1982.

[30] Shannon, A mathematical theory of communications, *Bell Systems Technical Journal,* 27, 379-423 and 623-656, 1948.

[31] Stanfill, C. & Waltz, D., The memory-based reasoning paradigm, in: *Case-based reasoning, Proceedings from a workshop*, 414-424, Morgan Kaufmann Publ., Clearwater Beach, Florida.

[32] Subbotin, I. Ya., Badkoobehi, H. & Bilotckii, N. N., Application of fuzzy logic to learning assessment, *Didactics of Mathematics: Problems and Investigations*, 22, 38-41, 2004.

[33] Taha, H. A., *Operations Research – An Introduction;* Second Edition, Collier Macmillan, N.Y., London, 1967.

[34] Umbers, G. & King, P. J., An analysis of human decision-making in cement kiln control and implications for automation, *Int. J. of Man-Machine Studies*, 12, 11-23, 1980.

[35] van Broekhoven, E. & De Baets, B., Fast and accurate centre of gravity defuzzification of fuzzy systems outputs defined on trapezoidal fuzzy partitions, *Fuzzy Sets and Systems,* 157(7), 904-918, 2006.

[36] Voskoglou, M. Gr., An application of Markov Chain to the process of Modelling; *International Journal of Mathematical Education in Science and Technology*, 25(4), 475-480, 1994.

[37] Voskoglou, M. Gr., An Application of Fuzzy Sets to the Process of Learning, *Heuristics and Didactics of Exact Sciences,* 10, 9-13, 1999.

[38] Voskoglou, M. Gr., A stochastic model for the modelling process, in Haines, C., Galbraith, P., Blum, W., Khan, S. (Eds.); *Mathematical*

Modelling: Education, Engineering and Economics (ICTMA 12), 149-157, 2006.

[39] Voskoglou, M. Gr., Problem solving from Polya to nowadays: A review and future perspectives, in Nata, R. V. (Ed.); *Progress in Education*, Nova Publishers: New York, N. Y.; Vol. 22, Chapter 4, pp. 65-82, 2011.

[40] Voskoglou, M. Gr., Case-Based Reasoning: History, Methodology and Development Trends, in Leeland, A. M. (Ed.); *Case-Based Reasoning: Processes, Suitability and Applications*. Nova Publishers: New York, N. Y.; Chapter 3, pp. 59-76, 2011.

[41] Voskoglou, M. Gr. & Buckley, S., Problem Solving and Computers in a Learning Environment, *Egyptian Computer Science Journal*, 36(4), 28-46, 2012.

[42] Voskoglou, M. Gr. & Salem A. B. M., Analogy-Based and Case-Based Reasoning: Two Sides of the Same Coin, *International Journal of Applications of Fuzzy Sets and Artificial Intelligence*, 4, 7-18, 2014.

[43] Voskoglou, M. Gr., Use of the Triangular Fuzzy Numbers for Student Assessment, *American Journal of Applied Mathematics and Statistics,* 3(4), 146-150, 2015.

[44] Voskoglou, M. Gr., Mathematical modelling as a teaching method of mathematics, *Journal for Research in Innovative Teaching* (National University, CA) 8(1), 35-50, 2015.

[45] Voskoglou, M. Gr., Teaching Mathematics or Mathematical Modelling? An Answer to a Comment, *ICTMA Newsletter*, 8(1), 12-13, 2015.

[46] Voskoglou, M. Gr., Computational Thinking and Fuzzy Logic in Problem Solving, in K. Newton (Ed.), *Problem- Solving: Strategies, Challenges and Outcomes*, Nova Publishers New York, N. Y., Chapter 1, 1-30, 2016.

[47] Voskoglou, M. Gr., *Finite Markov Chain and Fuzzy Logic Assessment Models: Emerging Research and Opportunities*, Columbia, SC, Createspace.com.– Amazon, 2017.

[48] Voskoglou, M. Gr. & Theodorou, Y., Application of Grey Numbers to Assessment Processes, *International Journal of Applications of Fuzzy Sets and Artificial Intelligence,* 7, 59-72, 2017.

[49] Voskoglou, M. Gr., Solving Systems of Equations with Grey Data, *International Journal of Applications of Fuzzy Sets and Artificial Intelligence*, 8, 103-111, 2018.

[50] Voskoglou, M. Gr., Solving Linear Programming Problems with Grey Data, *Oriental Journal of Physical Sciences*, 3(1), 17-23, 2018.

[51] Voskoglou, M. Gr., Fuzzy Linear Programming, *Egyptian Journal of Computer Science*, 42(3), 1-14, 2018.

[52] Voskoglou, M. Gr., Systems of Equations with Fuzzy Coefficients, *Journal of Physical Sciences,* 23, 77-88, 2018.

[53] Wikipedia, *Computational Thinking,* retrieved from http://en.wikipedia.org/wiki/Computational_thinking, 2017.

[54] Wing, J. M., Computational thinking, *Communications of the ACM*, 49, 33-35, 2006.

[55] Zadeh, L. A., Fuzzy Sets, *Information and Control*, 8, 338-353, 1965.

[56] Zadeh, L. A., Outline of a new approach to the analysis of complex systems and decision processes, *IEEE Transactions on Systems, Man and Cybernetics*, 3, 28–44, 1973.

[57] Zadeh, L. A., The role of fuzzy logic in the management of uncertainty of expert systems, *Memorandum No. UCB/ERL M83/41*, University of California, Berkeley, 1983.

[58] Zadeh, L. A., The Concept of a Linguistic Variable and its Application to Approximate Reasoning, Parts 1-3, *Information Science*, 8, 199-249, 301-357, and 9, 43-80, 1975.

[59] Zadeh, L. A., From Search Engines to Question-Answering Systems – The Role of Fuzzy Logic, *Progress in Informatics*, 1, 1-3, 2005.

[60] Zadeh, L. A., Toward a Generalized Theory of Uncertainty (GTU) – An Outline, *Information Sciences*, 172, 1-40, 2005.

In: An Essential Guide to Fuzzy Systems ISBN: 978-1-53616-128-1
Editor: Michael Gr. Voskoglou

Chapter 2

EVALUATION OF FUZZY DATA AND FUZZY RELATION EQUATIONS

Michael Gr. Voskoglou*
Graduate Technological Educational Institute of Western Greece
Koukouli, Patras, Greece

ABSTRACT

The Chapter at hand is focused on two targets. The first target is the presentation of a method for the management of fuzzy data and the second one is the use of Fuzzy Relation Equations as an assessment tool of a group's performance. The data obtained from the operation mechanisms of large and complex systems connected to science and real life applications are frequently characterized by uncertainty and fuzziness. A model is proposed in this work for evaluating such kind of data with the help of the corresponding membership degrees and possibilities. The general character of that model enables its application to a variety of human and machine activities. Examples are presented here for the process of learning a subject-matter in the classroom and for a market's research illustrating its usefulness in practical problems. On the other hand Fuzzy Relation Equations, which are associated with the

* Corresponding Author's Email: mvoskoglou@gmail.com.

composition of Fuzzy Binary Relations, have currently become a valuable tool for applications of fuzzy mathematics to many topics. Here, after presenting the headlines of the corresponding theory, an application is performed on the assessment of student mathematical modelling skills.

Keywords: Fuzzy Set (FS), Membership Degree, Possibility, Fuzzy Data, Fuzzy Variables, Fuzzy Binary Relation (FBR), Fuzzy Relation Equations (FRE), Mathematical Modelling (MM)

INTRODUCTION

In large and complex systems that frequently appear nowadays in science, technology and real life situations (like the socio – economic, the biological ones, etc.) many different and constantly changing factors are usually involved, the relationships among which are indeterminate. As a result the data obtained from their operation mechanisms cannot be easily determined precisely and in practice estimates of them are used. While 50-60 years ago the unique tool in the hands of the scientists for handling such kind of data and situations of uncertainty in general used to be the theory of *Probability*, nowadays the *Fuzzy Set (FS)* theory, initiated by Zadeh in 1965 [12], and its extensions and generalizations that followed in the recent years (see the author's Commentary in this book), have given a new dynamic to this field.

In the Chapter at hand a method is developed for evaluating a system's fuzzy data in terms of the corresponding fuzzy possibilities. Also Fuzzy Relation Equations are used as a tool for assessing student mathematical modeling skills. The rest of the Chapter is organized as follows: In the next Section a general model is presented for evaluating fuzzy data and an application is provided on learning mathematics illustrating the model's applicability in real situations. The general model is further extended for studying the combined results of the evaluation of fuzzy data obtained from two (or more) different sources and an example is presented on a market's research to emphasize the usefulness of this extension for tackling real life problems. In the third Section the headlines of the theory of Fuzzy

Relation Equations are introduced, a model is developed that uses them as a tool for assessing mathematical model building abilities, and examples are provided illustrating several special situations that they appear. The Chapter closes with the final conclusions stated in its last Section.

EVALUATION OF FUZZY DATA

The General Model

The reader is supposed to be familiar to the basics of the FS theory and the book [5] is proposed as a general reference on the subject.

Assume that one wants to study a system's behavior consisting of n components (objects), $n \geq 2$, during a process involving vagueness and/or uncertainty. Denote by S_i, i = 1,2,3 the main steps of that process[1] and by *a, b, c, d, e* the linguistic labels of very low, low, intermediate, high, and very high success respectively of the system components in each step.

Set $U = \{a, b, c, d, e\}$. A FS A_i in U will be associated to each step S_i, i = 1, 2, 3. For this, if n_{ia}, n_{ib}, n_{ic}, n_{id}, n_{ie} denote the numbers of the system components that faced very low, low, intermediate, high, and very high success respectively at stage S_i, we define the *membership degree* $m_{A_i}(x)$ of each x in U by

$$m_{A_i}(x) = \frac{n_{ix}}{n} \tag{1}$$

Then the FS A_i in U associated to S_i is of the form:

$$A_i = \{(x, m_{A_i}(x)): x \in U\}, \ i=1, 2, 3 \tag{2}$$

---

[1] For reasons of simplicity we consider here three steps, but the model holds in general for a finite number of steps.

In order to represent all possible *profiles (overall states)* of the system components during the corresponding process a *fuzzy relation*, say R, in U^3 (i.e., a FS in U^3) is considered of the form:

$$R= \{(s, m_R(s)): s=(x, y, z) \in U^3\} \tag{3}$$

Usually in practical applications the degree of success of each system's component in a certain step of the process depends upon the degree of its success in the previous step. Under this assumption and in order to define properly the membership function m_R, the following definition is given:

Definition: A profile $s=(x, y, z)$, with x, y, z in U, is said to be *well ordered* if x corresponds to a degree of success equal or greater than y and y corresponds to a degree of success equal or greater than z.

For example, (c, c, a) is a well ordered profile, while (b, a, c) is not. The membership degree of a well ordered profile s is defined now to be equal to the product

$$m_R(s) = m_{A_1}(x).\, m_{A_2}(y).\, m_{A_3}(z) \tag{4}$$

On the contrary, the degree of the profiles which are not well ordered is defined to be zero. In fact, if for example the profile (b, a, c) possessed a nonzero membership degree, then at least one of the system components demonstrating a very low performance at step S_2 would perform satisfactorily at the next step S_3, which is impossible to happen.

However, they are also real situations in which the performance of each component at each step does not depend on its performance in the previous steps (e.g., see Example 2). In such cases the membership degrees of all the profiles are defined by equation (4).

Next, for reasons of brevity we shall write m_s instead of $m_R(s)$. Then the *fuzzy probability* p_s of the profile s is defined by

$$P_s = \frac{m_s}{\sum_{s \in U^3} m_s} \tag{5}$$

However, according to the British economist Shackle [8] and many other researchers after him, the human behavior can be better studied by using the possibilities rather of the several profiles, than their probabilities. The *possibility* r_s of the profile *s* is defined by

$$r_s = \frac{m_s}{\max\{m_s\}} \tag{6}$$

In equation (6) *max* $\{m_s\}$ denotes the greatest value of m_s for all *s* in U^3. In other words the possibility of *s* expresses the “relative membership degree” of *s* with respect to *max* $\{m_s\}$.

The following application to the process of *learning* a subject matter in the classroom illustrates the applicability of the present model to real life situations:

Example 1: There is no doubt that learning is one of the fundamental components of human cognitive action. There are many different theories and models developed by psychologists, educators and other cognitive scientists for the description of the mechanisms of learning, Nevertheless, although the process of learning differs in details from person to person, it is in general accepted that it involves *representation* and *interpretation* of the input data in order to produce the new knowledge (step S_1), *generalization* of this knowledge to a variety of situations (step S_2) and *categorization* of the generalized knowledge by embodying it to the individual’s appropriate cognitive structures, widely termed as *schemas of knowledge* (step S_3). In this way the individual becomes able to derive from memory the suitable in each case piece of knowledge for facilitating the solution of related composite and complex problems (e.g., see [11]).

On the other hand the process of learning is usually connected with uncertainty and vagueness. In fact, the learner is in many cases not sure about the good understanding of a new concept or topic and also the teacher is in doubt about the degree of acquisition of a new subject matter by students. Consequently, the use of principles of the FS theory could be a valuable tool in the effort of a more effective description of the mechanisms of learning.

The following experiment took place some time ago at the Graduate Technological Educational Institute of Western Greece, in the city of Patras, during the teaching (in three teaching hours) of the definite integral to a group of 35 students of the School of Management and Economics.

In the instructor's short introduction during the first teaching hour the concept of the definite integral was introduced through the need of calculating the area between a curve and the x-axis, but the fundamental theorem of the integral calculus connecting the indefinite with the definite integral of a continuous in a closed interval function was stated without proof. Then the students were left to work alone on their papers and the instructor was inspecting their efforts and reactions giving to them from time to time the proper hints and instructions. His intension was to help students to understand the basic methods of calculating a definite integral in terms to the already known methods for the indefinite integral (step S_I of the model).

It was observed that 17, 8 and 10 students respectively achieved intermediate, high and very high understanding of the new subject. In other words, in terms of the model, one obtains that $n_{ia} = n_{ib} = 0$, n_{ic}=17, $n_{id} = 8$ and n_{ie} = 10. Therefore the step of representation-interpretation of the process of learning can be represented as a FS in U in the form

$$A_I = \{(a, 0), (b, 0), (c, \tfrac{17}{35}), (d, \tfrac{8}{35}), (e, \tfrac{10}{35})\}.$$

At the second teaching hour a series of exercises involving the calculation of improper integrals as limits of definite integrals and of the area under a curve (or among curves) was given to students for solution.

The target in that case was to help students to generalize the new knowledge to a variety of situations (step S_2 of the model). Working in the same way as above it was found that the step of generalization can be represented as a FS in U in the form

$$A_2 = \{(a, \frac{6}{35}), (b, \frac{6}{35}), (c, \frac{16}{35}), (d, \frac{7}{35}), (e, 0)\}.$$

At the third teaching hour a number of composite problems was forwarded to students for solution involving applications to economics, like the calculation of the present value in cash flows, of the consumer's and producer's surplus resulting from the change of prices of a given good, of probability density functions, etc. ([2], Chapter 17). The target this time was to help students to relate the new information to their existing schemas of knowledge (step S_3 of the model). In that case it was found that the step of categorization can be represented as a FS in U in the form

$$A_3 = \{(a, \frac{12}{35}), (b, \frac{10}{35}), (c, \frac{13}{35}), (d, 0), (e, 0)\}.$$

Then the membership degrees of all student profiles involved in the fuzzy relation (3) were calculated. For example, for $s = (c, b, a)$ one finds that $m_s = m_{A_1}(c).\ m_{A_2}(b).\ m_{A_3}(a) = \frac{17}{35} \text{x} \frac{6}{35} \text{x} \frac{12}{35} \approx 0.029$.

It turns out that the profile (c, c, c) possesses the greatest membership degree, which is equal to 0.082. Therefore the possibility of each profile s is calculated by $r_s = \frac{m_s}{0.082}$. For example, the possibility of (c, b, a) is equal to $\frac{0.029}{0.082} \approx 0.353$, while the possibility of (c, c, c) is equal to 1, etc.

The total number of the student profiles is obviously equal to the total number of the ordered samples with replacement of three objects taken from five, i.e., equal to 5^3. Among all those profiles the profiles possessing non zero membership degrees and their possibilities are presented in Table 1. All calculations have been made with accuracy up to the third decimal point. The fuzzy data of Table 1 give not only quantitative information, but also a qualitative view of the student behavior in the classroom during the learning process. This is obviously very useful to the instructor for organizing his/her future teaching plans.

Table 1. Student profiles with non zero membership degrees

A_1	A_2	A_3	m_s	r_s
c	*c*	*c*	0.082	1
c	*c*	*a*	0.076	0.927
c	*c*	*b*	0.063	0.768
c	*a*	*a*	0.028	0.341
c	*b*	*a*	0.028	0.341
c	*b*	*b*	0.024	0.293
d	*d*	*a*	0.016	0.195
d	*d*	*b*	0.013	0.159
d	*d*	*c*	0.021	0.256
d	*a*	*a*	0.013	0.159
d	*b*	*a*	0.013	0.159
d	*b*	*b*	0.011	0.134
d	*c*	*a*	0.031	0.378
d	*c*	*b*	0.026	0.317
d	*c*	*c*	0.034	0.415
e	*a*	*a*	0.017	0.207
e	*b*	*b*	0.014	0.171
e	*c*	*a*	0.039	0.476
e	*c*	*b*	0.033	0.402
e	*c*	*c*	0.042	0.512
e	*d*	*a*	0.025	0.305
e	*d*	*b*	0.021	0.256
e	*d*	*c*	0.027	0.329

Combined Results of Fuzzy Data

Frequently in practical applications it becomes necessary to study the *combined results* of the behavior of k different groups of a system's components, $k \geq 2$, during the same process (e.g., the combined performance of two or more student classes in solving the same problems).

For measuring the degree of evidence of the combined results of the *k* groups it is necessary to define the *combined probability p(s)* and the *combined possibility r(s)* of each profile *s* with respect to the membership degrees of *s* in all the groups involved. The values of *p(s)* and *r(s)* can be defined with respect to the *pseudo-frequency*

$$f(s) = \sum_{t=1}^{k} m_s(t) \tag{7}$$

and they are equal to

$$p(s) = \frac{f(s)}{\sum_{s \in U^3} f(s)} \tag{8}$$

and

$$r(s) = \frac{f(s)}{\max\{f(s)\}} \tag{9}$$

respectively, where *max{f(s)}* denotes the maximal pseudo-frequency.

Obviously the same procedure could be applied, if one wanted to study the combined results of the behavior of a single group during *k* different activities (e.g., the combined performance of a student class during the solution of two or more different problems).

The following example on a *market's research* illustrates the importance of the above procedure:

Example 2: A company performed a market's research about the degree of the consumer preference for its negotiable products, which was characterized by the above defined fuzzy linguistic labels *a, b, c, d, e*. The research was performed separately for men and women and for three different categories of age, namely C_1: 18-30 years, C_2: 31-50 years and C_3: over 50 years old.

Table 2. Customers' profiles with non zero pseudo-frequencies

A_1	A_2	A_3	$m_s(1)$	$m_s(2)$	$f(s)$	$r(s)$
b	*b*	*b*	0	0.016	0.016	0.092
b	*a*	*b*	0	0.012	0.012	0.069
b	*c*	*b*	0	0.032	0.032	0.184
b	*b*	*a*	0	0.021	0.021	0.121
b	*b*	*c*	0	0.016	0.016	0.092
b	*a*	*a*	0	0.016	0.016	0.092
b	*a*	*c*	0	0.012	0.012	0.069
b	*c*	*a*	0	0.042	0.042	0.241
b	*c*	*c*	0	0.032	0.032	0.184
c	*c*	*c*	0.072	0.080	0.152	0.874
c	*a*	*c*	0.082	0.030	0.112	0.644
c	*b*	*c*	0.031	0.040	0.071	0.408
c	*d*	*c*	0.046	0	0.046	0.264
c	*c*	*a*	0.067	0.107	0.174	1
c	*c*	*b*	0.056	0.008	0.064	0.368
c	*a*	*a*	0.028	0.040	0.068	0.391
c	*a*	*b*	0.024	0.030	0.054	0.310
c	*b*	*a*	0.028	0.053	0.081	0.466
c	*b*	*b*	0.024	0.040	0.064	0.368
c	*d*	*a*	0.043	0	0.043	0.247
c	*d*	*b*	0.036	0	0.036	0.207
d	*d*	*a*	0.020	0	0.020	0.115
d	*d*	*b*	0.017	0	0.017	0.098
d	*d*	*c*	0.022	0	0.022	0.126
d	*a*	*a*	0.013	0.024	0.037	0.213
d	*a*	*b*	0.011	0.018	0.029	0.167
d	*a*	*c*	0.015	0.018	0.033	0.190
d	*b*	*a*	0.013	0.032	0.045	0.259
d	*b*	*b*	0.011	0.024	0.035	0.201
d	*b*	*c*	0.014	0.024	0.038	0.218
d	*c*	*a*	0.031	0.064	0.095	0.546
d	*c*	*b*	0.026	0.048	0.074	0.425
d	*c*	*c*	0.034	0.048	0.082	0.471
e	*a*	*a*	0.017	0	0.017	0.098
e	*a*	*b*	0.014	0	0.014	0.080
e	*a*	*c*	0.018	0	0.018	0.103
e	*b*	*a*	0.017	0	0.017	0.098
e	*b*	*b*	0.014	0	0.014	0.080
e	*b*	*c*	0.018	0	0.018	0.103
e	*c*	*a*	0.039	0	0.039	0.224
e	*c*	*b*	0.033	0	0.033	0.190
e	*c*	*c*	0.042	0	0.042	0.241
e	*d*	*a*	0.025	0	0.025	0.144
e	*d*	*b*	0.021	0	0.021	0.121
e	*d*	*c*	0.027	0	0.027	0.155

Denote by A_1 (t), A_2 (t) and A_3 (t) respectively the FSs representing the consumers' degree of preference for each of the above three categories of age, where the variable t takes the values t = 1 for men and t = 2 for women. Such kind of FSs, whose entries depend on the values of a variable, are usually referred as *fuzzy variables.*

According to the collected data the FSs A_i (t), for i = 1, 2, 3 and t = 1, 2, were found to be the following:

$A_1(1) = \{(a, 0), (b, 0), (c, 0.486), (d, 0.228), (e, 0.286)\}$
$A_2(1) = \{(a, 0.171), (b, 0.171), (c, 0.4), (d, 0.257), (e, 0)\}$
$A_3(1) = \{(a, 0.343), (b, 0.0286), (c, 0.371), (d, 0), (e, 0)\}$
$A_1(2) = \{(a, 0), (b, 0.2), (c, 0.5), (d, 0.3), (e, 0)\}$
$A_2(2) = \{(a, 0.2), (b, 0.267), (c, 0.533), (d, 0), (e, 0)\}$
$A_3(2) = \{(a, 0.4), (b, 0.3), (c, 0.3), (d, 0), (e, 0)\}$

In this example the degree of the customer preferences in each age category does not depend on the previous categories. Therefore the calculation of the membership degrees of all the customer profiles is done by the product law of equation (4). For example, for the profile $s = (c, c, a)$ one finds that

$m_s(1) = 0.486 \text{ x } 0.4 \text{ x } 0.343 \approx 0.67$ and
$m_s(2) = 0.5 \text{ x } 0.5 \text{ x } 0.33 \approx 0.107.$

It turns out that the above profile has the greater pseudo-frequency $f(s) = 0.67 + 0.107 = 0.174$ and therefore its combined possibility is equal to 1, while the combined possibilities of all the other profiles are calculated by $r(s) = \frac{f(s)}{0.174}$.

The membership degrees, the pseudo-frequencies and the combined possibilities of all the customer profiles with nonzero pseudo-frequencies are presented in Table 2, where all calculations have been made with accuracy up to the third decimal point. The fuzzy data of Table 2 give to the managers of the company a detailed idea about the consumers' preferences for its products.

APPLICATION OF FUZZY RELATION EQUATIONS TO MATHEMATICAL MODELLING

Mathematical Modelling

Mathematical Modelling (MM) appears today as a dynamic tool for teaching and learning mathematics, because it connects mathematics with our everyday life problems giving the possibility to students to understand its usefulness in practice and therefore increasing their interest about it [9]. Details about the MM process are presented in Chapter 1 of this book. It is recalled here that the main steps of the MM process involve:

- S_1: Analysis of the problem (understanding the statement and recognizing the restrictions and requirements of the real system).
- S_2: Mathematization (formulation of the problem and construction of the model).
- S_3: Solution of the model.
- S_4: Validation (control) of the model and implementation of the final mathematical results to the real system, in order to give the "answer" to the real world problem.

In this Section we are going to apply the theory of *Fuzzy Relation Equations (FRE)* for assessing MM skills. Notice that the validation of the model and its implementation to the real system are considered by most authors as two distinct steps of the MM process, but we have joined them here (without loss of generality) in one step, in order to make our application simpler.

Fuzzy Relation Equations

FRE are associated with the composition of *Fuzzy Binary Relations (FBRs)*. A FBR is defined as follows:

Definition 1: Let X, Y be two crisp sets. Then a FBR R(X, Y) is a FS on the Cartesian product X x Y of the form: R(X, Y) = {(r, m_R(r): r = (x, y) $\in$ X x Y}, where m_R : X x Y $\rightarrow$ [0, 1] is the corresponding membership function.

When X = {x_1,………,x_n} and Y = {y_1,……, y_m}, then a FBR R(X, Y) can be represented by a n X m matrix of the form:

$$R = \begin{matrix} & \begin{matrix} y_1 & \ldots\ldots\ldots\ldots & y_m \end{matrix} \\ \begin{matrix} x_1 \\ \cdot \\ \cdot \\ \cdot \\ x_n \end{matrix} & \begin{pmatrix} r_{11} & \cdot & \cdot & \cdot & r_{1n} \\ \cdot & \cdot & \cdot & \cdot & \cdot \\ \cdot & \cdot & \cdot & \cdot & \cdot \\ \cdot & \cdot & \cdot & \cdot & \cdot \\ r_{m1} & \cdot & \cdot & \cdot & r_{mn} \end{pmatrix} \end{matrix} = [r_{ij}],$$

where r_{ij} = m_R (x_i, y_j), with i = 1,…, n and j =1,…, m. The matrix R is called the *membership matrix* of the FBR R(X, Y).

The basic ideas of fuzzy relations, which were introduced by Zadeh [13] and were further investigated by other researchers, are extensively covered in the book [4].

Definition 2: Consider two FBR P(X, Y) and Q(Y, Z) with a common set Y. Then, the standard composition of these relations, denoted by P(X, Y) $\circ$Q(Y, Z), produces a FBR R(X, Z) with membership function m_R defined by:

$$m_R(x_i, z_j) = \underset{y \in Y}{Max} \ \min [m_P(x_i, y), m_Q(y, z_j)] \tag{10}$$

for all i=1,…,n and all j=1,…,m. This composition is usually referred as *max-min composition.*

Compositions of FBRs are conveniently performed in terms of the membership matrices of them. In fact, if P=[p_{ik}] and Q=[q_{kj}] are the membership matrices of the FBRs P(X, Y) and Q(Y, Z) respectively, then by equation (10) we get that the membership matrix of R(X, Y) = P(X, Y) $\circ$ Q(Y, Z) is the matrix R = [r_{ij}], with

$$r_{ij} = \underset{k}{Max}\ \min(p_{ik}, q_{kj}) \qquad (11)$$

Example 3: If P = $\begin{array}{c} \\ x_1 \\ x_2 \\ x_2 \end{array}\begin{array}{c} \begin{array}{ccc} y_1 & y_2 & y_3 \end{array} \\ \begin{pmatrix} 0.2 & 0.4 & 0.8 \\ 0.1 & 0.5 & 1 \\ 0.4 & 0.7 & 0.3 \end{pmatrix} \end{array}$ and Q = $\begin{array}{c} \\ y_1 \\ y_2 \\ y_2 \end{array}\begin{array}{c} \begin{array}{cccc} z_1 & z_2 & z_3 & z_4 \end{array} \\ \begin{pmatrix} 0.2 & 0.7 & 0 & 0.4 \\ 0.8 & 0.1 & 0.5 & 0.6 \\ 1 & 0.3 & 0.2 & 0.9 \end{pmatrix} \end{array}$

are the membership matrices of the FBRs P(X, Y) and Q(Y, Z) respectively, then by equation (11) the membership matrix of R(X, Z) is the matrix

$$R = P \circ Q = \begin{array}{c} \\ x_1 \\ x_2 \\ x_2 \end{array}\begin{array}{c} \begin{array}{cccc} z_1 & z_2 & z_3 & z_4 \end{array} \\ \begin{pmatrix} 0.8 & 0.3 & 0.4 & 0.8 \\ 1 & 0.3 & 0.5 & 0.9 \\ 0.7 & 0.4 & 0.5 & 0.6 \end{pmatrix} \end{array}.$$

Observe that the same elements of P and Q are used in the calculation of m_R, as would be used in the regular multiplication of matrices, but the product and sum operations are here replaced with the min and max operations respectively.

Definition 3: Consider the FBRs P(X, Y), Q(Y, Z) and R(X, Z), defined on the sets

X = {x_i : i$\in$ N_n} , Y = {y_j : j $\in N_m$ }, Z= {z_k : k $\in$Ns}, where N_t = {1,2,…,t}, for t = n, m, k, and let P=[p_{ij}], Q=[q_{jk}] and R=[r_{ik}] be the membership matrices of P(X, Y), Q(Y, Z) and R(X, Z) respectively. Assume that the above three relations constrain each other in such a way that

$$P \circ Q = R \tag{12}$$

where $\circ$ denotes the max-min composition. This means that

$$\max_{j \in J} \min (p_{ij}, q_{jk}) = r_{ik} \tag{13},$$

for each i in N_n and each k in N_s. Therefore the matrix equation (12) encompasses nXs simultaneous equations of the form (13). When two of the components in each of the equations (13) are given and one is unknown, these equations are referred as FRE.

The notion of FRE was first proposed by Sanchez [7] and later was further investigated by other researchers ([1, 3, 6], etc.}.

A Study of MM Skills Using FRE

Let us consider the crisp sets X = {M}, Y = {A, B, C, D, F} and Z = $\{S_1, S_2, S_3, S_4\}$, where M denotes the imaginary notion of the *average student* of a class, A = Excellent, B = Very Good, C = Good, D = Fair and F = Failed are the linguistic grades used for the assessment of the student performance and S_1, S_2, S_3, S_4 are the steps of the MM process.

Further, let n be the total number of students of a class and let n_i be the numbers of students of the class who obtained the grade i assessing their performance, i $\in$ Y. Then one can represent the average student of the class as a FS on Y in the form

$$M = \{(i, \frac{n_i}{n}): i \in Y\}.$$

The FS M induces a FBR P(X, Y) with membership matrix

$$P = [\frac{n_A}{n} \ \frac{n_B}{n} \ \frac{n_C}{n} \ \frac{n_D}{n} \ \frac{n_F}{n}]$$

In an analogous way the average student of a class can be represented as a FS on Z in the form M′ = {(j, m(j): j∈ Z}, where m: Z→ [0, 1] is the corresponding membership function. In this case the FS M′ induces a FBR R(X, Z) with membership matrix

$$R = [m(S_1)\ m(S_2)\ m(S_3)\ m(S_4)].$$

We consider also the FBR Q(Y, Z) with membership matrix the 5X4 matrix Q = [q_{ij}], where q_{ij} = m_Q(i, j) with i∈ Y and j∈ Z and the FRE encompassed by the matrix equation $P \circ Q = R$.

When the matrix Q is fixed and the row-matrix P is known, then the equation (12) has always a unique solution with respect to R, which enables the representation of the average student of a class as a FS on the set of the steps of the MM process. This is useful for the instructor for designing his/her future teaching plans. On the contrary, when the matrices Q and R are known, then the equation (12) could have no solution, or could have more than one solution with respect to P, which makes the corresponding situation more complicated.

A Classroom Application and Other Examples

The model presented in the previous subsection will be illustrated here with a classroom application and other suitable examples.

The Classroom Application

The following experiment took place at the Graduate Technological Educational Institute of Western Greece, in the city of Patras, when I was teaching to a group of 60 students of the School of Technological

Applications (future engineers) the use of the derivative for the maximization and minimization of a function. A written test was performed after the end of the teaching process involving the two problems listed in the Appendix at the end of this paper. The results of the test are depicted in Table 3.

Table 3. Student Performance

Grade	Number of students
A	20
B	15
C	7
D	10
F	8
Total	60

Therefore, the average student M of the class can be represented as a FS on

Y = {A, B, C, D, F} by

$$M = \{(A, \tfrac{20}{60}), (B, \tfrac{15}{60}), (C, \tfrac{7}{60}), (D, \tfrac{10}{60}), (F, \tfrac{8}{60})\}$$

$$\approx \{(A, 0.33), (B, 0.25), (C, 0.12), (D, 0.17), (F, 0.13)\}$$

Thus M induces a FBR P(X, Y), where X = {M}, with membership matrix

$$P = [0.33\ 0.25\ 0.12\ 0.17\ 0.13].$$

Also, using statistical data of the last five academic years concerning the MM skills of the students of the School of Technological Applications of the Graduate Technological Educational Institute of Western Greece, we fixed the membership matrix Q of the binary fuzzy relation Q(Y, Z), where $Z = \{S_1, S_2, S_3, S_4\}$, in the form:

$$Q = \begin{array}{c} \\ A \\ B \\ C \\ D \\ F \end{array} \begin{array}{c} \begin{array}{cccc} S_1 & S_2 & S_3 & S_4 \end{array} \\ \begin{pmatrix} 0.7 & 0.5 & 0.3 & 0 \\ 0.4 & 0.6 & 0.3 & 0.1 \\ 0.2 & 0.7 & 0.6 & 0.2 \\ 0.1 & 0.5 & 0.7 & 0.5 \\ 0 & 0.1 & 0.5 & 0.8 \end{pmatrix} \end{array}$$

The statistical data were collected by the instructor who was inspecting the student reactions during the solution of several MM problems in the classroom.

Next, using the max-min composition of FBR one finds that the membership matrix of R(X, Z) = P(X, Y) o Q (Y, Z) is equal to R = P o Q = [0.33 0.33 0.3 0.17].

Therefore the average student of the class can be expressed as a fuzzy set on Z by

$$M = \{(S_1, 0.33), (S_2, 0.33), (S_3, 0.3), (S_4, 0.17)\}.$$

The conclusions obtained from the above expression of M are the following:

- Only the $\frac{1}{3}$ of the students of the class were ready to use contents of their memory (background knowledge, etc.) in order to facilitate the solution of the given problems.
- All the above students were able to design the model and almost all of them were able to execute the solutions of the given problems.
- On the contrary, half of the above students did not succeed to check the correctness of the solutions found and therefore to implement correctly the mathematical results to the real system.

The first conclusion was not surprising at all, since the majority of the students have the wrong habit to start studying the material of their courses the last month before the final exams. On the other hand, the second conclusion shows that the instructor's teaching procedure was successful enabling the diligent students to plan and execute successfully the solutions of the given problems. Finally, the last conclusion is due to the fact that students, when solving MM problems, they usually omit to check if their solutions are compatible to the restrictions imposed by the real world (for example see the two problems in the Appendix). Therefore, the instructor should emphasize during his/her lectures that the last step of the MM process (checking and implementation) is not a formality, but it has its own importance for preventing several mistakes.

Let us now consider the case where the membership matrices Q and R are known and we want to determine the matrix P representing the average student of the class as a FS on Y. This is a complicated case, because we may have more than one solution or no solution at all. The following two examples illustrate this situation:

Example 4: Consider the membership matrices Q and R of the previous application and set $P = [p_1\ p_2\ p_3\ p_4\ p_5]$. Then the matrix equation P o Q = R encompasses the following equations:

$$\max\{\min(p_1, 0.7), \min(p_2, 0.4), \min(p_3, 0.2),$$
$$\min(p_4, 0.1), (p_5, 0)\} = 0.33$$
$$\max\{\min(p_1, 0.5), \min(p_2, 0.6), \min(p_3, 0.7),$$
$$\min(p_4, 0.5), \min(p_5, 0.1)\} = 0.33$$
$$\max\{\min(p_1, 0.3), \min(p_2, 0.3), \min(p_3, 0.6),$$
$$\min(p_4, 0.7), (p_5, 0.5)\} = 0.3$$
$$\max\{\min(p_1, 0), \min(p_2, 0.1), \min(p_3, 0.2), \min(p_4, 0.5),$$
$$\min(p_5, 0.8)\} = 0.17$$

The first of the above equations is true if, and only if, $p_1 = 0.33$ or $p_2 = 0.33$, values that satisfy the second and third equations as well. Also, the fourth equation is true if, and only if, $p_3 = 0.17$ or $p_4 = 0.17$ or $p_5 = 0.17$. Therefore, any combination of values of p_1, p_2, p_3, p_4, p_5 in [0, 1] such that $p_1 = 0.33$ or $p_2 = 0.33$ and $p_3 = 0.17$ or $p_4 = 0.17$ or $p_5 = 0.17$ is a solution of P o Q = R.

Let S(Q, R) = {P: P o Q = R } be the set of all solutions of P o Q = R. Then one can define a partial ordering on S(Q, R) by $P \lesssim P' \Leftrightarrow p_i \lesssim p'_i$, $\forall \iota = 1, 2, 3, 4, 5$.

It is well established that whenever S(Q, R) is a non-empty set, it always contains a unique maximum solution and it may contain several minimal solutions [11]. It is further known that S(Q, R) is fully characterized by the maximum and minimal solutions in the sense that all its other elements are between the maximal and each of the minimal solutions [12]. A method of determining the maximal and minimal solutions of P o Q = R with respect to P is developed in [11].

Example 5: Let $Q = [q_{ij}]$, i = 1, 2, 3, 4, 5 and j = 1, 2, 3, 4 be as in Example 2 and let R = [1 0.33 0.3 0.17]. Then the first equation encompassed by the matrix equation P o Q = R is max {min (p_1, 0.7), min (p_2, 0.4), min (p_3, 0.2), min (p_4, 0.1), min (p_5, 0)}= 1.

In this case it is easy to observe that the above equation has no solution with respect to p_1, p_2, p_3, p_4, p_5, therefore P o Q = R has no solution with respect to P.

In general, writing R = {r_1 r_2 r_3 r_4}, it becomes evident that we have no solution if max $q_{ij} < r_j$.

Discussion and Conclusion

Based on what was reported in this Chapter, one can lead to the following conclusions:

- The management and evaluation of the fuzzy data obtained by the operation mechanisms of large and complex systems is very important for real life and science applications. A model has been developed for evaluating such kind of data in terms of the corresponding membership degrees and possibilities and examples were presented illustrating it. The general character of the proposed model enables its application to a variety of other human and machine activities (e.g., see the book [10] for a description of such kind of activities) and this is one of our main targets for future research.

- We have also used FRE for assessing student MM skills. In this way we have managed to express the "average student" of a class as a FS on the set of the steps of the MM process, which gives valuable information to the instructor for designing his/her future teaching plans. On the contrary, we have realized that the problem of representing the "average student" of a class as a FS on the set of the linguistic grades characterizing his performance using FRE is complicated, since it may have more than one solutions or no solution at all. In general, the use of FRE looks as being a powerful tool for the assessment of human skills. Therefore, our future research plans include the effort of using them in other human activities apart from the MM process, like learning, decision-making, etc.

APPENDIX

The MM problems used in the classroom application of the fourth Section of this Chapter are the following:

Problem 1: We want to construct a channel to run water by folding the two edges of a rectangle metallic leaf having sides of length 20 cm and 32 cm, in such a way that they will be perpendicular to the other parts of the leaf. Assuming that the flow of water is constant, how we can run the maximum possible quantity of the water through the channel?

Solution: Folding the two edges of the metallic leaf by length x across its longer side the vertical cut of the constructed channel forms a rectangle with sides x and 32-2x (Figure 1).

The area of the rectangle, which is equal to $E(x) = x(32-2x) = 32x-2x^2$, has to be maximized. Taking the derivative $E'(x)$, the equation $E'(x) = 32-4x = 0$ gives that $x = 8$ cm. But $E''(x) = -4 < 0$, therefore $E(8) = 128$ cm^2 is the maximum possible quantity of water to run through the channel.

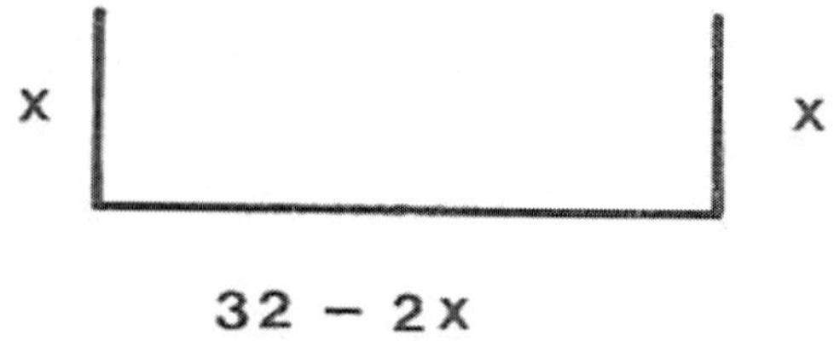

Figure 1. The vertical cut of the channel.

Remark: A number of students folded the edges of the other side of the leaf and they found that $E(x) = x(20-2x) = 20x-2x^2$. In this case the equation $E'(x) = 0$ gives that $x = 5$ cm, while $E(5) = 50\ cm^2$. Their solution was mathematically correct, but many of them failed to realize that it is not acceptable in practice (real world)

Problem 2: Among all the cylindrical towers having a total surface of $180\pi\ m^2$, which one has the maximal volume?

Solution: Let R be the radius of the basement of the tower and let h be its height. Then its total surface is equal to $2\pi Rh+2\pi R^2=180\pi \Rightarrow h = \frac{90-R^2}{R}$. Therefore the volume of the tower as a function of R is equal to $V(R) = \pi R^2 \frac{90-R^2}{R} = 90\pi R-\pi R^3$. But $V'(R) = 90\pi-3\pi R^2 = 0$ gives that $R = \sqrt{30}$ m, while $V''(R) = -6\pi R < 0$. Thus, the maximal volume of the tower is equal to $V(\sqrt{30}) = 90\pi\sqrt{30} - \pi(\sqrt{30})^3 = 60\sqrt{30}\pi \approx 1032\, m^3$

Remark: A number of students considered the total surface of the tower as being equal to $2\pi Rh$, not including to it the areas of its basement and its roof. In this case they found that $h=\frac{90}{R}$, $V(R) = 90\pi R$ *and* $V'(R) = 90\pi > 0$, which means that under these conditions there is no tower having a maximal volume. However, some of them failed to correct their model in order to find the existing solution of the real problem (unsuccessful transition from the model to the real world).

REFERENCES

[1] Dowling E. T., *Mathematics for Economists,* Schaum's Outline Series, New York, Mc Graw – Hill, 1980.

[2] Czogala, E., Drewniak, J. & Pedryz, W., Fuzzy relation equations on a finite set, *Fuzzy Sets and Systems*, 7, 89-101, 1982.

[3] Higashi, M. & Klir, G.J., Resolution of finite fuzzy relation equations, *Fuzzy Sets and Systems*, 13, 65-82, 1984.

[4] Kaufmann, A., *Introduction to the Theory of Fuzzy Subsets,* New York: Academic Press, 1975

[5] Klir, G. J. & Folger, T. A., *Fuzzy Sets, Uncertainty and Information*, Prentice-Hall, London, 1988.

[6] Prevot, M., Algorithm for the solution of fuzzy relations, *Fuzzy Sets and Systems,* 5, 319-322, 1981.

[7] Sanchez, E., Resolution of Composite Fuzzy Relation Equations, *Information and Control*, 30, 38-43, 1976.

[8] Shackle, G. L. S., Decision, Order and Time in Human Affairs, Cambridge University Press, Cambridge, UK, 1961.

[9] Voskoglou, M. Gr., Mathematical modelling as a teaching method of mathematics, *Journal for Research in Innovative Teaching* (National University, CA) 8(1), 35-50, 2015.

[10] Voskoglou, M. Gr., *Finite Markov Chain and Fuzzy Logic Assessment Models: Emerging Research and Opportunities*, Createspace.com. – Amazon, Columbia, SC, USA, 2017.

[11] Voss, J. F., Learning and transfer in subject matter learning: A problem solving model, *International Journal of Educational Research*, 1, 607-622, 1987.

[12] Zadeh, L. A., Fuzzy Sets, *Information and Control*, 8, 338-353, 1965.

[13] Zadeh, L. A., Similarity relations and fuzzy orderings, *Information Sciences*, 3, 177- 200, 1971.

In: An Essential Guide to Fuzzy Systems
Editor: Michael Gr. Voskoglou
ISBN: 978-1-53616-128-1

Chapter 3

AN INTRODUCTION TO FUZZY GRAPH THEORY

Madhumangal Pal*

Department of Applied Mathematics with Oceanology and Computer Programming,Vidyasagar University, Midnapore, India

ABSTRACT

Now-a-days fuzzy logic has become very useful tool in modeling huge number of real life problems where certain types of uncertaintities are presents. At the same time a graph is a mathematical structure which is also used regorously in modeling real life problems. So, a fuzzy graph being a combination of the two, can be used to solve almost all sets of problems where uncertainty is present. In this chapter, fuzzy graph theory is introduced. Different types of fuzzy graphs, fundamental terminologies and concepts are explained with appropriate examples. One special type and widely used graph is planar graph. Here fuzzy planar graph is introduced and several results are presented with examples. Also, the fuzzy tolerance graph and fuzzy competition graph are defined. Many interesting properties of these two types of fuzzy graphs are given. Lastly, some other related works are highlighted.

*Corresponding Author's Email: mmpalvu@gmail.com.

Keywords:fuzzy graph, fuzzy planar graph, fuzzy tolerance graph, fuzzy competition graph

1. INTRODUCTION

A graph is a very useful tool that helps to model many problems of practical importance. For example, a network of cities, can berepresented by vertices, and connections between the cities by roads, trains, ferries, flights, etc. by edges. The travel time and/or distance between the cities represents the weight of the edge. The well-known travelling salesman problem determines the shortest possible path to all the cities exactly once and come back to the starting city. The concept of graph theory is very old and it was used for the first time in 1730's. Leonhard Euler solved the famous long standing problem of Seven Bridges of Koningsberg in 1736 using graphs and thereby marked the beginning of graph theory. And the journey of graph theory begins. Furthermore, the large part of graph theory has been motivated by the study of games and recreational mathematics. Graphs are very convenient tools to represent the relationships between objects, which are represented by vertices. The relationships among the vertices are represented by connections called edges. In general, any mathematical structure involving points and connections among them can be called a graph or a hypergraph. A graph can be represented by a diagraph on a plane paper and mathematically with two sets: one is called set of vertices and other is called set of edges. Also, it can be represented by different types of matrices. A huge number of problems can be modelled by graphs and by solving the problems on graphs one can find the solution of the original problem. Few applications of graphs are database management, ecosystems, organic molecules, physical networks, signal-flow graphs, map coloring, tracing mazes, flow of a computer program, etc. At present, graph theory has taken the shape of a dynamic field in both theory and applications.

The following abbreviations are used throughout the chapter.

Table 1. Some abbreviations

Abbreviation	Description
FG	fuzzy graph
FHG	fuzzy hypergraph
FMG	fuzzy multigraph
FPG	fuzzy planar graph
FTG	fuzzy tolerance graph
FCG	fuzzy competition graph

2. Fuzzy Graphs

Today's world is full of uncertainty. So the classical graph called the crisp graph in contrast of FG cannot model all real problems completely and properly. For example, a social network, viz. facebook, whatsapp, etc. are modelled as a graph whose vertices are accounts (facebook ID, whatsapp ID, etc.) (i.e., person, institution, etc.) and edges are the relation between the accounts. The frequency of occurrence of an account in a social network varies such as some accounts are very active in the networks, while some other accounts are less active and some are rarely active, etc. So depending on the activeness we can assign the membership value to an account (vertex). Again, the frequency of communication between any pair of accounts are not same. So same as vertices one can assign different membership values to the edges. And hence the entire network becomes a graph with varied membership values on the vertices and edges. That is, the appearance or occurrence or existence of vertices and edges in the network is uncertain. This kind of graph is known as FG. Rosenfeld [46] first introduced the concept of FGs. In[25], Koczy used FGs in the evaluation and optimization of some problems of uncertain networks. After this work FG theory has developed into a vast research area. FGs are applied in several areas of computer science that include data mining, image segmentation, image capturing, clustering, communication, networking, scheduling, planning, etc. and other areas of science, engineering, social sciences, medical sciences,as well.

Some fundamental definitions and terminologies on FGs are given below.

Definition 1. [46] *A fuzzy graph* $\xi = (V, \sigma, \mu)$ *is a non-empty set* V *together with a pair of functions* $\sigma: V \to [0,1]$ *and* $\mu: V \times V \to [0,1]$ *such that for all* $x, y \in V$, $\mu(x,y) \leq min\{\sigma(x), \sigma(y)\}$, *where* $\sigma(x)$ *and* $\mu(x,y)$ *represent the membership values of the vertex* x *and of the edge* (x,y) *in* ξ *respectively.*

An edge is non-trivial if $\mu(x,y) \neq 0$. If there is a loop at a vertex x in a FG, then $\mu(x,x) \neq 0$.

A FG is illustrated in Figure 1.

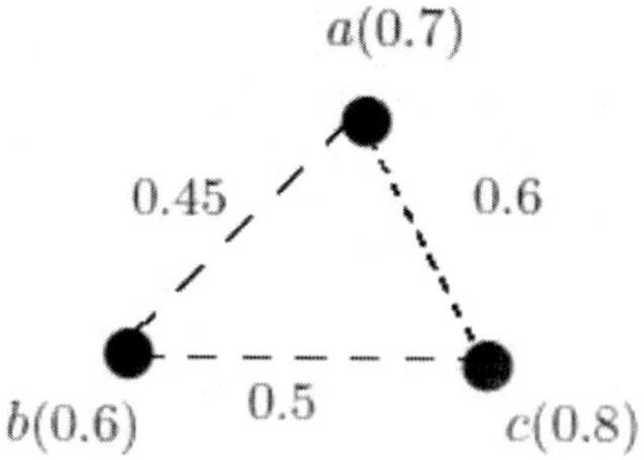

Figure1. An example of fuzzy graph.

In Figure 1, a, b, c are the vertices and their membership values are $0.7, 0.6, 0.8$ respectively. This graph has 3 edges $(a,b), (b,c)$ and (c,a). Their membership values are $0.45, 0.50$ and 0.60.

The uncertainties shown in a FG may occure in many different ways. Here five possible cases are discussed which are cited by Blue in [6].

Type I: Crisp vertex set and fuzzy edge set.

Type II: Crisp vertices and edges with fuzzy Connectivity.

Type III: Fuzzy vertex set and crisp edge set.
Type IV: Crisp graph with fuzzy weights.
Type V: Fuzzy set of crisp graphs.

2.1. Type 1 Fuzzy Graphs

Suppose in a graph the vertices are known (certain), but their edges are unknown (uncertain). In this case the vertices are crisp and the edges are fuzzy.

Example 1. Suppose we would like go from my own city to another city by the shortest route. Unfortunately, a lot of road construction work is going on, making quite a few roads fully or partially closed; but I do not know for certaint which roads are affected. This type of situation leads to a FG of Type I.

2.2. Type 2 Fuzzy Graphs

In contrast with Type I fuzziness, it may happen that the graph has known (certain) vertices and edges, but unknown (uncertain) edge connectivity.

Example 2. Let us consider the same example as Example 1. Let us suppose no road is under construction. Instead, I may look upon few routes that involve ferry crossings over a large body of water andunfortunately, the ferry schedule is vague as to which drop-off points correspond to which pick-up points which leads to uncertain edge connectivity.

2.3. Type 3 Fuzzy Graphs

Another type of fuzziness in contrast to Type I FG is based on the following condition: it may happen that a graph has known (certain) edges but unknown (uncertain) vertices.

Example 3. Suppose I need to give lectures at several conferences (occuring at different places) consecutively. Naturally, I would be interested in finding the most cost-effective travel plan for attending the conferences. The conference committees, however, have not declared the conference venues yet.

2.4. Type 4 Fuzzy Graphs

This kind of graph with fuzziness occurs when the graph has known (certain) vertices and edges, but unknown (uncertain) weights (or capacities) on the edges.

Example 4. Suppose I need to plan the quickest bus route from one city to another. It is well know that the map provides the distances, not travel times, so the exact travel time through any particular road segment is not known in advance.

2.5. Type 5 Fuzzy Graphs

A trivial type of graph fuzziness arises from considering a fuzzy set G of crisp graphs G_i: $G = \frac{G_1}{\mu_1} + \frac{G_2}{\mu_2} + \cdots + \frac{G_n}{\mu_n}$. From the analitical point of view, this type of fuzziness is really not very interesting unless the graphs G_i have some vertices or edges in common.

Lot of important works have been done on fuzzy graphs. Interested readers may read [30-39]to have some idea.

3. Some Terminologiesy of Fuzzy Graphs

The FG $\xi' = (V', \tau, \nu)$ is called a *fuzzy subgraph* [30]of ξ if $\tau(x) \leq \sigma(x)$ for all $x \in V'$ and $\nu(x, y) \leq \mu(x, y)$ for all $x, y \in V'$ where $V' \subset V$.

Definition 2. *A FG* $\xi = (V, \sigma, \mu)$ *is called complete if* $\mu(x,y) = min\{\sigma(x), \sigma(y)\}$ *for all* $x, y \in V$.

Unlike crisp graph, several types of edges are available in literature. One important type of edge is the strong edge. But, several type of strong edges are defined by the researchers. Among them, the definition given in [14] is more acceptable and widely used which is given below:

An edge $(x, y) \in \xi$ is called *strong* [14] if

$$\frac{1}{2}\min\{\sigma(x), \sigma(y)\} \leq \mu(x,y)$$

and weak otherwise. The strength of the fuzzy edge (x, y) is measured by

$$\frac{\mu(x,y)}{\min\{\sigma(x), \sigma(y)\}}.$$

Other different types of arcs are defined in [4,5,26,27].

Definition 3. *A FG* $\xi = (V, \sigma, \mu)$ *is called bipartite if the vertex set* V *can be partitioned into two nonempty disjoint sets* V_1 *and* V_2 *such that* $\mu(v_1, v_2) = 0$ *if* $v_1, v_2 \in V_1$ *or* $v_1, v_2 \in V_2$. *Further, if* $\mu(v_1, v_2) = min\{\sigma(v_1), \sigma(v_2)\}$ *for all* $v_1 \in V_1$ *and* $v_2 \in V_2$, *then* ξ *is called a complete bipartite FG.*

If the edge (x, y) satisfies the condition $\mu(x,y) = \min\{\sigma(x), \sigma(y)\}$, then it is called effective edge [33]. Two vertices are said to be effective adjacent if they are the end vertices of the same effective edge. Then the effective incident degree of a FG is defined as the number of effective incident edges on a vertex v. If all the edges of a FG are effective, then the FG becomes complete FG. A pendent vertex in a FG is defined as a vertex of an effective incident degree one. A fuzzy edge is called a fuzzy *pendant edge* if one end vertex is fuzzy *pendant vertex*. The membership value of the pendant edge is the minimum among the membership values of the end vertices.

A FG $\xi = (V, \sigma, \mu)$ is called *regular* [32] if all the vertices have equal degree. If each vertex of ξ has same total degree k, then ξ is said to be a*totally regular* FG. A FG is said to be *irregular* [34], if there is a vertex which is adjacent to vertices with distinct degrees. A FG is said to be *neighbourly irregular* [34], if every pair of adjacent vertices have different degrees. A FG is said to be*totally irregular*, if there is a vertex which is adjacent to vertices with distinct total degrees. If every two adjacent vertices have distinct total degrees is a FG then it is said to be *neighbourly total irregular* [34]. A FG is said to be *highly irregular* [34] if every vertex is adjacent to vertices with distinct degrees.

Definition 4. *The underlying crisp graph of the FG ξ is denoted as $\xi^* = (V, \sigma^*, \mu^*)$ where $\sigma^* = \{u \in V | \sigma(u) > 0\}$ and $\mu^* = \{(u, v) \in V \times V | \mu(u, v) > 0\}$.*

Theoretically, every element of $V \times V$ is an edge of a FG $\xi = (V, \sigma, \mu)$. But, normally we consider such tuple (x, y) as an edge of ξ if $\mu(x, y) > 0$. With this very simple assumption, the FG (ξ) and underlying crisp graph (ξ^*) are structurally same. In underlying crisp graph (ξ^*), all the edges and vertices are certain, i.e., their degree of membership are all same and is equal to 1. On the other hand, the vertices and edges of ξ are uncertain and hence their degree of membership are different.

Definition 5. *A directed FG (fuzzy digraph) $\overleftarrow{\xi} = (V, \sigma, \overleftarrow{\mu})$ is a non-empty set V together with a pair of functions $\sigma: V \to [0,1]$ and $\overleftarrow{\mu}: V \times V \to [0,1]$ such that for all $x, y \in V$, $\overleftarrow{\mu}(x, y) \le \sigma(x) \wedge \sigma(y)$.*

Since the function $\overleftarrow{\mu}$ is well defined, then in a fuzzy digraph, between any two vertices there is at most two directed edges. $\overleftarrow{\mu}(u, v)$ represent the membership value of the edge $\overleftarrow{(u, v)}$. If there is a loop at a vertex x, then $\overleftarrow{\mu}(x, x) \neq 0$. In a fuzzy digraph, the function $\overleftarrow{\mu}$ is, in general, not symmetric because $\overleftarrow{\mu}(x, y)$ and $\overleftarrow{\mu}(y, x)$ may have distinct values. The *underlying directed crisp graph of directed FG* is similar as underlying FG with directed edges.

Now, we introduce another kind of FG called fuzzy hypergraph [19]. The formal definition of fuzzy hypergraph is given below:

Definition 6. [30] *Let X be a finite set and let $\mathcal{E}$ be a finite family of nontrivial fuzzy sets on X (or subsets of X) such that $X = \cup \{suppA|A \in \mathcal{E}\}$. Then the pair $\mathcal{H} = (X, \mathcal{E})$ is a fuzzy hypergraph on X.*

Here, X and $\mathcal{E}$ are respectively vertex set and fuzzy edge set of $\mathcal{H}$. The height of $\mathcal{H}$, $h(\mathcal{H})$, is defined by $h(\mathcal{H}) = max\{h(A)|A \in \mathcal{E}\}$. A fuzzy hypergraph is*simple* if $\mathcal{E}$ has no repeated fuzzy edges and whenever $A, B \in \mathcal{E}$ and $A \subseteq B$, then $A = B$. A fuzzy hypergraph $\mathcal{H} = (X, \mathcal{E})$ is *support simple* if whenever $A, B \in \mathcal{E}$, $A \subseteq B$ and $supp(A) = supp(B)$, then $A = B$. Suppose $A = (X_1, \mu) \in \xi, X_1 \subseteq X$ and $c \in (0,1]$. The c −cut of A, A^c, is defined by $A^c = \{x \in X|\mu(x) \geq c\}$. Also, $\mathcal{E}^c = \{A^c| \in \mathcal{E}/\{\phi\}\}$ and $X^c = \cup \{A^c|A \in \mathcal{E}\}$. If $\mathcal{E}^c \neq \phi$, then the (crisp) hypergraph $H^c = (X^c, \mathcal{E}^c)$ is the *c-level hypergraph* of $\mathcal{H}$.

Suppose $\mathcal{H}_1 = (X, \mathcal{E}_1)$ and $\mathcal{H}_2 = (X, \mathcal{E}_2)$ are fuzzy hypergraphs. Then $\mathcal{H}_1$ is partial hypergraph of $\mathcal{H}_2$ if $\mathcal{E}_1 \subseteq \mathcal{E}_2$. A fuzzy set $A = (X, \mu)$ with $\mu: X \rightarrow [0,1]$ is an *elementary fuzzy set* if μ is constant function or μ has range $\{0, a\}, 0 \neq a$. An *elementary fuzzy hypergraph* is a fuzzy hypergraph in which all fuzzy edges are elementary.

A fuzzy hypergraph $\mathcal{H} = (X, \mathcal{E})$ is an m *tempered fuzzy hypergraph* of a crisp hypergraph $H^* = (X, E)$ if there exists a fuzzy set $A = (X, m)$ such that $m: X \rightarrow (0,1]$ and $\mathcal{E} = \{\gamma_{E_i}|E_i \in E\}$ where

$$\gamma_{E_i}(x) = \begin{cases} min\{m(e)|e \in E_i\}; & if x \in E_i \\ 0; & otherwise \end{cases}$$

A *fuzzy transversal* $\mathcal{T} = (X, \tau)$ of $\mathcal{H}$ is a fuzzy set defined on X with the property that $\tau_{h(A)} \cap \mu_{h(A)} \neq \phi$ for each $A \in \mathcal{E}$ (recall that $h(A)$ is the height of A). A *minimal fuzzy transversal* $\mathcal{T}$ for $\mathcal{H}$ is a transversal of $\mathcal{H}$ with the property that if $T_1 < T$, then T_1 is not a fuzzy transversal of $\mathcal{H}$.

Many works on fuzzy hypergraphs are available in [19,20,21].

4. Fuzzy Multigraph

In this section, fuzzy multigraph (FMG) is defined and is illustrated by an example.

Definition 7. *Let V be a non-empty set and $\sigma: V \to [0,1]$ be a mapping. Also, let $E = \{((x,y),(x,y)_{\mu^j}), j = 1,2,\dots,p_{xy} | (x,y) \in V \times V\}$ be a fuzzy multiset of $V \times V$ such that $(x,y)_{\mu^j} \leq min\{\sigma(x),\sigma(y)\}$ for all $j = 1,2,\dots,p_{xy}$, where $p_{xy} = max\{j | (x,y)_{\mu^j} \neq 0\}$. Then $\psi = (V,\sigma,E)$ is denoted by FMG.*

In FMG, there may exist more than one edge between any pair of vertices x and y. $(x,y)_{\mu^j}$ denotes the membership value of the j-th edge between the vertices x and y. The number of edges between the vertices x and y is denoted by p_{xy}.

Also, for a FMG, $\psi = (V,\sigma,E)$, E is said to be fuzzy multi-edge set. Following is an example of a FMG.

Example 5. *Let $V = \{u,x,y,z\}$ and $\sigma: V \to [0,1]$ such that $\sigma(u) = 0.6, \sigma(x) = 0.7, \sigma(y) = 0.65, \sigma(z) = 0.8$.*

Let

$$E = \{((u,x),0.5),((u,x),0.55),((u,x),0.6),((x,y),0.6),((x,z),0.7).$$

Obviously, (V,σ,E) is a FMG and it is depicted in Figure 2.

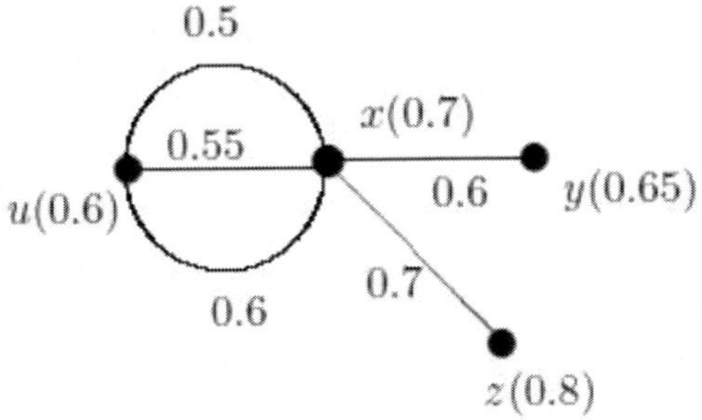

Figure 2. Example of fuzzy multigraph.

The degree of a vertex in a FMG is defined below:

Definition 8. *Let* $\psi = (V, \sigma, E)$ *be a FMG where* $E = \{((x,y),(x,y)_{\mu^j}), j = 1,2,\ldots,p_{xy} | (x,y) \in V \times V\}$ *and* $p_{xy} = max\{j | (x,y)_{\mu^j} \neq 0\}$. *The degree of a vertex* $v \in V$ *is denoted by* $d(v)$ *and is defined by* $d(v) = \sum_{j=1}^{p_{xy}} (u,v)_{\mu^j}$ *for all* $u \in V$.

The multi graphs are useful for any design where multi edges exist. Similarly, FMG have many important uses in FG theory. Theoretically, fuzzy planar graph could not be defined without the concept of FMG.

5. Fuzzy Planar Graphs

The planar graph is a very important topic of graph theory. It has many applications in designing problems based on circuits, utility lines, subways design, etc. The crossing of two connections generally means that the communication lines must run at different layers. This is not a big issue for electrical circuit design, but it generates extra cost for certain types of communication wires. For example, circuits are easier to manufacture if their connections can be constructed in fewer layers. The concept of planar graph is highly used to design of circuits with less number of layers. Circuit where crossing of wires is necessary, can not be represented by planar graphs. Huge number of computational challenges including shape matching or image segmentation, etc. can be solved by means of cuts of planar graph.

The subway tunnels, pipelines, metro lines, etc. are essential for city planning. Due to overlapping, there remains some chance of accident on the road. Also, the construction cost of underground junction is high. But, underground routes reduce traffic jam and can avoid collision, too. In a city planning, routes without crossing are preferable for safety. But, due to lack of space and other factor crossing roads are constructed. It is easy to observe that the crossing between one congested and one non-congested route is better than the crossing between two congested routes. The term "congested" has no specific meaning, it depends on public habit, mentality, places of residence, number of commuters, etc. We generally use the terms "congested,""very congested," "highly congested," etc. for routes. These terms are used as linguistic variables in fuzzy setup and they have some membership values. A congested route may be refer to a strong route and a low congested route may be called as a weak route. Thus inter crossing between strong route and weak route is more safe than the inter crossing between two strong routes. That is, inter crossing between strong route and weak route may be constructed with certain amount of safety with proper traffic system. The terms "strong route" and "weak route" resemble strong edge and weak edge of a FG as defined earlier. Use of inter crossing between strong and weak edge leads to a new kind of planar graph called fuzzy planar graph (FPG).

With this analogy Samanta and Pal [56] introduced a new type of FPG.

Normally, in planar graph no edge cross each other. But, in FPG we assume that some edges may cross each other. The reason for this assumption is that in FG, the membership values of some edges may be very small. The significance for such edges in the graph is negligible. So, the importance of crossing of such edges is very less.

A (crisp) graph is called non-planar graph if at least two edges are crossed for all possible geometrical representations of the graph. Suppose G is a (crisp) graph and (a, b) and (c, d) be two edges. It is assumed that they cross in a particular representation of the graph. In terms of fuzzy setup, we say that the membership values if these edges are 1. See that if the edge (c, d) is removed, the graph becomes planar. In fuzzy sense, we can say that the membership values of the edges (a, b) and (c, d) are 1 and 0 respectively.

Let $\xi = (V, \sigma, \mu)$ be a FG and for simplicity we assume that there is only one crossing in the graph and this crossing happens between the fuzzy edges $((w, x), \mu(w, x))$ and $((y, z), \mu(y, z))$. If $\mu(w, x) = 1$ and $\mu(y, z) = 0$, then we say that the FG has no crossing. Similarly, if the value of $\mu(w, x)$ is near to 1 and that of $\mu(w, x)$ near to 0, then in this case the crossing is not important for planarity. If $\mu(w, x)$ is near to 1 and $\mu(w, x)$is close to 1, then, in this case the crossing is highly significant for testing the planarity.

Before going to the main definition, some related terms are discussed below.

6. Intersecting Value between Two Edges

In FMG, when two edges are intersected at a point, a numerical value is defined to that point as follows:

Let in a FMG, $\psi = (V, \sigma, E)$, $((a, b), (a, b)_{\mu^k})$ and $((c, d), (c, d)_{\mu^l})$ be two edges which intersect at a point P, where k and l are fixed integers.

The strength $(I_{(a,b)})$ of the fuzzy edge (a, b) is calculated by the following formula:

$$I_{(a,b)} = \frac{(a, b)_{\mu^k}}{\min\{\sigma(a), \sigma(b)\}}.$$

If $I_{(a,b)} \geq 0.5$, then the fuzzy edge is called strong otherwise weak.

Let us define another parameter called intersecting value at the intersecting point. The intersecting value at the point P is defined by

$$\mathcal{J}_P = \frac{I_{(a,b)} + I_{(c,d)}}{2}.$$

If the number of points of intersection in a FMG increases, planarity decreases. So for FMG, $\mathcal{J}_P$ is inversely proportional to the degree of planarity. Based on this concept, a new terminology is introduced below for a FPG.

Definition 9. *Let ψ be a FMG and for a certain geometrical representation $P_1, P_2, \dots, P_z$ be the points of intersection between the edges. ψ is said to be FPG with degree of planarity f, where*

$$f = \frac{1}{1 + \{\mathcal{J}_{P_1} + \mathcal{J}_{P_2} + \cdots + \mathcal{J}_{P_z}\}}.$$

From definition it follows that f is bounded and $0 < f \leq 1$.

If there is no point of intersection for a certain geometrical representation of a FG, then the degree of planarity is 1. In this case, the underlying crisp graph of this FG is the (crisp) planar graph. If f decreases, then the number of points of intersection between the edges increases and obviously the nature of planarity decreases. From this analogy, one can say that every FG is a FPG with certain degree of planarity.

Example 6. Let us consider the FPG of Figure 3. Here $V = \{a, b, c, d, e\}$ is the vertex set and let the membership values of the vertices be $\sigma(a) = 0.8, \sigma(b) = 0.7, \sigma(c) = 0.6, \sigma(d) = 0.9, \sigma(e) = 0.9$ and that of edges be $(a, b)_{\mu^1} = 0.6, (a, c)_{\mu^1} = 0.5, (a, d)_{\mu^1} = 0.45,$ $(a, d)_{\mu^2} = 0.3, (b, c)_{\mu^1} = 0.4, (b, d)_{\mu^1} = 0.55, (c, d)_{\mu^1} = 0.6, (a, e)_{\mu^1} = 0.8,$ $(c, e)_{\mu^1} = 0.5, (e, d)_{\mu^1} = 0.8, (e, b)_{\mu^1} = 0.7$. In this FMG, there are two points of intersection P_1 and P_2. P_1 is a point between the edges $((a, d), 0.45)$ and $((b, c), 0.4)$ and P_2 is between $((a, d), 0.3)$ and $((b, c), 0.4)$. For the edge $((a, d), 0.45)$, $I_{(a,d)} = \frac{0.45}{0.8} = 0.5625$ and for the edge $((b, c), 0.4)$, $I_{(b,c)} = \frac{0.4}{0.6} = 0.6667$. For the first point of intersection

P_1, intersecting value $\mathcal{I}_{P_1}$ is 0.6146 and that for the second one is $\mathcal{I}_{P_2}$ = 0.5208. Therefore, the degree of planarity for the FMG shown in Figure 3 is 0.4683.

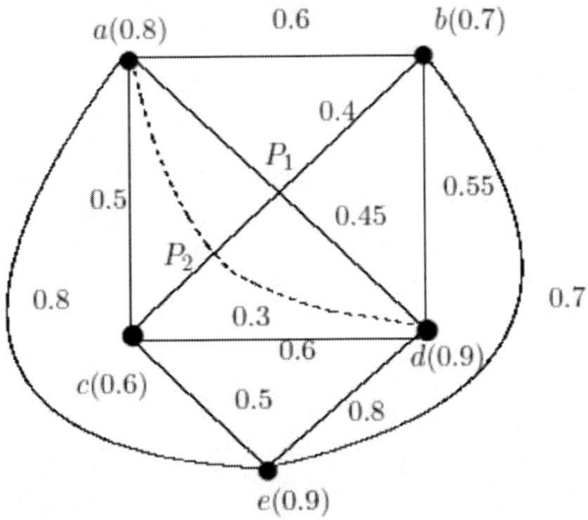

Figure 3. Example of fuzzy planar graph with fuzzy planarity value 0.4683.

Let ψ be a fuzzy complete multigraph. Then the degree of planarity f of ψ is given by

$$f = \frac{1}{1 + N_p},$$

where N_p is the number of points of intersection between the edges in ψ.

Definition 10. *A FPG ψ is called strong FPG if the degree of planarity of the graph is greater than* 0.5.

The FPG of Example 5 is not strong FPG as its degree of planarity is less than 0.5.

Thus, depending on the degree of planarity, the fuzzy planar graphs are divided into two groups, viz. strong FPG and weak FPG.

Theorem 1. *If ψ be a strong FPG, then the number of points of intersection between strong edges in ψ is at most one.*

Proof. Let $\psi = (V, \sigma, E)$ be a strong FPG. If possible letψ has at least two point of intersection P_1 and P_2 between two strong edges in ψ.

For any strong edge $((a,b),(a,b)_{\mu^j})$, $(a,b)_{\mu^j} \geq \frac{1}{2}\min\{\sigma(a),\sigma(b)\}$. So $I_{(a,b)} \geq 0.5$.

Thus for two intersecting strong edges $((a,b),(a,b)_{\mu^j})$ and $((c,d),(c,d)_{\mu^i})$, $\frac{I_{(a,b)}+I_{(c,d)}}{2} \geq 0.5$, that is, $\mathcal{I}_{P_1} \geq 0.5$. Similarly, $\mathcal{I}_{P_2} \geq 0.5$. Then $1 + \mathcal{I}_{P_1} + \mathcal{I}_{P_2} \geq 2$. Therefore, $f = \frac{1}{1+\mathcal{I}_{P_1}+\mathcal{I}_{P_2}} \leq 0.5$. It contradicts the fact that the FG is a strong fuzzy planar graph.

So the number of points of intersection between strong edges cannot be two. It is clear that if the number of points of intersection of strong fuzzy edges increases, the degree of planarity decreases. Similarly, if the number of points of intersection of strong edges is one, then the degree of planarity is greater than 0.5.

Any FPG without any crossing between edges is a strong FPG. Thus, we conclude that the maximum number of points of intersection between the strong edges in ψ is one.

Also, the above theorem is verified by the following illustration.

Example 7. *Let us consider two FPGs depicted in Figure 4. In Figure 4(a), there is one crossing between two strong edges* (a,d) *and* (b,e). *Let the membership values of the vertices and edges be* $\sigma(a) = \sigma(b) = \sigma(d) = \sigma(e) = 1$ *and* $\mu(a,d) = 0.99, \mu(b,e) = 0.99$. *The degree of planarity of this graph is* 0.50251 *and hence this FPG is strong and the number of points of intersection is* 1. *In Figure 4(b), there are two crossings between the strong edges* $(a,d),(b,e)$ *and* $(a,d),(c,e)$. *For this FG, let the membership values of vertices and edges be* $\sigma(a) = \sigma(b) = \sigma(c) = \sigma(d) = \sigma(e) = 1$ *and* $(a,d)_{\mu^1} = 0.5, (b,e)_{\mu^1} = 0.5, (a,d)_{\mu^2} = 0.5, (c,e)_{\mu^1} = 0.5$. *The degree of planarity of this graph is* 0.5, *and hence this FPG is not strong. Also, it is easy to observe that if there is no* *crossing in the FG, then the FPG must be strong. This observation support the statement of Theorem* 2.

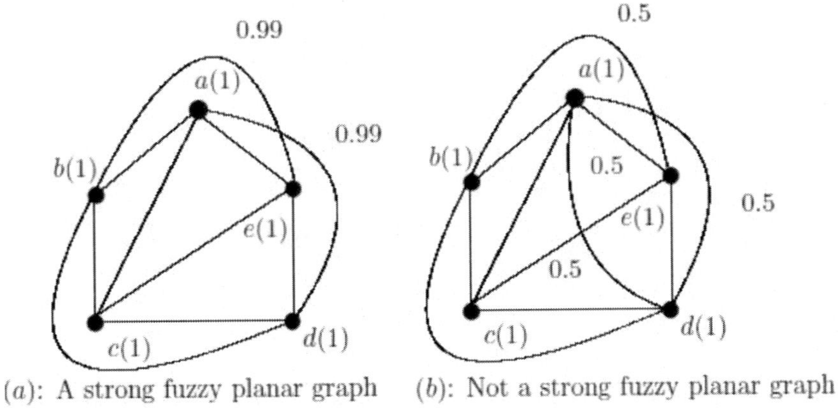

Figure 4. Strong fuzzy planar graph with at most one crossing between strong edges.

The strength of a fuzzy edge has an important role to play in modeling some real life problems. If the strength of an edge is very small, then the edge has very less significance in the problem and sometimes this kind of edge is removed from the graph. The edges with good strengths are considered to model such type of problems. These types of edges are called considerable edges and is defined below:

Definition 11. *Let* (V,σ,μ) *be a FG and* c *be a rational number such that* $0 < c < 0.5$. *An edge* (x,y) *is called considerable edge if* $\frac{\mu(x,y)}{min\{\sigma(x),\sigma(y)\}} \geq c$. *If an edge is not considerable, it is called non-considerable edge. For FMG* (V,σ,E), *a multi-edge* (x,y) *is called considerable edge if* $\frac{(x,y)_{\mu^j}}{min\{\sigma(x),\sigma(y)\}} \geq c$ *for all* $j = 1,2,\ldots,p_{xy}$.

In crisp sense, K_5 and $K_{3,3}$ are non-planar graphs and cannot be drawn without crossing of edges. So any graph containing K_5 or $K_{3,3}$ is non-planar, in crisp sense.

Note 1. *A fuzzy complete graph of five vertices $\mathcal{K}_5$ or fuzzy complete bipartite graph $\mathcal{K}_{3,3}$ are not strong FPG. This statement is justified below.*

Let $\psi = (V, \sigma, E)$ be a fuzzy complete graph of five vertices where $V = \{a, b, c, d, e\}$ and $E = \{((x,y),(x,y)_\mu) | (x,y) \in V \times V\}$.

$\mu(x,y) = \min\{\sigma(x), \sigma(y)\}$ for all $x, y \in V$.

We have mentioned that the degree of planarity of a fuzzy complete graph is $f = \frac{1}{1+N_p}$, where N_p is the number of points of intersection of the edges in ψ. We know that the fuzzy complete graph of five vertices is non-planar. If we remove one edge then the graph becomes planar. So, there is at least one crossing between the edges in a particular geometrical representation and hence the degree of planarity of this FG is $f = \frac{1}{1+1} = 0.5$. Hence ψ is not strong FPG.

Similarly, it can be shown that a complete bipartite FG $\mathcal{K}_{3,3}$ is not a strong FPG.

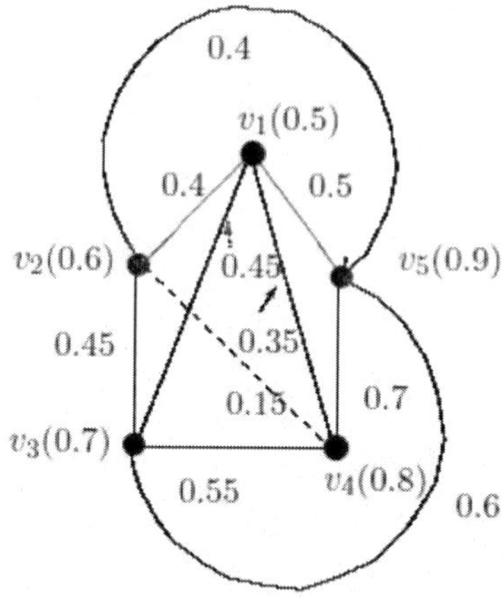

Figure 5. Example of a strong fuzzy planar graph.

A FG with five vertices and each pair of vertices connected by a fuzzy edge may or may not be strong FPG. In Figure 5, there is one point of intersection between two fuzzy edges $((v_2, v_4), 0.15), ((v_1, v_3), 0.45)$. This is an example of strong FPG with degree of planarity 0.63492, which has five vertices and each pair of vertices are connected by a fuzzy edge. Also, a fuzzy complete graph of five vertices is not a strong FPG.

A fuzzy bipartite graph of six vertices can be partitioned in two subsets containing three vertices each and all pairs of vertices between two partitioned sets are connected by fuzzy edges may or may not be strong FPG. The graph of Figure 6 with intersection at a point between two edges $((a,f)$,0.5) and $((c,d)$,0.2)is strong FPG with degree of planarity 0.64122. Also, we know that a complete bipartite FG of six vertices is not strong FPG. Hence this conclusion is true.

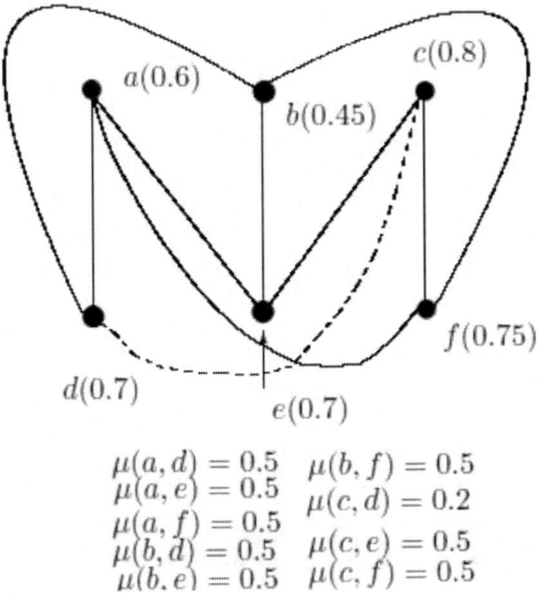

Figure 6. Example of strong fuzzy planar graph with degree of planarity 0.588.

Theorem 2. Let ψ be a FPG with the degree of planarity f. If $f \geq 0.67$, then no two strong edges will intersect in the graph.

Proof. Let $\psi = (V, \sigma, E)$ be a FPG with the degree of planarity $f \geq 0.67$. Let if possible, P be a point of intersection between two strong fuzzy edges $((a,b),(a,b)_{\mu^j})$ and $((c,d),(c,d)_{\mu^j})$.

For any strong edge $((a,b),(a,b)_{\mu^j})$, $(a,b)_{\mu^j} \geq \frac{1}{2}\min\{\sigma(a),\sigma(b)\}$. Therefore, $I_{(a,b)} \geq 0.5$. For the minimum value of $I_{(a,b)}$ and $I_{(c,d)}$, $\mathcal{I}_P = 0.5$ and $f = \frac{1}{1+0.5} < 0.67$. A contradiction arises. Hence, there is no point of intersection in ψ.

Motivated from this result, a new type of FPG is defined called 0.67-FPG, whose degree of planarity is more than or equal to 0.67. As mentioned earlier, if the degree of planarity is 1, then the geometrical representation of FPG is similar to the crisp planar graph. It is shown in Theorem 2, if the degree of planarity is 0.67, then there is no crossing between any pair of strong edges. For this case, if there is any point of intersection between edges in the graph, then it is definitely the crossing between the weak edge and any other edge. Again, the significance of a weak edge is less compared to strong edges. Thus, 0.67-FPG is more significant. If the degree of planarity increases, then the geometrical structure of planar graph tends to a crisp planar graph.

Any FG without any point of intersection of fuzzy edges is a FPG with the degree of planarity 1. Therefore, it is a 0.67-FPG.

7. Fuzzy Tolerance Graph

The intersection graph [28] is a very special and useful subclass of graphs. For every intersection graph there is a family of sets. The family of sets represents the vertices of the graph and if two sets have non-empty intersection then the corresponding vertices have an edge. Depending on the type of family of sets, different types of intersection graphs can be derived. Some common intersection graphs are interval graph, circular-arc graph, permutation graph [45], trapezoidal graph, etc.

Tolerance graphs [9,22,29] are generalization of interval graphs in which each vertex can be represented by an interval and a tolerance value such that an edge occurs if and only if the overlap of corresponding intervals is at least as large as the tolerance associated with one of the vertices. Hence a graph $G = (V, E)$ is a tolerance graph [22]if there is a set

$I = \{I_v : v \in V\}$ of closed real intervals and a set $\{T_v : v \in V\}$ of positive real numbers such that $(x, y) \in E$ if $|I_x \cap I_y| \geq min\{T_x, T_y\}$. The collection $\langle I, T \rangle$ of intervals and tolerances is called tolerance representation of the graph G.

Tolerance graphs were introduced by Golumbic to generalize some widely use applications of interval graphs. The main aim was to model resource allocation and certain scheduling problems, in which resources such as rooms and vehicles, can tolerate sharing among users. Tolerance graphs are widely used to solve huge number of problems occurred in biology and bio-informatics. The tolerance graphs have other several applications in constraint-based temporal reasoning, scheduling aircraft and crews, data transmission as well as contributing to genetic analysis and studies of the brain.

Due to the uncertain nature of real life problems and available data, fuzzy tolerance graphs are very useful and are natural extension of crisp tolerance graphs. These graphs are defined below.

Definition 12. *Fuzzy tolerance of a fuzzy interval is defined by an arbitrary fuzzy interval whose core length is a positive real number and is denoted by* $\mathcal{T}$. *Let* $i_{k-1}, i_k \in R$. *If such real number is L and* $|i_k - i_{k-1}| = L$ *then fuzzy tolerance is a fuzzy set of the interval* $[i_{k-1}, i_k]$.

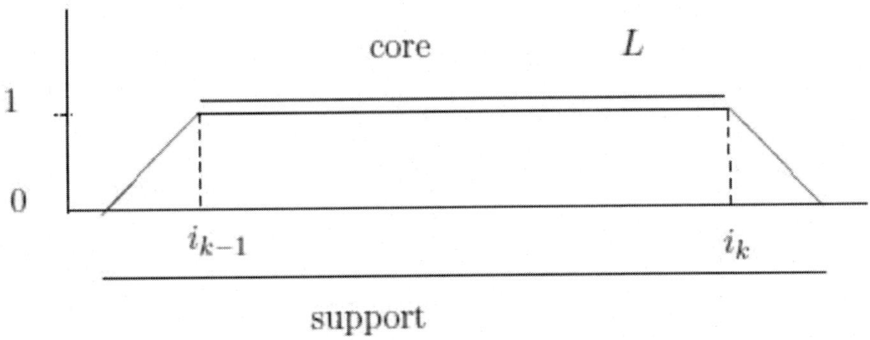

Figure 7. Fuzzy tolerance.

The core and support of a fuzzy tolerance are defined as tolerance core and tolerance support respectively. The tolerance core and tolerance support of fuzzy tolerance $\mathcal{T}$ are denoted by $c(\mathcal{T})$ and $s(\mathcal{T})$ respectively. Fuzzy tolerance may be a fuzzy number. The fuzzy tolerance is shown in Figure 7.

The fuzzy tolerance graph (FTG) is the extension of fuzzy interval graph defined below:

Definition 13. *Let* $\mathfrak{I} = \{\mathfrak{I}_1, \mathfrak{I}_2, \dots, \mathfrak{I}_n\}$ *be a finite family of fuzzy intervals on the real line and* $\mathcal{T} = \{\mathcal{T}_1, \mathcal{T}_2, \dots, \mathcal{T}_n\}$ *be the corresponding fuzzy tolerances. Suppose* $\mathfrak{I}$ *as a crisp vertex set* $V = \{v_1, v_2, \dots, v_n\}$. *The FTG is the FG* (V, σ, μ) *where* $\sigma: V \to [0,1]$ *is defined by* $\sigma(v_i) = h(\mathfrak{I}_i) = 1$ *for all* $v_i \in V$ *and* $\mu: V \times V \to [0,1]$ *is defined by*

$$\mu(v_i, v_j) = \begin{cases} 1, & if\, c(\mathfrak{I}_i \cap \mathfrak{I}_j) \geq min\{c(\mathcal{T}_i), c(\mathcal{T}_j)\} \\ \dfrac{s_{ij} - min\{s(\mathcal{T}_i), s(\mathcal{T}_j)\}}{s_{ij}} h_{ij}, & otherwise\, if\, s_{ij} \geq min\{s(\mathcal{T}_i), s(\mathcal{T}_j)\} \\ 0, & otherwise. \end{cases}$$

where $c(\mathfrak{I}_i \cap \mathfrak{I}_j)$ is the core of the intersection of the intervals $\mathfrak{I}_i$ and $\mathfrak{I}_j$.

$min\{c(\mathcal{T}_i), c(\mathcal{T}_j)\}$ is the minimum of the cores of the corresponding tolerances $\mathcal{T}_i$ and $\mathcal{T}_j$. Also $h_{ij} = h(\mathfrak{I}_i \cap \mathfrak{I}_j)$, $s_{ij} = s(\mathfrak{I}_i \cap \mathfrak{I}_j)$ and $s(\mathfrak{I}_i \cap \mathfrak{I}_j)$, $min\{s(\mathcal{T}_i), s(\mathcal{T}_j)\}$ are the support of the intersection of intervals $\mathfrak{I}_i$, $\mathfrak{I}_j$ and minimum support of the tolerances $\mathcal{T}_i$, $\mathcal{T}_j$ respectively.

If the (fuzzy) tolerances are not assigned to the fuzzy intervals, then the corresponding graph becomes fuzzy interval graph. The FTGs are depicted in Figures 8, 9, 10 for three different cases. Let $\mathfrak{I}_i, \mathfrak{I}_j$ be two fuzzy intervals and $\mathcal{T}_i, \mathcal{T}_j$ are the fuzzy tolerances. In the first case minimum of two tolerance cores $min\{c(\mathcal{T}_i), c(\mathcal{T}_j)\}$ is less than the intersection core $c(\mathfrak{I}_i \cap \mathfrak{I}_j)$ of the intervals $\mathfrak{I}_i$ and $\mathfrak{I}_j$. In this case, edge membership value is 1. If $min\{c(\mathcal{T}_i), c(\mathcal{T}_j)\}$ is greater than $c(\mathfrak{I}_i \cap \mathfrak{I}_j)$ then second case arises where membership value of the edge is less than 1. Here minimum of two tolerance supports $min\{s(\mathcal{T}_i), s(\mathcal{T}_j)\}$ is less than the intersection support

$s(\mathcal{I}_i \cap \mathcal{I}_j)$ of the intervals $\mathcal{I}_i$, $\mathcal{I}_j$. Otherwise, the third case arises where edge membership value is 0.

If $\langle I, T\rangle$ is a tolerance representation of a crisp graph G and v is a point instead of an interval then the corresponding vertex is an isolated vertex in G. But, it may not be true in the fuzzy sense. If $\langle \mathcal{I}, \mathcal{T}\rangle$ is a fuzzy tolerance representation of ξ and f is a fuzzy number instead of fuzzy interval, then the corresponding vertex may not be isolated. Intersection of supports of a fuzzy interval and a fuzzy number may be greater than the minimum support of their corresponding tolerances. So there is an edge of some positive strength. If it is not a strong arc then the fuzzy number is isolated otherwise it is not isolated.

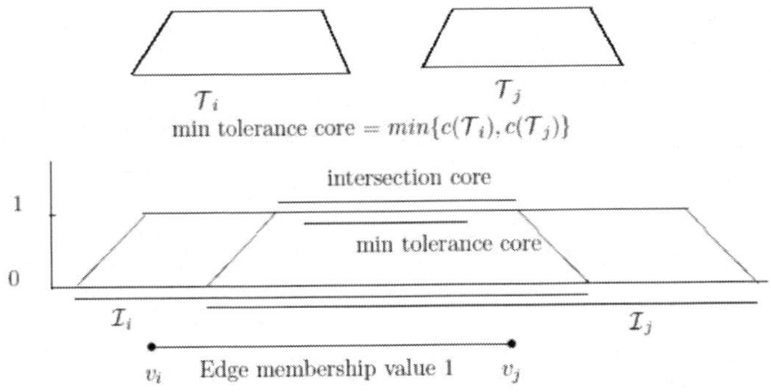

Figure 8. Fuzzy tolerance graph with edge membership value 1.

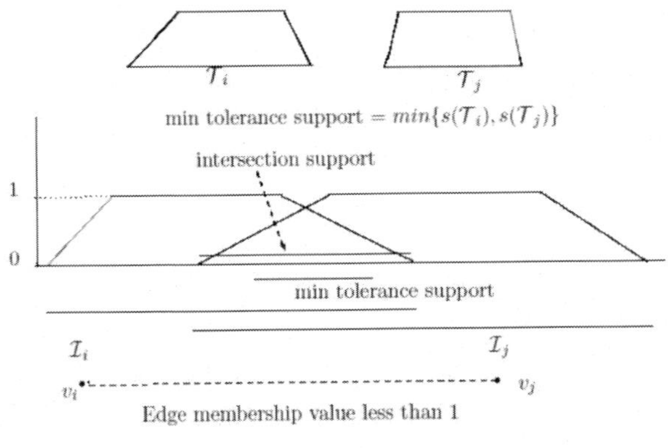

Figure 9. Fuzzy tolerance graph with edge membership value less than 1.

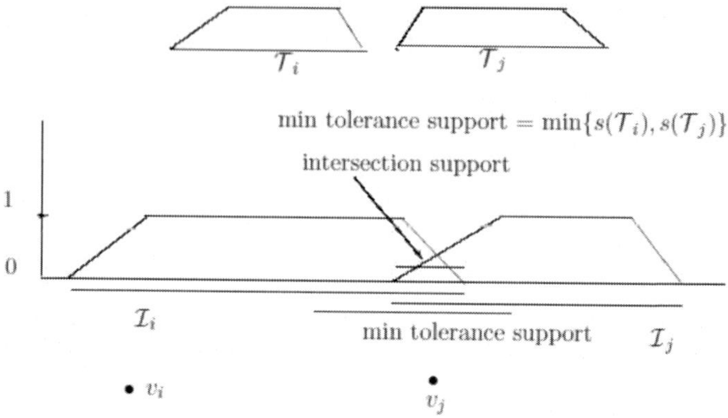

Figure 10. Fuzzy tolerance graph with edge membership value 0.

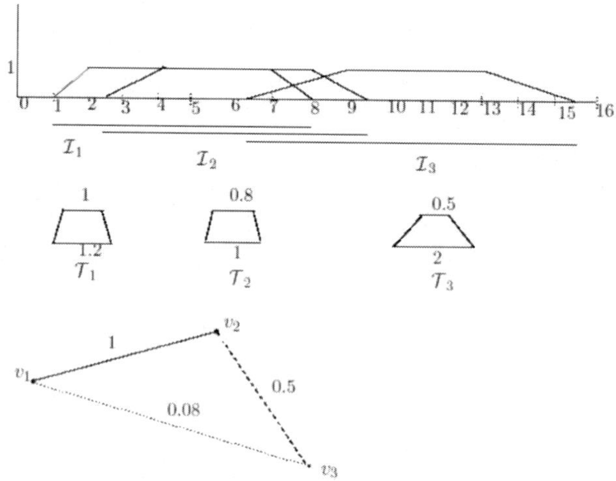

Figure 11. Example of fuzzy tolerance graph.

Example 8. Let us consider an example of three fuzzy intervals $\mathcal{I}_1$, $\mathcal{I}_2$, $\mathcal{I}_3$. Also, let $\mathcal{T}_1$, $\mathcal{T}_2$,$\mathcal{T}_3$ be the corresponding fuzzy tolerances. Let the core and support of $\mathcal{I}_1$ be [2,7] and [1,8]; that of $\mathcal{I}_2$ are [4,8] and [2.5,9.5]; that of $\mathcal{I}_3$ are [9,13] and [6.5,15.5]. The core and support lengths of $\mathcal{T}_1$, $\mathcal{T}_2$,$\mathcal{T}_3$ be respectively {1,1.2}, {0.8,1}, {0.5,2}. So $h(\mathcal{I}_1 \cap \mathcal{I}_2) = 1$, $h(\mathcal{I}_2 \cap \mathcal{I}_3) = 0.75$ and $h(\mathcal{I}_1 \cap \mathcal{I}_3) = 0.43$. Clearly, the edge membership values of the edges (v_1, v_2), (v_2, v_3) and (v_1, v_3) are obtained as 1, 0.5 and 0.08 respectively. The corresponding graph is shown in Figure 11.

8. Fuzzy Bounded Tolerance Graph

Let G be a crisp graph and its tolerance representation be $\langle I, T\rangle$, where $T_j \leq |I_j|$ for every $j = 1,2, \dots, n$ then G is called *bounded tolerance graph*[29].

Following is the formal definition of fuzzy bounded tolerance.

Definition 14. *Let $\mathcal{I} = \{\mathcal{I}_1, \mathcal{I}_2, \dots, \mathcal{I}_n\}$ be a finite family of fuzzy intervals on a real line and $\mathcal{T} = \{\mathcal{T}_1, \mathcal{T}_2, \dots, \mathcal{T}_n\}$ be the corresponding fuzzy tolerances. If the fuzzy interval $\mathcal{I}_i$ with core $c(\mathcal{I}_i)$ and support $s(\mathcal{I}_i)$ has tolerance with core $c(\mathcal{T}_i)$ and support $s(\mathcal{T}_i)$ such that $c(\mathcal{I}_i) \geq c(\mathcal{T}_i)$ and $s(\mathcal{I}_i) \geq s(\mathcal{T}_i)$ of a FTG then the corresponding representation is called fuzzy bounded tolerance representation and the FG is called fuzzy bounded tolerance graph.*

It may be noted that if $\langle \mathcal{I}, \mathcal{T}\rangle$ is a fuzzy bounded tolerance representation then no fuzzy interval $\mathcal{I}_v, v \in V$ is fuzzy number.

Theorem 3. *If ξ is a fuzzy tolerance graph with constant core and constant support a of tolerances then ξ is fuzzy bounded tolerance graph.*

Proof. Let $\mathcal{I} = \{\mathcal{I}_1, \mathcal{I}_2, \dots, \mathcal{I}_n\}$ be a finite family of fuzzy intervals on real line and $\mathcal{T} = \{\mathcal{T}_1, \mathcal{T}_2, \dots, \mathcal{T}_n\}$ be the corresponding fuzzy tolerances. Let ξ be a FTG with $\mathcal{I}$ and $\mathcal{T}$ be the corresponding fuzzy interval and fuzzy tolerance representation. Let k_1 and k_2 be two positive real numbers and suppose that $c(\mathcal{T}_i) = k_1$ and $s(\mathcal{T}_i) = k_2$ for $i = 1,2, \dots, n$.

If $c(\mathcal{I}_i) \geq k_1$ and $s(\mathcal{I}_i) \geq k_2$ for all $i = 1,2, \dots, n$ then ξ is a fuzzy bounded tolerance graph. If $c(\mathcal{I}_i) < k_1$ for any i and $s(\mathcal{I}_j) < k_2$ for any j then let $c(\mathcal{I}_j) = k_1$ and $s(\mathcal{I}_j) = k_2$, to make ξ be bounded. In this situation, the neighbour of all vertices remain unchanged.

Theorem 4. *Let ξ be a FTG. For each $t \in [0,1]$, the cut level graphs are tolerance graphs.*

Proof. Let ξ be a fuzzy tolerance graph and let $\mathcal{I} = \{\mathcal{I}_1, \mathcal{I}_2, \dots, \mathcal{I}_n\}$ be a finite family of fuzzy intervals on real line and $\mathcal{T} = \{\mathcal{T}_1, \mathcal{T}_2, \dots, \mathcal{T}_n\}$ be the corresponding fuzzy tolerances. For each $t \in [0,1]$, $\mathcal{I}_i^t \in \mathcal{I}^t$ for all $i = 1,2,\dots,n$. So it is a crisp interval. Also $t \in [0,1]$ then $\mathcal{T}_i^t \in \mathcal{T}^t$ is a crisp length, that is, a crisp tolerance.

So crisp intervals together with crisp tolerances represent a tolerance representation. Hence ξ^t, $t \in [0,1]$ is a tolerance graph.

Weakly chordal graphs [22] are those graphs with no induced subgraphs isomorphic to C_n or $\bar{C}_n$ for $n \geq 5$. A graph G is called *alternately orientable* graphs[22] if this is an orientation of G which is transitive on every chordless cycle of length greater than or equal to 4 i.e., the directions of the oriented edges must alternate. A graph G is*perfect* if for all induced subgraphs H of G, the chromatic number of H equals the number of vertices in the largest clique in H. Now we prove the following result.

Theorem 5. *If ξ is a FTG then each cut level graph is (i) weakly chordal, (ii) alternately orientable, (iii) perfect graph.*

Proof. It is known that each cut level graph of FTG is a tolerance graph. Also, we know that tolerance graphs are (i) weakly chordal, (ii) alternately orientable, (iii) perfect graph [22]. So each cut level graph of FTG is (i) weakly chordal, (ii) alternately orientable, (iii) perfect graph.

9. Fuzzy Interval Containment Graph

An interval containment graph $G = (V, E)$ is one that can be represented by a set of real intervals $I = \{I_v : v \in V(G)\}$ so that $(k, l) \in E(G)$ precisely when one of I_k, I_l contains the other. Such a representation is called an *interval containment representation*.

Fuzzy interval containment graph is defined below:

Definition 15. *Let* $\mathcal{J} = \{\mathcal{J}_1, \mathcal{J}_2, \dots \mathcal{J}_n\}$ *be a finite family of fuzzy intervals on real line. We consider the crisp vertex set* $V = \{v_1, v_2, \dots, v_n\}$. *The fuzzy interval containment graph is the FG* (V, σ, μ) *where* $\sigma: V \to [0,1]$ *is defined by* $\sigma(v_i) = h(\mathcal{J}_i) = 1$ *for all* $i = 1,2,\dots,n$ *and* $\mu: V \times V \to [0,1]$ *is defined by*

$$\mu(v_i, v_j) = \begin{cases} 1, if core and support of one of \mathcal{J}_i, \mathcal{J}_j contain the other \\ \frac{1}{2}\left(\frac{c(\mathcal{J}_i \cap \mathcal{J}_j)}{min\{c(\mathcal{J}_i), c(\mathcal{J}_j)\}} + \frac{s(\mathcal{J}_i \cap \mathcal{J}_j)}{min\{s(\mathcal{J}_i), s(\mathcal{J}_j)\}}\right) h(\mathcal{J}_i \cap \mathcal{J}_j), otherwise. \end{cases}$$

where $c(\mathcal{J}_i \cap \mathcal{J}_j)$, $s(\mathcal{J}_i \cap \mathcal{J}_j)$ are the core and support of the intersection of the intervals $\mathcal{J}_i$, $\mathcal{J}_j$ and $min\{c(\mathcal{J}_i), c(\mathcal{J}_j)\}$, $min\{s(\mathcal{J}_i), s(\mathcal{J}_j)\}$ are the minimum of core and support of the intervals $\mathcal{J}_i$ and $\mathcal{J}_j$.

Example 9. *Let us consider the four fuzzy intervals* $\mathcal{J}_1, \mathcal{J}_2, \mathcal{J}_3, \mathcal{J}_4$. *Let the core and the support of* $\mathcal{J}_1$ *be* $[2,7]$ *and* $[1,8]$; *that of* $\mathcal{J}_2$ *be* $[4,6]$ *and* $[3,7]$; *that of* $\mathcal{J}_3$ *be* $[9,13]$ *and* $[7.5,14.5]$; *that of* $\mathcal{J}_4$ *be* $[12,16]$ *and* $[11,17]$. *So* $h(I_1 \cap I_2) = 1$, $h(I_3 \cap I_4) = 1$ *and* $h(I_1 \cap I_3) = 0.2$. *Clearly, the edge membership values of the edges* (v_1, v_2), (v_1, v_3) *and* (v_3, v_4) *are respectively* 1, 0.007 *and* 0.42.

10. Regular Representation of Fuzzy Tolerance Graph

A FTG is said to have regular representation if it satisfies the following three properties:

1) Any fuzzy tolerance core and support lengths larger than the core and support lengths of the corresponding fuzzy interval can be set to infinity without changing the adjacencies of vertices in the fuzzy tolerance graph.
2) All cores and supports of fuzzy tolerances are distinct.
3) The cores and supports of two different fuzzy intervals do not share an end point.

Theorem 6. *Every FTG has a regular representation.*

Proof. Let $\mathcal{J} = \{\mathcal{J}_1, \mathcal{J}_2, \dots, \mathcal{J}_n\}$ be a finite family of fuzzy intervals on real line and $\mathcal{T} = \{\mathcal{T}_1, \mathcal{T}_2, \dots, \mathcal{T}_n\}$ be the corresponding fuzzy tolerances. Assuming $\mathcal{J}$ as a crisp vertex set $V = \{v_1, v_2, \dots, v_n\}$, let $\xi = (V, \sigma, \mu)$ be a FTG and $< I, T >$ be the tolerance representation of it. Let $L(c(\mathcal{J}_v))$ and $R(c(\mathcal{J}_v))$ be the left and right end points of the core $c(\mathcal{J}_v)$ and that of support $s(\mathcal{J}_v)$ be $L(s(\mathcal{J}_v))$ and $R(s(\mathcal{J}_v))$ respectively for each $v \in V$.

For a vertex v_i, let $c(\mathcal{T}_i) > c(\mathcal{J}_i)$ and $s(\mathcal{T}_i) > s(\mathcal{J}_i)$. There exists an edge (v_i, v_j) in ξ if $c(\mathcal{J}_i \cap \mathcal{J}_j) \geq min\{c(\mathcal{T}_i), c(\mathcal{T}_j)\} = c(\mathcal{T}_j)$ otherwise if $s(\mathcal{J}_i \cap \mathcal{J}_j) > min\{s(\mathcal{T}_i), s(\mathcal{T}_j)\} = s(\mathcal{T}_j)$. So tolerance core and support for v_i can be set to infinity without changing the set of neighbouring vertices in the FTG. Thus, any fuzzy tolerance core and support length larger than the core and support length of the corresponding fuzzy interval can be set to infinity without changing the adjacency of vertices in FTG.

Let ε be the smallest positive number appearing in the union of the sets (i) $\{|L(c(\mathcal{J}_x)) - L(c(\mathcal{J}_y))|\}$, (ii) $\{|R(c(\mathcal{J}_x)) - R(c(\mathcal{J}_y))|\}$, (iii) $\{|L(c(\mathcal{J}_x)) - R(c(\mathcal{J}_y))|\}$, (iv) $\{c(\mathcal{T}_x)\}$, (v) $\{|c(\mathcal{T}_x) - c(\mathcal{T}_y)|\}$, (vi) $\{c(\mathcal{T}_x) - |c(\mathcal{J}_x) \cap c(\mathcal{J}_y)|\}$.

If x and y be two distinct vertices with $c(\mathcal{T}_x) = c(\mathcal{T}_y)$, choosing one of them, say x, we replace$c(\mathcal{T}_x)$ by $c(\mathcal{T}_x') = c(\mathcal{T}_x) - \frac{\varepsilon}{2}$, and we leave $c(\mathcal{T}_y)$ unchanged. We show that this gives a representation of ξ with one fewer repeated tolerance core. If $(x, z) \in$ E, then $|c(\mathcal{J}_x) \cap c(\mathcal{J}_z)| \geq$ min $\{c(\mathcal{T}_x), c(\mathcal{T}_z)\} \geq$ min $\{c(\mathcal{T}_x'), c(\mathcal{T}_z)\}$ is one possible case. If $(x, z) \notin$ E, then $|c(\mathcal{J}_x) \cap c(\mathcal{J}_z)| <$ min $\{c(\mathcal{T}_x), c(\mathcal{T}_z)\}$ is a possible case and by our choice of ε we know min $\{c(\mathcal{T}_x), c(\mathcal{T}_z)\} - |c(\mathcal{J}_x) \cap c(\mathcal{J}_z)| \geq \varepsilon$. Thus $|c(\mathcal{J}_x) \cap$

$c(\mathcal{I}_z)| \leq min\{c(\mathcal{T}_x), c(\mathcal{T}_z)\} - \varepsilon < min\{c(\mathcal{T}_x'), c(\mathcal{T}_z)\}$ as desired. If necessary, we recompute ε and repeat the process until all tolerance cores are distinct. Similarly, it can be done for the case of tolerance support.

Now suppose two different fuzzy intervals share an end point of core. Let $S = \{L(c(\mathcal{I}_v)), R(c(\mathcal{I}_v)) \forall v \in V\}$ be the set of core end points. Let x and y be the distinct elements of V for which p is an end point of $c(\mathcal{I}_x)$ and $c(\mathcal{I}_y)$. If there exist $x \in V$ with $R(c(\mathcal{I}_x)) = p$, picking the one whose length $c(\mathcal{I}_x)$ is longest and replacing $c(\mathcal{I}_x)$ by $c(\mathcal{I}_x') = [L(c_x) + \frac{\varepsilon}{2}, R(c_x) + \frac{\varepsilon}{2}]$ otherwise, we pick $x \in V$ with $L(c(\mathcal{I}_x)) = p$ and $|c(\mathcal{I}_x)|$ as large as possible and replace $c(\mathcal{I}_x)$ by $c(\mathcal{I}_x') = [L(c(\mathcal{I}_x)) + \frac{\varepsilon}{2}, R(c(\mathcal{I}_x)) + \frac{\varepsilon}{2}]$. It is not hard to see that this gives a representation of ξ with one fewer pair of elements sharing an end point. All tolerances are still distinct. If necessary, we recompute ε and repeat the process until all end points of core are distinct. The similar process can be applied to support also.

11. Fuzzy Competition Graphs

In [12], Cohen introduced the concept of competition graphs in connection with the problem of ecology. Cohen showed that any food web in ecology can be represented by a diagraph. A related graph is defined in [10]. Let the digraph $\overleftarrow{D} = (V, \overleftarrow{E})$ represents a food web. A vertex $x \in V(\overleftarrow{D})$ represents a species in the food web and there is a directed arc $\overleftarrow{(x,s)} \in \overleftarrow{E}(\overleftarrow{D})$ from the vertex x to the vertex s if x preys on the species s. If two species x and y have a common prey s, this means x and y both compete for the prey s. Based on this concept, Cohen defined a graph that represents the relations of competition among the species in ecology, particularly in the food web. The competition graph $C(\overleftarrow{D})$ of a digraph $\overleftarrow{D} = (V, \overleftarrow{E})$ is an undirected graph $G = (V, E)$ which has the same set of vertices V and there is an edge between two distinct vertices $x, y \in V$ if there is a vertex $s \in V$ and arcs $\overleftarrow{(x,s)}, \overleftarrow{(y,s)} \in \overleftarrow{E}(\overleftarrow{D})$. The competition graph is not only used in foodweb, it is also applicable in channel assignment, coding, energy systems, modelling of complex economic, etc. This idea can be extended for more vertices. For example, if the vertices x and y have, say p, common neighbours, then the vertices are joined by an edge. If a graph is constructed from the digraph $\overleftarrow{G}$ based on this idea, then the resulted graph $C(\overleftarrow{G})$ is called p-competition graph. There is another technique to construct other type of competition graph. If two vertices x and v are adjacent, i.e., they are neighbours then x and v are called 1-step neighbours. If v is 1-step neighbour of both x and y, i.e.,x and v are adjacent, and also, y and v are adjacent, then v is called 1-step common neighbour of x and y. If x and v are connected through a path containing $m+1$ vertices (including end vertices), then x and v are said to be m-step neighbours. If both x and y are m-step neighbours of v, then v is called m-step common neighbour of x and y. 2-competition graph is studied in [23].

Like crisp graph, fuzzy out-neighbourhood and fuzzy in-neighbourhood of a vertex in directed FG are defined below.

Definition 16. *Fuzzy out-neighbourhood of a vertex v of a directed FG $\overleftarrow{\xi} = (V, \sigma, \overleftarrow{\mu})$ is the fuzzy set $\mathcal{N}^+(v) = (X_v^+, m_v^+)$, where $X_v^+ = \{u | \overleftarrow{\mu}(v,u) > 0\}$ and $m_v^+ : X_v^+ \to [0,1]$ defined by $m_v^+(u) = \overleftarrow{\mu}(v,u)$. Similarly, fuzzy in-neighbourhood of a vertex v of a directed FG $\overleftarrow{\xi} = (V, \sigma, \overleftarrow{\mu})$ is the fuzzy set $\mathcal{N}^-(v) = (X_v^-, m_v^-)$, where $X_v^- = \{u | \overleftarrow{\mu}(u,v) > 0\}$ and $m_v^- : X_v^- \to [0,1]$ defined by $m_v^-(u) = \overleftarrow{\mu}(u,v)$.*

Example 10. Let $\overleftarrow{\xi}$ be a directed FG. Let $\{a,b,c,d\}$ be the set of vertices with membership values $\sigma(a) = 0.4$, $\sigma(b) = 0.5$, $\sigma(c) = 0.6$, $\sigma(d) = 0.55$. Let the membership values of directed arcs be $\overleftarrow{\mu}(b,a) = 0.3$, $\overleftarrow{\mu}(d,a) = 0.35$, $\overleftarrow{\mu}(b,c) = 0.5$, and $\overleftarrow{\mu}(d,c) = 0.4$. Thus, $\mathcal{N}^+(d) = \{(a, 0.35), (c, 0.4)\}$.$\mathcal{N}^-(c) = \{(b, 0.5), (d, 0.4)\}$. (Note that $(a, \sigma(a))$ represents the vertex a with membership value $\sigma(a)$). It is shown in Figure 12.

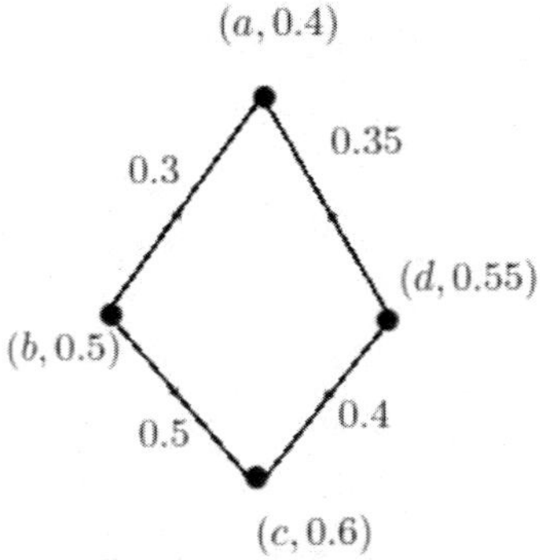

$$\mathcal{N}^+(d) = \{(a, 0.35), (c, 0.4)\}$$

$$\mathcal{N}^-(c) = \{(b, 0.5), (d, 0.4)\}$$

Figure 12. Example of fuzzy out-neighbourhood and in-neighbourhood of a vertex.

The fuzzy competition graph is defined by Samanta and Pal[53] and they also defined other extension of FCG called *m*-step competition graph [55]. Following is the definition of the fuzzy competition graph (FCG).

Definition 17. *The FCG* $\mathcal{C}(\overleftarrow{\xi})$ *of a fuzzy digraph* $\overleftarrow{\xi} = (V, \sigma, \overleftarrow{\mu})$ *is an undirected FG* $\xi = (V, \sigma, \mu)$ *which has the same set of fuzzy vertices as in* $\overleftarrow{\xi}$ *and has a fuzzy edge between two vertices* $x, y \in V$ *in* $\mathcal{C}(\overleftarrow{\xi})$ *if and only if* $\mathcal{N}^+(x) \cap \mathcal{N}^+(y)$ *is non-empty fuzzy set in* $\overleftarrow{\xi}$ *and the membership value of the edge* (x, y) *in* $\mathcal{C}(\overleftarrow{\xi})$ *is* $\mu(x, y) = (\sigma(x) \wedge \sigma(y))h(\mathcal{N}^+(x) \cap \mathcal{N}^+(y))$.

Example 11. Let $\overleftarrow{\xi}$ be a directed FG. Let the vertices be $\overleftarrow{\xi}$ be $(a, 0.3)$, $(b, 0.6)$, $(c, 0.4)$, $(d, 0.5)$, $(e, 0.4)$ and the arcs be $\overleftarrow{\mu}(b, a) = 0.2$, $\overleftarrow{\mu}(c, b) = 0.35$, $\overleftarrow{\mu}(d, c) = 0.2$, $\overleftarrow{\mu}(e, d) = 0.2$, $\overleftarrow{\mu}(e, a) = 0.25$, $\overleftarrow{\mu}(a, d) = 0.3$. It is shown in Figure 13(a). The corresponding FCG is shown in Figure 13(b).

Edges of fuzzy competition graphs represent that two vertices (species) compete for at least one prey. So the strengths of the edges are important to the nature of competitions.

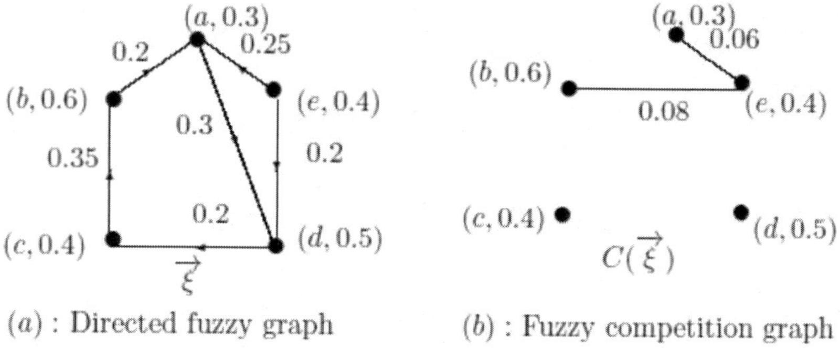

Figure 13. Example of fuzzy competition graph.

Now an extension of fuzzy competition graph, called fuzzy k-competition graph is defined in the following manner.

Definition 18. [24]*Let k be a non-negative number. The fuzzy k-competition graph $\mathcal{C}_k(\overleftarrow{\xi})$ of a fuzzy digraph $\overleftarrow{\xi} = (V, \sigma, \overleftarrow{\mu})$ is an undirected FG $\xi = (V, \sigma, \mu)$ which has the same fuzzy vertex set as $\overleftarrow{\xi}$ and has a fuzzy edge between two vertices x, $y \in V$ in $\mathcal{C}_k(\overleftarrow{\xi})$ if and only if $|\mathcal{N}^+(x) \cap \mathcal{N}^+(y)| > k$. The edge membership value between x and y in $\mathcal{C}_k(\overleftarrow{\xi})$ is $\mu(x,y) = \frac{(k'-k)}{k'}[\sigma(x) \wedge \sigma(y)]h(\mathcal{N}^+(x) \cap \mathcal{N}^+(y))$, where $k' = |\mathcal{N}^+(x) \cap \mathcal{N}^+(y)|$.*

So fuzzy k-competition graph is simply fuzzy competition graph when $k = 0$. An example of fuzzy 0.2-competition graph is given below.

Example 12. Let $\overleftarrow{\xi}$ be a directed FG. Let vertices with membership values of $\overleftarrow{\xi}$ be $(x, 0.4)$, $(y, 0.6)$, $(a, 0.6)$, $(b, 0.7)$, $(c, 0.8)$, $(d, 0.65)$ and the membership values of arcs be $\overleftarrow{\mu}(x,a) = 0.3$, $\overleftarrow{\mu}(x,b) = 0.35$, $\overleftarrow{\mu}(x,c) = 0.36$, $\overleftarrow{\mu}(x,d) = 0.4$, $\overleftarrow{\mu}(y,a) = 0.6$, $\overleftarrow{\mu}(y,b) = 0.5$, $\overleftarrow{\mu}(y,c) = 0.45$, $\overleftarrow{\mu}(y,d) = 0.35$. It is shown in Figure 14(a). The corresponding fuzzy 0.2-competition graph is shown in the Figure 14(b).

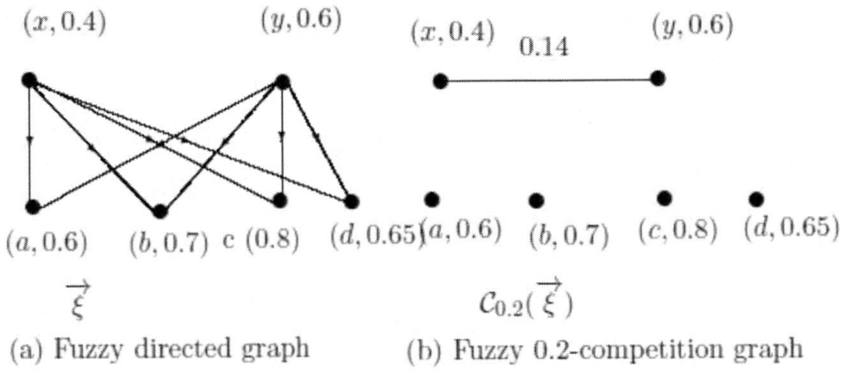

Figure 14. Example of fuzzy 0.2-competition graph.

Theorem 7. Let $\overleftarrow{\xi} = (V, \sigma, \overleftarrow{\mu})$ be a fuzzy digraph. If $\mathcal{N}^+(x) \cap \mathcal{N}^+(y)$ has only one element of $\overleftarrow{\xi}$, then the edge (x, y) of $C(\overleftarrow{\xi})$ is strong if and only if $|\mathcal{N}^+(x) \cap \mathcal{N}^+(y)| > 0.5$.

Proof. Here $\overleftarrow{\xi} = (V, \sigma, \overleftarrow{\mu})$ is a fuzzy digraph. Let $\mathcal{N}^+(x) \cap \mathcal{N}^+(y) = \{(a, m)\}$, where m is the membership value of the element a. Here, $|\mathcal{N}^+(x) \cap \mathcal{N}^+(y)| = m = h(\mathcal{N}^+(x) \cap \mathcal{N}^+(y))$. So, $\mu(x, y) = m \times \sigma(x) \wedge \sigma(y)$. Hence, the edge (x, y) in $C(\overleftarrow{\xi})$ is strong if and only if $m > 0.5$.

If all the edges of a fuzzy digraph are strong, then all the edges of the corresponding FCG may not be strong. This result is illustrated below:

Suppose x, y are two vertices in a fuzzy digraph and $\sigma(x) = 0.3, \sigma(y) = 0.4$. Assumed that the vertices have a common prey z with $\sigma(z) = 0.2$. Let $\overleftarrow{\mu}(x, z) = 0.2, \overleftarrow{\mu}(y, z) = 0.15$. Clearly, the edges $\overleftarrow{(x, z)}$and $\overleftarrow{(y, z)}$ are strong. But, the membership value of the edge (x, y) in corresponding competition graph is $0.3 \times 0.15 = 0.045$. Hence, the edge is not strong as $\frac{0.045}{0.3} = 0.15 < 0.5$.

If all the edges are strong in a fuzzy digraph, then a different result is proved in the following theorem.

Theorem 8. If all the edges of a fuzzy digraph $\overleftarrow{\xi} = (V, \sigma, \overleftarrow{\mu})$ are strong, then $\frac{\mu(x,y)}{(\sigma(x) \wedge \sigma(y))^2} > 0.5$ for all edges (x, y) in $C(\overleftarrow{\xi})$.

Proof. Let $\overleftarrow{\xi} = (V, \sigma, \overleftarrow{\mu})$ be a fuzzy digraph and also let every edge of $\overleftarrow{\xi}$ be strong i.e., $\frac{\overleftarrow{\mu}(x,y)}{\sigma(x) \wedge \sigma(y)} > 0.5$ for all edge (x, y) in $\overleftarrow{\xi}$. Let the associated FCG be $C(\overleftarrow{\xi}) = (V, \sigma, \mu)$.

Case 1.

Let $\mathcal{N}^+(x) \cap \mathcal{N}^+(y)$ be the null set for all $x, y \in V$. Then there is no edge between x and y in $C(\overleftarrow{\xi})$.

Case 2.

$\mathcal{N}^+(x) \cap \mathcal{N}^+(y)$ is not a null set. Let $\mathcal{N}^+(x) \cap \mathcal{N}^+(y) = \{(a_1, m_1), (a_2, m_2), \ldots, (a_z, m_z)\}$, where $m_i, i = 1,2, \ldots, z$ are the membership values of$a_i, i = 1,2, \ldots, z$, respectively. So $m_i = min\{\overleftarrow{\mu}(x, a_i), \overleftarrow{\mu}(y, a_i)\}, i = 1,2, \ldots, z.$ Let $h(\mathcal{N}^+(x) \cap \mathcal{N}^+(y)) = max\{m_i, i = 1,2, \ldots, z\} = m_{max}.$

$\mu(x, y) = (\sigma(x) \wedge \sigma(y)) h(\mathcal{N}^+(x) \cap \mathcal{N}^+(y)) = m_{max} \times \sigma(x) \wedge \sigma(y)$. Hence, $\frac{\mu(x,y)}{(\sigma(x) \wedge \sigma(y))^2} = \frac{m_{max}}{\sigma(x) \wedge \sigma(y)} > 0.5.$

It is observed that if the height of intersection between two out neighbourhoods of two vertices of a fuzzy digraph is greater than 0.5, the edge between two vertices in the associated FCG is strong. But, this result is not valid in the corresponding fuzzy k-competition graph. A related result is proved below.

Theorem 9. Let $\overleftarrow{\xi} = (V, \sigma, \overleftarrow{\mu})$ be a fuzzy digraph. If $h(\mathcal{N}^+(x) \cap \mathcal{N}^+(y)) = 1$ and $|\mathcal{N}^+(x) \cap \mathcal{N}^+(y)| > 2k$, then the edge (x, y) is strong in $C_k(\overleftarrow{\xi})$.

Proof. Let $\overleftarrow{\xi} = (V, \sigma, \overleftarrow{\mu})$ be a fuzzy digraph and $C_k(\overleftarrow{\xi}) = (V, \sigma, \mu)$ be the corresponding fuzzy k-competition graph. Also let, $h(\mathcal{N}^+(x) \cap \mathcal{N}^+(y)) = 1$ and $|\mathcal{N}^+(x) \cap \mathcal{N}^+(y)| > 2k$.

Now, $\mu(x,y) = \frac{k'-k}{k'}\sigma(x) \wedge \sigma(y)h(\mathcal{N}^+(x) \cap \mathcal{N}^+(y))$, where $k' = |\mathcal{N}^+(x) \cap \mathcal{N}^+(y)|$. So, $\mu(x,y) = \frac{k'-k}{k'}\sigma(x) \wedge \sigma(y)$. Hence, $\frac{\mu(x,y)}{\sigma(x)\wedge\sigma(y)} = \frac{k'-k}{k'} > 0.5$ as $k' > 2k$ i.e., the edge (x,y) is strong.

11.1. Fuzzy Neighbourhood Graphs

The fuzzy open neighbourhood and fuzzy closed neighbourhood of a vertex in FG are defined below.

Definition 19. *The fuzzy open neighbourhood of a vertex v of a FG $\xi = (V,\sigma,\mu)$ is the fuzzy set $\mathcal{N}(v) = (X_v, m_v)$, where $X_v = \{u|\mu(v,u) > 0\}$ and $m_v: X_v \to [0,1]$ defined by $m_v(u) = \mu(v,u)$. For each vertex $v \in V$, there is a fuzzy singleton set, $A_v = (\{v\},\sigma')$ such that $\sigma': \{v\} \to [0,1]$ defined by $\sigma'(v) = \sigma(v)$. The fuzzy closed neighbourhood of a vertex v is $\mathcal{N}[v] = \mathcal{N}(v) \cup A_v$.*

In this section, fuzzy open neighbourhood and fuzzy closed neighbourhood graphs are defined. Based on these FGs fuzzy k-neighbourhood graphs of open and closed types are introduced.

Definition 20. *Let $\xi = (V,\sigma,\mu)$ be a FG. The fuzzy open neighbourhood graph of ξ is a FG $\mathcal{N}(\xi) = (V,\sigma,\mu')$ whose fuzzy vertex set is same as ξ and there is a fuzzy edge between the vertices x and $y \in V$ in $\mathcal{N}(\xi)$ if and only if $\mathcal{N}(x) \cap \mathcal{N}(y)$ is non-empty fuzzy set in ξ and $\mu': V \times V \to [0,1]$ such that $\mu'(x,y) = [\sigma(x) \wedge \sigma(y)]h(\mathcal{N}(x) \cap \mathcal{N}(y))$.*

Definition 21. *Let $\xi = (V,\sigma,\mu)$ be a FG. The fuzzy closed neighbourhood graph of ξ is a FG $\mathcal{N}[\xi] = (V,\sigma,\mu')$ whose fuzzy vertex set is same as ξ and there is a fuzzy edge between two vertices x and $y \in V$ in $\mathcal{N}[\xi]$ if and only if $\mathcal{N}[x] \cap \mathcal{N}[y]$ is non-empty fuzzy set in ξ and $\mu': V \times V \to [0,1]$ such that $\mu'(x,y) = [\sigma(x) \wedge \sigma(y)]h(\mathcal{N}[x] \cap \mathcal{N}[y])$.*

Definition 22. *Let $\xi = (V, \sigma, \mu)$ be a FG. Fuzzy (k)-neighbourhood graph (read as open fuzzy k-neighbourhood graph) of ξ is a FG $\mathcal{N}_{\mathrm{k}}(\xi) = (V, \sigma, \mu')$ whose vertex set is same as ξ and two vertices x and $y \in V$ are connected by an edge in $\mathcal{N}_{\mathrm{k}}(\xi)$ if and only if $|\mathcal{N}(x) \cap \mathcal{N}(y)| > k$ in ξ and $\mu': V \times V \to [0,1]$ such that $\mu'(x,y) = \frac{(k'-k)}{k'}[\sigma(x) \wedge \sigma(y)]h(\mathcal{N}(x) \cap \mathcal{N}(y))$, where $k' = |\mathcal{N}(x) \cap \mathcal{N}(y)|$.*

Definition 23. *Let $\xi = (V, \sigma, \mu)$ be a FG. Fuzzy $[k]$-neighbourhood graph (read as fuzzy closed k-neighbourhood graph) of ξ is a FG $\mathcal{N}_{\mathrm{k}}[\xi] = (V, \sigma, \mu')$ whose fuzzy vertex set is same as ξ and two vertices x and $y \in V$ are connected by an edge in $\mathcal{N}_{\mathrm{k}}[\xi]$ if and only if $|\mathcal{N}[x] \cap \mathcal{N}[y]| > k$ in ξ and $\mu': V \times V \to [0,1]$ such that $\mu'(x,y) = \frac{(k'-k)}{k'}[\sigma(x) \wedge \sigma(y)]h(\mathcal{N}[x] \cap \mathcal{N}[y])$, where $k' = |\mathcal{N}[x] \cap \mathcal{N}[y]|$.*

The examples of fuzzy (0.2)-neighbourhood and fuzzy [0.2]-neighbourhood graphs are considered below.

Example 13. Let $\xi = (V, \sigma, \mu)$ be a FG. Let the vertices with membership values in ξ be $(a, 0.3)$, $(b, 0.6)$, $(c, 0.4)$, $(d, 0.5)$, $(e, 0.4)$ and edges with their membership values be $\mu(a,b) = 0.3, \mu(b,c) = 0.4, \mu(c,d) = 0.4, \mu(d,e) = 0.4, \mu(e,a) = 0.3, \mu(a,d) = 0.3$. The membership values of other edges are taken as 0. The fuzzy open neighbourhood graph $\mathcal{N}(\xi)$, the fuzzy closed neighbourhood graph $\mathcal{N}[\xi]$, the fuzzy (0.2)-neighbourhood graph $\mathcal{N}_{0.2}(\xi)$ and the fuzzy [0.2]-neighbourhood graph $\mathcal{N}_{0.2}[\xi]$ of the FG ξ are shown in Figure 15.

Theorem 10. For every edge of a FG ξ, there exists one edge in $\mathcal{N}[\xi]$.

Proof. Let (x, y) be an edge of a FG $\xi = (V, \sigma, \mu)$. Let the corresponding closed neighbourhood graph be $\mathcal{N}[\xi] = (V, \sigma, \nu)$. Then $x, y \in \mathcal{N}[x]$ and $x, y \in \mathcal{N}[y]$. So $x, y \in \mathcal{N}[x] \cap \mathcal{N}[y]$. Hence, $h(\mathcal{N}[x] \cap \mathcal{N}[y]) \neq 0$. Now, $\nu(x,y) = \sigma(x) \wedge \sigma(y)h(\mathcal{N}[x] \cap \mathcal{N}[y]) \neq 0$. So for every edge (x, y) in ξ, there exists an edge (x, y) in $\mathcal{N}[\xi]$. W

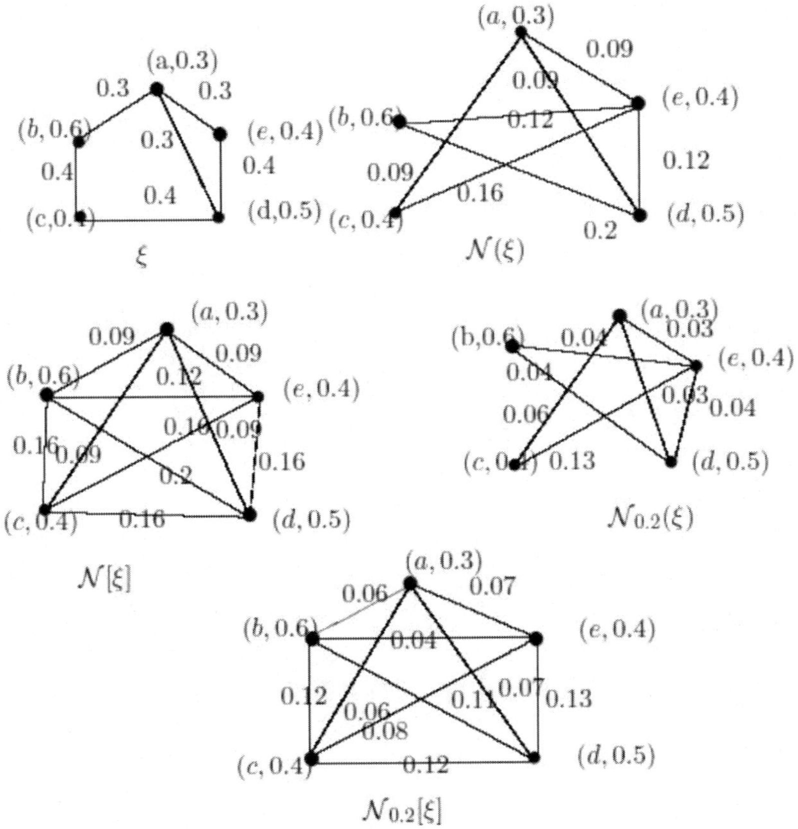

Figure 15. Example of fuzzy neighbourhood graphs.

Definition 24. Let $\overleftarrow{\xi} = (V, \sigma, \overleftarrow{\mu})$ be a fuzzy digraph. The underlying FG of $\overleftarrow{\xi}$ is denoted by $\mathcal{U}(\xi)$ and is defined as $\mathcal{U}(\xi) = (V, \sigma, \mu)$, where $\mu(u, v) = \min\{\overleftarrow{\mu}(u, v), \overleftarrow{\mu}(v, u)\}$ for all $u, v \in V$.

A relation between fuzzy (k)-neighbourhood graph and fuzzy k-competition graph is established below.

Theorem 11. If the symmetric fuzzy digraph $\overleftarrow{\xi}$ is loop less, then $\mathcal{C}_k(\overleftarrow{\xi}) = \mathcal{N}_k(\mathcal{U}(\xi))$, where $\mathcal{U}(\xi)$ is the FG underlying $\overleftarrow{\xi}$.

Proof. Let a directed FG be $\overleftarrow{\xi} = (V, \sigma, \overleftarrow{\mu})$ and the corresponding underlying FG be $\mathcal{U}(\xi) = (V, \sigma, \mu)$. Also let $\mathcal{C}_k(\overleftarrow{\xi}) = (V, \sigma, \nu)$ and

$\mathcal{N}_k(\mathcal{U}(\xi)) = (V,\sigma,\nu')$. The fuzzy vertex set of $\overleftarrow{\xi}$ is equal to $\mathcal{C}_k(\overleftarrow{\xi})$. Also, an underlying FG has the same fuzzy vertex set as the directed FG. Hence, $\mathcal{N}_k(\mathcal{U}(\xi))$ has the same fuzzy vertex set as $\overleftarrow{\xi}$. Now we need to show that $\nu(x,y) = \nu'(x,y)$ for all $x,y \in V$. If $\nu(x,y) = 0$ in $\mathcal{C}_k(\overleftarrow{\xi})$ then $|\mathcal{N}^+(x) \cap \mathcal{N}^+(y)| \le k$. As $\overleftarrow{\xi}$ is symmetric fuzzy set, $|\mathcal{N}[x] \cap \mathcal{N}[y]| \le k$ in $\mathcal{U}(\xi)$. So $\nu'(x,y) = 0$.

If $|\mathcal{N}^+(x) \cap \mathcal{N}^+(y)| > k$ then $\nu(x,y) > 0$ in $\mathcal{C}_k(\overleftarrow{\xi})$.

So $\nu(x,y) = \frac{(k'-k)}{k'}[\sigma(x) \wedge \sigma(y)]h(\mathcal{N}^+(x) \cap \mathcal{N}^+(y))$,

where $k' = |\mathcal{N}^+(x) \cap \mathcal{N}^+(y)|$. As $\overleftarrow{\xi}$ is symmetric fuzzy digraph, $|\mathcal{N}[x] \cap \mathcal{N}[y]| > k$ in $\mathcal{U}(\xi)$. So $\nu' = \frac{(k''-k)}{k''}[\sigma(x) \wedge \sigma(y)]h(\mathcal{N}[x] \cap \mathcal{N}[y])$, where $k'' = |\mathcal{N}[x] \cap \mathcal{N}[y]|$. It is clear that $h(\mathcal{N}^+(x) \cap \mathcal{N}^+(y))$ in $\overleftarrow{\xi}$ is equal to $h(\mathcal{N}[x] \cap \mathcal{N}[y])$ in $\mathcal{U}(\xi)$ as $\overleftarrow{\xi}$ is symmetric. $k' = k''$ for similar reason. Hence, $\nu(x,y) = \nu'(x,y)$ for all $x,y \in V$.

Similarly, a relation between fuzzy $[k]$-neighbourhood graph and fuzzy k-competition graph is established below.

Theorem 12. If the symmetric fuzzy digraph $\overleftarrow{\xi}$ has loop at every vertex, then $\mathcal{C}_k(\overleftarrow{\xi}) = \mathcal{N}_k[\mathcal{U}(\xi)]$, where $\mathcal{U}(\xi)$ is the loop less FG underlying $\overleftarrow{\xi}$.

Proof. Let a directed FG be $\overleftarrow{\xi} = (V,\sigma,\overleftarrow{\mu})$ and the corresponding underlying loop less graph be $\mathcal{U}(\xi) = (V,\sigma,\mu)$. Also let $\mathcal{C}_k(\overleftarrow{\xi}) = (V,\sigma,\nu)$ and $\mathcal{N}_k[\mathcal{U}(\xi)] = (V,\sigma,\nu')$. The fuzzy vertex set of $\overleftarrow{\xi}$ is equal to $\mathcal{C}_k(\overleftarrow{\xi})$. Also, an underlying FG has the same fuzzy vertex set as the directed FG. Hence, $\mathcal{N}_k[\mathcal{U}(\xi)]$ has the same fuzzy vertex set as $\overleftarrow{\xi}$. Now, we have to prove that $\nu(x,y) = \nu'(x,y)$ for all $x,y \in V$. As the directed FG $\overleftarrow{\xi}$ has a loop at every vertex, out-neighbourhood of each vertex contains the vertex itself. Hence, if $\nu(x,y) = 0$ in $\mathcal{C}_k(\overleftarrow{\xi})$ then $|\mathcal{N}^+(x) \cap \mathcal{N}^+(y)| \le k$. As $\overleftarrow{\xi}$ is symmetric fuzzy set, $|\mathcal{N}(x) \cap \mathcal{N}(y)| \le k$ in $\mathcal{U}(\xi)$. So $\nu'(x,y) = 0$.

If $|\mathcal{N}^+(x) \cap \mathcal{N}^+(y)| > k$ then $\nu(x,y) > 0$ in $\mathcal{C}_k(\overleftarrow{\xi})$.

So $\nu(x,y) = \frac{(k'-k)}{k'}[\sigma(x) \wedge \sigma(y)]h(\mathcal{N}^+(x) \cap \mathcal{N}^+(y))$,

where $k' = |\mathcal{N}^+(x) \cap \mathcal{N}^+(y)|$. As $\overleftarrow{\xi}$ is a symmetric fuzzy digraph, $|\mathcal{N}(x) \cap \mathcal{N}(y)| > k$ in $\mathcal{U}(\xi)$. So $\nu' = \frac{(k''-k)}{k''}[\sigma(x) \wedge \sigma(y)]h(\mathcal{N}(x) \cap \mathcal{N}(v))$, where $k'' = |\mathcal{N}(x) \cap \mathcal{N}(y)|$. It is clear that $h(\mathcal{N}^+(x) \cap \mathcal{N}^+(y))$ in $\overleftarrow{\xi}$ is equal to $h(\mathcal{N}(x) \cap \mathcal{N}(v))$ in $\mathcal{U}(\xi)$ and $k' = k$"as $\overleftarrow{\xi}$ is symmetric. Hence, $\nu(x,y) = \nu'(x,y)$ for all $x, y \in V$.

For any positive integer p, we define a new kind of FG known as p-competition FG, defined below:

Definition 25. Let p be a positive integer. The p-competition FG $\mathcal{C}^p(\overleftarrow{\xi})$ of a fuzzy digraph $\overleftarrow{\xi} = (V, \sigma, \overleftarrow{\mu})$ is an undirected FG $\xi = (V, \sigma, \mu)$ which has same set of fuzzy vertices as in $\overleftarrow{\xi}$ and two vertices x and $y \in V$ are adjacent in $\mathcal{C}^p(\overleftarrow{\xi})$ if and only if $|supp(\mathcal{N}^+(x) \cap \mathcal{N}^+(y))| \geq p$. The membership value of the edge (x,y) in $\mathcal{C}^p(\overleftarrow{\xi})$ is $\mu(x,y) = \frac{(n-p)+1}{n}[\sigma(x) \wedge \sigma(y)]h(\mathcal{N}^+(x) \cap \mathcal{N}^+(y))$, where $n = |supp(\mathcal{N}^+(x) \cap \mathcal{N}^+(y))|$.

The following example illustrates a 2-competition FG.

Example 14. Let us consider the FG shown in Figure 16(a) with set of vertices be $\{x, y, z, a, b, c\}$. Let $\sigma(x) = 0.7, \sigma(y) = 0.8, \sigma(z) = 0.9, \sigma(a) = 0.75, \sigma(b) = 0.85, \sigma(c) = 0.95$ and membership values of edges be $\overleftarrow{\mu}(\overleftarrow{x,a}) = 0.7, \overleftarrow{\mu}(\overleftarrow{x,b}) = 0.7$, $\overleftarrow{\mu}(\overleftarrow{y,a}) = 0.6$, $\overleftarrow{\mu}(\overleftarrow{y,b}) = 0.8$, $\overleftarrow{\mu}(\overleftarrow{y,c}) = 0.8$, $\overleftarrow{\mu}(\overleftarrow{z,c}) = 0.9$. The corresponding 2-competition FG is shown in the Figure 16(b).

Fuzzy k-competition FG and p-competition FG are two generalizations of fuzzy competition graph, where in the first FG a real positive number k is related to the cardinality of a fuzzy set. In the second FG, i.e.,p-competition FG, p is related to the cardinality of a crisp set.

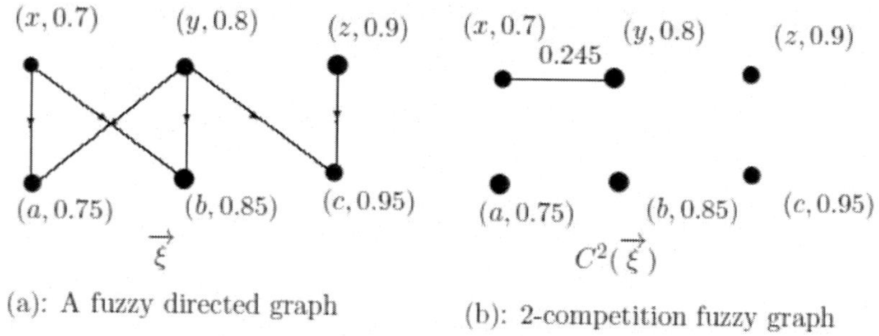

Figure 16. Example of 2-competition fuzzy graph.

Theorem 13. *Let* $\overleftarrow{\xi} = (V, \sigma, \overleftarrow{\mu})$ *be a fuzzy digraph. If* $h(\mathcal{N}^+(x) \cap \mathcal{N}^+(y)) = 1$ *in* $C^{[\frac{n}{2}]}(\overleftarrow{\xi})$, *then the edge* (x, y) *is strong, where* $n = |supp(\mathcal{N}^+(x) \cap \mathcal{N}^+(y))|$. *(*$[x]$ *= greatest integer not exceeding* x.*)*

Proof. Here $\overleftarrow{\xi} = (V, \sigma, \overleftarrow{\mu})$ is a fuzzy digraph. Let the corresponding $[\frac{n}{2}]$-fuzzy competition graph be $\xi = (V, \sigma, \mu)$, where $n = |supp(\mathcal{N}^+(x) \cap \mathcal{N}^+(y))|$. Also, assumed that $h(\mathcal{N}^+(x) \cap \mathcal{N}^+(y)) = 1, x, y \in V$. Now, $\mu(x, y) = \frac{n-[\frac{n}{2}]+1}{n} \sigma(x) \wedge \sigma(y)$. This gives the result, $\frac{\mu(x,y)}{\sigma(x) \wedge \sigma(y)} = \frac{n-[\frac{n}{2}]+1}{n} > 0.5$. Hence, the edge (x, y) is strong.

Another variety of competition graph is the m-step competition graph [11,12,15].

12. Other Related Works

After the development of fuzzy graph theory using several extension of fuzzy sets many generalization of fuzzy graphs have been made by researchers. The intuitionistic fuzzy graphs [3,47,48], interval-valued fuzzy graphs [1,40,41,42,60], bipolar fuzzy graphs [2,43,44,51,52,54], m-polar fuzzy graphs [15,16,17,18], etc. are some of the useful and applicable classes of fuzzy graphs. Many researchers are working on these types of graphs and are applying these theories in different areas of science and technology.

REFERENCES

[1] Akram, M., and W. A. Dudek."Interval-valued fuzzy graphs." *Computers and Mathematics with Applications* 61(2): (2011) 289-299.

[2] Akram, M., and W.A. Dudek."Regular bipolar fuzzy graphs." *Neural Computing and Applications*, 21 (suppl 1): (2012) 197-205.

[3] Akram, M, and W. A. Dudek. "Intuitionistic fuzzy hypergraphs with applications." *Information Sciences* 218: (2013) 182-193.

[4] Bhutani, K. R.,and A. Battou."On M-strong fuzzy graphs." *Information Sciences* 155: (2003) 103-109.

[5] Bhutani, K. R., and A. Rosenfeld."Strong arcs in fuzzy graphs." *Information Sciences* 152: (2003) 319-322.

[6] Blue, M., B. Bush, and J. Puckett."Unified approach to fuzzy graph problems." *Fuzzy Sets and Systems* 125: (2002) 355-368.

[7] Burt, R. S. *Structural holes*. (Harvard University Press, 1992).

[8] Burt, R. S. "The contingent value of social capital." *Administrative Science Quarterly* 42: (1997) 339-365.

[9] Brigham, R. C., F. R. McMorris, and R. P. Vitray. "Tolerance competition graphs." *Linear Algebra and its Application* 217: (1995) 41-52.

[10] Cable, C., K. F. Jones, J. R. Lundgren, and S. Seager."Niche graphs." *Discrete Apllied Mathematics* 23(3): (1989) 231-241.

[11] Cho, H. H., S. R. Kim, and Y. Nam."The m-step competition graph of a diagraph."*Discrete Applied Mathematics* 105(1-3):(2000) 115-127.

[12] Cohen, J. E. "Interval graphs and food webs: a finding and a problem."(Document 17696-PR, RAND Corporation, Santa Monica, CA, 1968).

[13] Craine, W. L. "Characterizations of fuzzy interval graphs."*Fuzzy Sets and Systems* 68: (1994) 181-193.

[14] Eslahchi, C., and B. N. Onaghe. "Vertex strength of fuzzy graphs."*International Journal of Mathematics and Mathematical Sciences.* Volume 2006, Article ID 43614, Pages 1-9.

[15] Ghorai, G., and M. Pal."A study on m-polar fuzzy planar graphs." *Int. J. Computing Science and Mathematics* 7, no. 3: (2016) 283-292.

[16] Ghorai, G., and M. Pal. "Some isomorphic properties of m-polar fuzzy graphs with applications."*SpringerPlus* 5: (2016) 1-21.

[17] Ghorai, G., and M. Pal. "Regular product vague graphs and product vague line graphs."*Cogent Mathematics* 3(1): (2016) 1213214.

[18] Ghosh, P., K. Kundu, and D. Sarkar. "Fuzzy graph representation of a fuzzy concept lattice." *Fuzzy Sets and Systems* 161 no. 12: (2010) 1669-1675.

[19] Goetschel, R. H. "Introduction to fuzzy hypergraphs and Hebbian structures." *Fuzzy Sets and Systems* 76: (1995) 113-130.

[20] Goetschel, R. H. "Fuzzy colorings of fuzzy hypergraphs."*Fuzzy Sets and Systems* 94: (1998) 185-204

[21] Goetschel, R. H., and W. Voxman. "Intersecting fuzzy hypergraphs." *Fuzzy Sets and Systems* 99: (1998) 81 – 96.

[22] Gulumbic, M. C., and A. Trenk. *Tolerance Graphs.*(Cambridge University Press, 2004).

[23] Isaak, G., S. R. Kim, T. A. McKee, F. R. McMorris, and F. S. Roberts, "2-competition graphs." *SIAM J. Disc. Math.* 5 no. 4: (1992) 524-538.

[24] Kim, S. R., T. A. McKee, F. R. McMorris, and F. S. Roberts."p-competition graphs." *Linear Algebra and its Application* 217: (1995) 167-178.

[25] Koczy, L. T. "Fuzzy graphs in the evaluation and optimization of networks." *Fuzzy Sets and Systems* 46: (1992) 307-319.

[26] Mathew, S., and M. S. Sunitha."Types of arcs in a fuzzy graph." *Information Sciences* 179: (2009) 1760-1768.

[27] Mathew, S., and M. S. Sunitha."Node connectivity and arc connectivity of a fuzzy graph." *Information Sciences* 180 no. 4: (2010) 519-531.

[28] McAllister, M. L. N. "Fuzzy intersection graphs." *Comput. Math. Applic.*15 no. 10: (1988)871 – 886.

[29] Mertzios, G. B., and I. S. S. Zaks."The recognition of tolerance and bounded tolerance graphs."*Symposiam on Theoritial Aspects of Computer Science,* (2000) 585-596.

[30] Mordeson, J. N., and P. S. Nair. *Fuzzy graphs and hypergraphs.* (Physica Verlag, 2000).

[31] Munoz, S., M. T. Ortuno, J. Rami'rez, and J. Yanez. "Coloring fuzzy graphs." *Omega* 33 no. 3: (2005) 211-221.

[32] Nagoorgani, A., and K. Radha. "On regular fuzzy graphs." *Journal of Physical Sciences* 12: (2008) 33-40.

[33] Nagoorgani, A., and R. J. Hussain. "Fuzzy effective distance k-dominating sets and their applications." *International Journal of Algorithms, Computing and Mathematics* 2(3): (2009) 25-36.

[34] Nagoorgani, A., and A. Latha. "On irregular fuzzy graphs." *Applied Mathematical Sciences* 6 no. 11: (2012) 517 - 523.

[35] Nair, P. S., and S. C. Cheng. "Cliques and fuzzy cliques in fuzzy graphs." *IFSA World Congress and 20th NAFIPS International Conference*, 4: (2001) 2277 - 2280.

[36] Nair, P. S. "Perfect and precisely perfect fuzzy graphs." *Fuzzy Information Processing Society* (2008) 19-22.

[37] Narayanam, R., "A game theoretic approach to identify critical components in networked systems." In *Annual SRII Global Conference (SRII)*, 2012, IEEE.

[38] Natarajan, C., and S. K. Ayyasawamy."On strong (weak) domination in fuzzy graphs." *World Academy of Science, Engineering and Technology* 67:(2010) 247-249.

[39] Nirmala, G., and K. Dhanabal. "Special planar fuzzy graph configurations." *International Journal of Scientific and Research Publications* 2 no. 7: (2012) 1-4.

[40] Rashmanlou, H., and M. Pal. "Balanced interval-valued fuzzy graphs." *Journal of Physical Sciences* 17: (2013) 43-57.

[41] Rashmanlou, H., and M. Pal. "Antipodal interval-valued fuzzy graphs." *International Journal of Applications of Fuzzy Sets and Artificial Intelligence* 3: (2013) 107-130.

[42] Rashmanlou, H., and M. Pal. "Some properties of highly irregular interval valued fuzzy graphs." *World Applied Sciences Journal* 27 no. 12: (2013) 1756-1773.

[43] Rashmanlou, H., S. Samanta, M. Pal, and R.A.Borzooei. "A study on bipolar fuzzy graphs." *Journal of Intelligent and Fuzzy Systems* 28: (2015) 571-580.

[44] Rashmanlou, H., S. Samanta, M. Pal, and R. A. Borzooei. "Bipolar fuzzy graphs with Categorical properties." *The International Journal of Computational Intelligence Systems* 8 no. 5: (2015) 808-818.

[45] Raut, S., M. Pal, and G. Ghorai. "Fuzzy permutation graph and its complements." *Journal of Intelligent & Fuzzy Systems* 35: (2018) 2199-2213.

[46] Rosenfeld, A. "Fuzzy graphs." in: L.A. Zadeh, K.S. Fu, M. Shimura (Eds.), *Fuzzy Sets and Their Applications.* (Academic Press, New York, 1975) 77-95.

[47] Sahoo, S., and M. Pal."Intuitionistic fuzzy competition graphs."*J. Appl. Math. Comput.,* 52 no. 1: (2016) 37-57.

[48] Sahoo, S., M. Pal, H.Rashmanlou, and R. A. Borzooei."Covering and paired domination in intuitionistic fuzzy graphs." *Journal of Intelligent and Fuzzy Systems* 33 no. 6: (2017) 4007-4015.

[49] Samanta, S., and M. Pal. "Fuzzy tolerance graphs." *International Journal of Latest Trends in Mathematics* 1 no. 2: (2011) 57-67.

[50] Samanta, S., and M. Pal. "Fuzzy threshold graphs." *CIIT International Journal of Fuzzy Systems* 3 no. 12: (2011) 360-364.

[51] Samanta, S., and M. Pal. "Irregular bipolar fuzzy graphs." *International Journal of Applications of Fuzzy Sets* 2: (2012) 91-102.

[52] Samanta, S., and M. Pal. "Bipolar fuzzy hypergraphs."*International Journal of Fuzzy Logic Systems* 2 no. 1: (2012) 17-28.

[53] Samanta, S., and M. Pal. "Fuzzy k-competition graphs and p-competition fuzzy graphs." *Fuzzy Inf. Eng*. 5 no. 2: (2013) 191-204.

[54] Samanta, S., and M. Pal. "Some more results on bipolar fuzzy sets and bipolar fuzzy intersection graphs." *The Journal of Fuzzy Mathematics* 22 no. 2: (2014) 253-262.

[55] Samanta, S., M. Akram, and M. Pal."*m*-step fuzzy compitition graphs." *Journal of Applied Mathematics and Computing* 47: (2015) 461-472.

[56] Samanta, S., and M. Pal. "Fuzzy planar graphs."*IEEE Transaction on Fuzzy System* 23 no. 6: (2015) 1936 - 1942.

[57] Samanta, S., B. Sarkar, D. Shin, and M. Pal. "Completeness and regularity of generalized fuzzy graphs."*SpringerPlus*, 5: (2016) 1979.

[58] Samanta, S., T. Pramanik, and M. Pal."Fuzzy colouring of fuzzy graphs." *Afrika Matematika* 27 no. 1-2: (2016) 37-50.

[59] Samanta, S., H. Rashmanlou, M. Pal, and R. A. Borzooei. "A study on vague graphs." *SpringerPlus*, 5: (2016) 1234. DOI: 10.1186/s40064-016-2892-z.

[60] Pramanik, T., S. Samanta, M. Pal, S. Mondal, and B. Sarkar. "Interval-valued fuzzy ϕ-tolerance competition graphs." *SpringerPlus*5: (2016) 1981. https://doi.org/10.1186/s40064-016-3463-z.

Biographical Sketch

Dr. Madhumangal Pal
Professor of Applied Mathematics
Department of Applied Mathematics with Oceanology and Computer Programming, Vidyasagar Univerity, Midnapore, India
Email: mmpalvu@gmail.com

Education:

- BSc, Mathematics (Honours), Vidyasagar University, Midnapore, India, 1988.
- MSc, Applied Mathematics, Vidyasagar University, Midnapore, India, 1990.
- PhD, Indian Institute of Technology Kharagpur, India, 1996.

Address: Department of Applied Mathematics with Oceanology and Computer Programming, Vidyasagar University, Midnapore-721102, India
Email: mmpalvu@gmail.com

Research and Professional Experience: Research experience 29 yrs. Professional experience 24 yrs.

Honors:

- Received University Silver Medal for rank second in B.Sc. (Honours) in the year 1988.
- Received University Gold Medal for rank first in M.Sc. in the year 1990.
- Received Computer Division medal from Institution of Engineers (India) in the year 1996 for the best research work published in the Institution journal jointly with Prof. G.P.Bhattacharjee
- Received Bharat Jyoti Award from International Friendship Association, New Delhi, 2012.
- Received 5th Faculty Branding Awards 2017, International Achievement Award, from Educationexpo, Research and Branding Company, on 23.07.2017.

Memberships: American Mathematical Society, USA; Calcutta Mathematical Society, India; Academy of Discrete Mathematics and Applications, India; Neutrosophic Science International Association, USA.

Links:
http://www.vidyasagar.ac.in/Faculty/Profile?fac_u_id=Fac-MATH-18

Scopus ID : 7101848756
ORCHID ID : orcid.org/0000-0002-6709-836X
Google Scholar ID: ZN2dqfoAAAAJ

Recent Publications (2016 – 2019)

1. A. Rana, A. Pal, M. Pal, An efficient algorithm to solve the distance *k*-domination problem on permutation graphs, *Journal of Discrete Mathematical Sciences and Cryptography,* 19(2) (2016) 241-255.
2. Sk. Amanathulla and M. Pal, L (0,1) and L(1,1) labeling problems on circular arc graphs, *International Journal of Soft Computing*, 11(6) (2016) 343-350.
3. S. Paul, A. Pa and M. Pal, A new method to compute square of interval graphs and their colouring in linear time, *AKCE International Journal of Graphs and Combinatorics*, 13(1) (2016) 54-64.
4. Sk. Amanathulla and M. Pal, L (3,2,1) and L(4,3,2,1) labeling problems on circular arc graphs, *Inter. J. Control Theory and Applications*, 9(34) (2016) 869-884.
5. T. Pramanik, S. Samanta and M. Pal, Interval-valued fuzzy planar graphs, *Int. J. Mach. Learn. & Cyber.*, 7(4) (2016) 653-664, Springer, DOI 10.1007/s13042-014-0284-7
6. Rajab Ali Borzooei, Hossein Rashmanlou, Sovan Samanta, Madhumangal Pal, A Study on Fuzzy Labeling Graphs, *Journal of Intelligent and Fuzzy Systems*, 30 (2016) 3349–3355.
7. Rajab Ali Borzooei, Hossein Rashmanlou, Madhumangal Pal, Regularity of vague graphs, *Journal of Intelligent and Fuzzy Systems,* 30 (2016) 3681–3689, DOI: 10.3233/IFS-162114
8. Sovan Samanta, Madhumangal Pal, Rajab Ali Borzooei, Hossein Rashmanlou, Vague graphs and strengths, *Journal of Intelligent and Fuzzy Systems*, 30 (2016) 3675–3680, DOI: 10.3233/IFS-162113
9. Hossein Rashmanlou, Sovan Samanta, Madhumangal Pal, Rajab Ali Borzooei, Product of bipolar fuzzy graphs and their degree, *International Journal of General Systems*, 45(1) (2016) 1-14.

10. Rajab Ali Borzooei, Hossein Rashmanlou, S. Samanta, Madhumangal Pal, New concept of vague competition graphs, *Journal of Intelligent and Fuzzy Systems*, 31 (2016) 69-75.
11. G. Ghorai, M. Pal, Faces and dual of m-polar fuzzy planar graphs, *Journal of Intelligent and Fuzzy System,* 31 (2016) 2041-2049.
12. S. Samanta, T. Pramanik and M. Pal, Fuzzy colouring of fuzzy graphs, *Afrika Matematika*, 27(1-2) (2016) 37-50 DOI 10.1007/s13370-015-0317-8
13. G. Ghorai, M. Pal**,** A study on m-polar fuzzy planar graphs, *Int. J. Computing Science and Mathematics*, 7(3) (2016) 283-292.
14. G. Ghorai, M. Pal**,** Some isomorphic properties of m-polar fuzzy graphs with applications, *SpringerPlus*, 5:1-21 (2016), DOI: 10.1186/s40064-016-3783-z
15. T. Pramanik, M. Pal and S. Mondal, Interval-valued fuzzy threshold graph, *Pacific Science Review A: Natural Science and Engineering*, 18 (2016) 66-71.
16. S.Samanta, H.Rashmanlou; M. Pal, R. A. Borzooei, A study on vague graphs, *SpringerPlus,* (2016) 5:1234**.** DOI: 10.1186/s40064-016-2892-z
17. Sankar Sahoo, M. Pal, Intuitionistic fuzzy competition graphs, *J. Appl. Math. Comput.*, 52(1) (2016) 37–57. DOI 10.1007/s12190-015-0928-0
18. G. Ghorai, M. Pal, Regular product vague graphs and product vague line graphs, *Cogent Mathematics,* 3(1) (2016) 1213214. https://doi.org/10.1080/23311835.2016.1213214
19. Hossein Rashmanlou, Sovan Samanta, Madhumangal Pal, Rajab Ali Borzooei, Properties of interval valued intuitionistic (S,T) – Fuzzy graphs, *Pacific Science Review A: Natural Science and Engineering*, 18(1) (2016) 30-37. doi:10.1016/j.psra.2016.06.003.

20. Sovan Samanta, Biswajit Sarkar, Dongmin Shin, Madhumangal Pal, Completeness and regularity of generalized fuzzy graphs, *SpringerPlus*, (2016) 5: 1979.
21. T. Pramanik, S. Samanta, M. Pal, S Mondal, B. Sarkar, Interval-valued fuzzy φ-tolerance competition graphs, *Springer Plus* (2016) 5: 1981. https://doi.org/10.1186/s40064-016-3463-z
22. G. Ghorai, M. Pal, Some properties of m-polar fuzzy graphs, *Pacific science review A: Natural science and engineering*, 18 (1) (2016) 38-46.
23. Sanjib Mondal and M. Pal, Similarity relations, eigenvalues and eigenvectors of bipolar fuzzy matrix, *Journal of Intelligent and Fuzzy Systems*, 30(2016) 2297-2307, DOI: 10.3233/IFS-151780
24. Sanjib Mondal and Madhumangal Pal, Rank of interval-valued fuzzy matrices, *Afrika Matematika*, 27 (1-2) (2016) 97-114 DOI: 10.1007/s13370-015-0325-8
25. M Pal, Fuzzy matrices with fuzzy rows and columns, *Journal of Intelligent & Fuzzy Systems*, 30 (1) (2016) 561-573.
26. Tapan Senapati, Monoranjan Bhowmik, Madhumangal Pal, Triangular norm based fuzzy BG-algebras, *Afrika Matematika*, 27 (1-2) (2016) 187-199 DOI 10.1007/s13370-015-0330-y
27. A. Dey, M. Pal, Multi-fuzzy complex nilpotent matrices, *International Journal of Fuzzy System Applications*, 5 (4) (2016) 52-76.
28. C. Jana, T. Senapati, M. Pal, (∈,∈∨ q)-intuitionistic fuzzy BCI-subalgebras of a BCI-algebra, *Journal of Intelligent & Fuzzy Systems*, 31 (1) (2016) 613-621
29. C Jana, M. Pal, Applications of new soft intersection set on groups, *Ann. Fuzzy Math. Inform.*, 11 (6) (2016) 923-944.
30. Sk. Amanathulla and M. Pal, L(3,2,1)-labeling problems on permutation graphs, *Transylvanian Review*, XXV (14) (2017) 3939-3953.
31. Sk. Amanathulla and M. Pal, L(h1, h2,…, hm)-labeling problems on interval graphs, *International Journal of Control Theory and Applications*, 10(1) (2017) 467-479.

32. M. Pal and A. Pal, Scheduling Algorithm to Select Optimal Programme Slots in Television Channels: A Graph Theoretic Approach, *International Journal of Applied and Computational Mathematics*, 3(3) (2017) 1931–1950, doi:10.1007/s40819-016-0216-1
33. Sk. Amanathulla and M. Pal, L(3,2,1)- and L(4,3,2,1)-labeling problems on interval graphs, *AKCE International Journal of Graphs and Combinatorics*. 14(3) (2017) 205-215
34. B. Jana, S. Mondal and M. Pal, Computation of inverse 1-centre location problem on the weighted interval graphs, *Int. J. Computing Science and Mathematics*, 8(6) (2017) 533-541
35. G. Ghorai, M. Pal, H. Rashmanlou and R. A. Borzooei, New concept of regularity in product m-polar fuzzy graphs, *Intern. Journal of Mathematics and Computation*, 28(4) (2017) 9-20.
36. S. Sahoo and M. Pal, Product of intuitionistic fuzzy graphs and their degree, *Journal of Intelligent and Fuzzy Systems*, 32(1) (2017) 1059-1067
37. M. Akram, S. Samanta and M. Pal, Cayley vague graphs, *The Journal of Fuzzy Mathematics*, 25(2) (2017) 449-462.
38. A. Nagoor Gani and K. Prasanna Devi and M. Pal, Reduction of Domination Parameter in Fuzzy Graphs, *Global Journal of Pure and Applied Mathematics*, 13(7) (2017) 3307–3315.
39. T. Pramanik, S. Samanta, B. Sarkar and M. Pal, Fuzzy φ-tolerance competition graphs, *Soft Computing*, 21 (2017) 3723–3734DOI 10.1007/s00500-015-2026-5, 2016
40. S. Sahoo and M. Pal, Intuitionistic fuzzy tolerance graphs with application, *J. Appl. Math. Comput.*, 55 (2017) 495–511.
41. G. Ghorai, M. Pal, Novel Concepts of Strongly Edge Irregular m-Polar Fuzzy Graphs, *International Journal of Applied and Computational Mathematics*, 3(4) (2017) 3321–3332.
42. G. Ghorai and M. Pal, Certain types of product bipolar fuzzy graphs, *Intern. J. Applied and Computational Mathematics*, 3(2) (2017) 605–619.
43. G. Ghorai, M. Pal, On Degrees of m-polar Fuzzy Graphs with Application, *Journal of Uncertain Systems*, 11(4) (2017) 294-305.

44. M. Akram, S. Samanta, M. Pal, Application of bipolar fuzzy sets in planar graphs, *International Journal of Applied and Computational Mathematics*, 3 (2) (2017) 773-785.
45. Sankar Sahoo, M. Pal, H.Rashmanlou and R. A Borzooei, Covering and paired domination in intuitionistic fuzzy graphs, *Journal of Intelligent and Fuzzy Systems*, 33 (6) (2017) 4007-4015.
46. R. Pradhan and M. Pal, Transitive and strongly transitive intuitionistic fuzzy matrices, *Annals of Fuzzy Mathematics and Informatics*, 13(4) (2017) 485-498.
47. Sankar Sahoo and Madhumangal Pal, Modular and homomorphic product of intuitionistic fuzzy graphs and their degree, *Int. J. Computing Science and Mathematics*, 8(5) (2017) 395-404.
48. Chiranjibe Jana and Madhumangal Pal, Application of (α, β)-soft intersectional sets on BCK/BCI-algebras, *Int. J. Intelligent Systems Technologies and Applications*. 16(3)(2017) 269-288
49. Chiranjibe Jana and Madhumangal Pal, On (∈α, ∈α ∨qβ)-fuzzy soft bci-algebras, *Missouri Journal of Mathematical Sciences*, 29(2) (2017) 197-215.
50. Chiranjibe Jana, Madhumangal Pal and Arsham Borumand Saied, (∈,∈∨ q)-bipolar fuzzy BCK/BCI-algebras, *Missouri Journal of Mathematical Sciences*, 29 (2) (2017) 139-160.
51. Chiranjibe Jana, Madhumangal Pal, Generalized intuitionistic fuzzy ideals of BCK/BCI-algebras based on 3-valued logic and its computational study, *Fuzzy Information and Engineering*, 9 (2017) 455-478
52. Madhumangal Pal and Rajkumar Pradhan, Triangular Fuzzy Matrix Norm and Its Properties, *The Journal of Fuzzy Mathematics*, 25(4) (2017) 823-834.
53. Mijanur Rahaman Seikh, Prasun Kumar Nayak, M. Pal, Aspiration level approach to solve matrix games with I-fuzzy goals and I-fuzzy pay-offs, *Pacific Science Review A: Natural Science and Engineering*, 18(1) (2016) 5-13, doi:10.1016/j.psra.2016.03.001
54. S. Ruidas, M. R. Seikh, P. K. Nayak and M. Pal, An interval valued EPQ model in imperfect production system with rework of regular

production, shortages and sales return via particle swarm optimization, *International Journal of Pure and Applied Mathematics*, 113(6) (2017) 375 – 384.

55. S Paul, M Pal, A Pal, L (0, 1)-Labelling of Trapezoid Graphs, *International Journal of Applied and Computational Mathematics*, 3 (1) (2017) 599-610.
56. Sk. Amanathulla and Madhumangal Pal, Surjective L(2, 1)-labeling of cycles and circular-arc graphs, *Journal of Intelligent & Fuzzy Systems*, 35 (2018) 739-748
57. Sankar Sahoo and Madhumangal Pal, Certain types of edge irregular intuitionistic fuzzy graphs, *Journal of Intelligent and Fuzzy Systems*, 34 (2018) 295–305
58. H. Rashmanlou, S. Sahoo, R. A. Borzooei, M. Pal , A. Lakdashti, Computation of Shortest Path in a Vague Network by Euclidean Distance, *Journal of Multiple-Valued Logic and Soft Computing*, 30 (2018) 115-123.
59. Sonia Mandal, Sankar Sahoo, Ganesh Ghorai, Madhumangal Pal, Genus value of m-polar fuzzy graphs, *Journal of Intelligent and Fuzzy Systems*, 34 (2018) 1947–1957.
60. Sankar Sahoo, M. Saravanan, R. Sujatha, R. Sundareswaran and Madhumangal Pal, Concept of integrity and its value of fuzzy graphs, *Journal of Intelligent and Fuzzy Systems*, 34 (2018) 2429–2439.
61. Kousik Das, Sovan Samanta, Madhumangal Pal, Study on centrality measures in social networks: a survey, *Social Network Analysis and Mining*, (2018) 8: 13. https://doi.org/10.1007/s13278-018-0493-2
62. Tarasankar Pramanik, Sukumar Mondal, Sovan Samanta, M. Pal, A study on bipolar fuzzy planar graph and its application in image shrinking, *Journal of Intelligent and Fuzzy Systems*, 34 (3), 1863-1874
63. Sankar Sahoo, M. Pal, Certain types of edge irregular intuitionistic fuzzy graphs, *Journal of Intelligent and Fuzzy Systems*, 34 (2018) 295–305. DOI:10.3233/JIFS-171187
64. Sankar Sahoo, Ganesh Ghorai and Madhumangal Pal, Embedding and Genus of Intuitionistic Fuzzy Graphs on Spheres, *J. of Mult.-Valued Logic & Soft Computing*, 31(1/2) (2018) 139-154.

65. G. Ghorai and M. Pal, A note on "Regular bipolar fuzzy graphs" Neural Computing and Applications 21(1) (2012) 197-205", *Neural Computing & Applications*, 30 (2018) 1569–1572
66. Sreenanda Raut, Madhumangal Pal and Ganesh Ghorai, Fuzzy permutation graph and its complements, *Journal of Intelligent & Fuzzy Systems*, 35 (2018) 2199-2213
67. Tarasankar Pramanik, Madhumangal Pala, Sukumar Mondal and Sovan Samanta, A study on bipolar fuzzy planar graph and its application in image shrinking, *Journal of Intelligent & Fuzzy Systems*, 34 (2018) 1863–1874.
68. R. Sumathi, R. Sujatha, R. Sundareswaran, Sankar Sahoo, Madhumangal Pal and Anita Pal, Nodes in fuzzy block with application, *Journal of Intelligent & Fuzzy Systems*, 35 (2018) 4785–4793
69. Chiranjibe Jana, Tapan Senapati & Madhumangal Pal, t-derivations on complicated subtraction algebras, *Journal of Discrete Mathematical Sciences and Cryptography*, 20 (8) (2017) 1583-1595
70. Ganeshsree Selvachandran, Madhumangal Pal, Tahani A. Abdusalam Alhawari, Abdul Razak Salleh, Interval-Valued Complex Fuzzy Sets and Its Application to the Malaysian Economy, *International Journal of Fuzzy System Applications*, 7(1) (2018) 22-31.
71. Chiranjibe Jana, Madhumangal Pal, Application of bipolar intuitionistic fuzzy soft sets in decision making problem, *International Journal of Fuzzy System Applications*, 7 (3) (2018) 32-55.
72. Chiranjibe Jana, Madhumangal Pal, Application of bipolar intuitionistic fuzzy soft sets in decision making problem, *International Journal of Fuzzy Systems Applications*, 7(3) (2018) 32-55.
73. T. Senapati, M. Bhowmik, M. Pal and K. P. Shum, Characterization of Intuitionistic Fuzzy BG-subalgebras of BG-algebras, *Journal of Discrete Mathematical Sciences & Cryptography*, 21(7-8) (2018) 1549-1558.
74. Anjana Das and M. Pal, A mathematical study of an imprecise SIR epidemic model with treatment control, *Journal of Applied*

Mathematics and Computing, 56 (1-2) (2018) 477-500. doi:10.1007/s12190-017-1083-6

75. A. Das and M. Pal, Modeling and Analysis of an Imprecise Epidemic System with Optimal Treatment and Vaccination Control, *Biophysical Reviews and Letters*, 13 (2) (2018) 37-60.
76. A. Das and M. Pal, An Imprecise Eco-Epidemic Model with Pesticide in Relevance to Agricultural Pest Control, *Biophysical Reviews and Letters*, 13(1) (2018) 1-23.
77. Chiranjibe Jana, Tapan Senapati, Madhumangal Pal, Ronald R. Yager, Picture fuzzy Dombi aggregation operators: Application to MADM Process, *Applied Soft Computing*, 74 (2019) 99–109.
78. Chiranjibe Jana and Madhumangal Pal, A Robust Single-Valued Neutrosophic Soft Aggregation Operators in Multi-Criteria Decision Making, *Symmetry*, 2019, 11, 110; doi:10.3390/sym11010110
79. Chiranjibe Jana and Madhumangal Pal, Assessment of Enterprise Performance Based on Picture Fuzzy Hamacher Aggregation Operators, *Symmetry*, 2019, 11, 75; doi:10.3390/sym11010075.
80. Tapan Senapati, Chiranjibe Jana, Madhumangal Pal and Young Bae Jun, Cubic Intuitionistic q-Ideals of BCI-Algebras, *Symmetry*, doi:10.3390/sym10120752.

In: An Essential Guide to Fuzzy Systems ISBN: 978-1-53616-128-1
Editor: Michael Gr. Voskoglou

Chapter 4

Controllability of Fuzzy Fractional Stochastic Differential Systems

P. Balasubramaniam **and** ***J. Priyadharsini***
Department of Mathematics,
The Gandhigram Rural Institute (Deemed to be University)
Gandhigram, Dindigul, Tamil Nadu, India

Abstract

In this chapter, first we recall the set-valued and fuzzy set-valued stochastic processes and the Lebesgue integral of a fuzzy set-valued stochastic process with respect to time t. Secondly, we shall discuss some basic results about fuzzy fractional calculus. Finally, we prove controllability for the nonlinear fuzzy fractional stochastic differential system by using Banach fixed point theorem.

Keywords: set-valued, fuzzy set-valued, fuzzy fractional, controllability

1. Introduction

In 1965 Zadeh first introduced the Fuzzy set theory [1]. The theory of metric spaces in a fuzzy set has been established by [2]. Due to uncertainty, the theory

*Corresponding Author's Email: balugru@gmail.com

of fuzzy differential equations have become popular and it provides a model for dynamical systems. Existence and uniqueness of fuzzy solution for semilinear fuzzy integrodifferential equations with nonlocal conditions have been derived in [3]. Fuzzy stochastic equation with fuzzy set-valued stochastic Lebesgue integral have been studied in [4, 5, 6]. Li and Wang proved the existence and uniqueness of solution of fuzzy set-valued stochastic process in [7]. Fuzzy and set-valued stochastic differential equations with local lipschitz condition have been studied in [8]. Aumann type Lebesgue integral of a fuzzy set-valued stochastic process with respect to time has been discussed in [9].

Controllability plays a major role in the development of modern mathematical control theory, it enables many applications in modern engineering. Balasubramaniam has derived the controllability for the nonlinear fuzzy neutral functional differential equations in [10]. Park et al. [11] studied the controllability for the semilinear fuzzy integrodifferential equations with nonlocal conditions. Controllability of a nonlinear fuzzy control system with nonlocal initial conditions in n-dimensional fuzzy vector space E_n^N have been studied in [12]. Lee et al. [13] examined the exact controllability for abstract fuzzy differential equations in credibility space.

Agarwal et al. [14] have established the concept of fuzzy Riemann-Liouville fractional differential equations and to solve uncertain fractional differential equations they have considered Riemann-Liouville differentiability notation based on the Hukuhara differentiability. Explicit solutions of uncertain fractional differential equations under Riemann-Liouville H-differentiability using Mittag-Leffler functions have been proved in [15].

However, till now no work has been reported in the literature about controllability of fuzzy fractional stochastic differential equation and also by considering the existence of solutions and controllability results of fuzzy fractional order stochastic differential system in finite dimensional space.

Motivated by the above-mentioned works, our aim in this chapter is to fill this gap. The purpose of this chapter is to show the controllability of fuzzy fractional order stochastic systems. The existence and uniqueness are acquired via Banach fixed point theorem, where the controllability of fuzzy mapping is defined by using Mittag-Leffler matrix function. In particular, the α-level controllability results are investigated for the nonlinear fuzzy fractional stochastic differential system.

2. Set-Valued Stochastic Process

Let N be the set of all natural numbers, R be the set of all real numbers, $\mathbb{R}^d$ be the d-dimensional Euclidean space with the usual norm $\|.\|$, R^+ be the set of all nonnegative numbers. Let $(\Omega, \mathcal{F}, P)$ be a complete probability space, $\{\mathcal{F}_t, t \in J := [0,T]\}$ a σ-field filtration satisfying the usual conditions such that $\mathcal{F}_0$ includes all P-null sets in $\mathcal{F}$ and non-decreasing and right continuous. Let $\mathcal{B}(E)$ be the Borel field of the metric space E. Let $f = \{f(t), \mathcal{F}_t : t \in J\}$ be a $\mathbb{R}^d$-valued adapted stochastic process. If f is progressively measurable if for any $t \in J$, the mapping $(s,\omega) \mapsto f(s,\omega)$ from $J \times \Omega$ to $\mathbb{R}^d$ is $\mathcal{B}(J) \times \mathcal{F}_t$-measurable. Each right continuous (left continuous) adapted process is progressively measurable. Let $\mathcal{C} = \{A \subseteq J \times \Omega : \forall t \in J, A \cap (J \times \Omega) \in \mathcal{B}(J) \times \mathcal{F}_t\}$ then f is progressively measurable if and only if f is $\mathcal{C}$ measurable.

Let $L^p(\mathbb{R}^d)$ denote the set of $\mathbb{R}^d$-valued stochastic processes $f = \{f(t), \mathcal{F}_t : t \in J\}$ such that f satisfies

(i) f is progressively measurable and

(ii) $|||f|||_p = \left[\mathbb{E}\left(\int_0^t \|f(t,\omega)\|^p\right)\right]^{\frac{1}{p}} < \infty.$

Let $f, f' \in L^p(\mathbb{R}^d)$, $f = f'$ if and only if $|||f - f'||| = 0$. Then $L^p(\mathbb{R}^d, |||\cdot|||_p)$ is complete.

Let $\mathcal{K}(\mathbb{R}^d)$ is the family of all non-empty, closed convex subsets of $\mathbb{R}^d$, which is also denoted as $\mathcal{K}_c(\mathbb{R}^d)$ or $\mathcal{K}_k(\mathbb{R}^d)$, A is a nonempty subset of $\mathbb{R}^d$, define the distance of x and A as $d(x,A) = inf_{y \in A}\|x - y\|$. The Hausdorff metric on $\mathcal{K}(\mathbb{R}^d)$ is defined as $d_H(A,B) = \max\{\sup_{a \in A} d(a,B), \sup_{b \in B} d(b,A)\}$ for $A, B \in \mathcal{K}(\mathbb{R}^d)$.

For a set-valued random variable F, define the set $S_F^p = \{f \in L^p[\Omega; \mathbb{R}^d] : f(\omega) \in F(\omega) \text{ for a.e. } \omega \in \Omega\}$ where $L^p[\Omega; \mathbb{R}^d]$ is the set of all $\mathbb{R}^d$-valued random variables f such that $\|f\|_p = [E(\|f\|^p)]^{\frac{1}{p}} < \infty$, and constant $p \geq 1$.

Definition 2..1 *[16] A set-valued random variable $F : \Omega \to \mathcal{K}(\mathbb{R}^d)$ is called integrable if $\int_\infty \|F(\omega)\|_{\mathcal{K}} d\mu < \infty$. Let $L^p[\Omega; \mathcal{K}(\mathbb{R}^d)]$ denote the family of $\mathcal{K}(\mathbb{R}^d)$-valued L^p-bounded random variables F such that $\|F(\cdot)\|_{\mathcal{K}} \in L^p[\Omega; \mathbb{R}]$.*

For any two set-valued random variables $F_1, F_2 \in L^p[\Omega; \mathcal{K}(\mathbb{R}^d)]$, *define*

$$\Delta_p(F_1, F_2) = \Big(\int_\Omega d_H^p(F_1(\omega), F_2(\omega))d\mu\Big)^{\frac{1}{p}},$$

then $(L^p[\Omega; \mathcal{K}(\mathbb{R}^d)], \Delta_p)$ *is a complete space.* $F = \{F(t) : t \in J\}$ *is called a set-valued stochastic process if* $F : J \times \Omega \to \mathcal{K}(\mathbb{R}^d)$ *is a set-valued function such that for any fixed* $t \in J$, $F(t, \cdot)$ *is a set-valued random variable.*

Definition 2..2 *[17] Let* X *be a non-empty set. A fuzzy set* ν *in* X *is characterized by its membership function* $\nu : X \to [0, 1]$. *Then for each* $x \in X$, *we interpret* $\nu(x)$ *as the degree of membership of the element* x *in the fuzzy set* $\nu : u(x) = 0$ *corresponded to nonmembership,* $0 < \nu(x) < 1$ *to partial membership and* $u(x) = 1$ *to full membership.*

Definition 2..3 *[17] Let* $F(\mathbb{R}^d)$ *denote the class of fuzzy subsets of the real axis* $\nu : \mathbb{R}^d \to [0, 1]$ *which satisfy the following conditions:*

(1) ν *is normal (i.e) there exists* $x_0 \in \mathbb{R}^d$ *with* $\nu(x_0) = 1$

(2) ν *is a convex fuzzy set (i.e)* $\nu(\lambda x + (1 - \lambda)y) \geq \min\{\nu(x) + \nu(y)\}$ *for all* $\lambda \in [0, 1]$ *and* $x, y \in \mathbb{R}^d$

(3) ν *is upper semicontinuous on* $\mathbb{R}^d$ *(i.e) for each* $a \in [0, 1]$, $\{x : \nu(x) < a\}$ *is open in* $\mathbb{R}^d$

(4) $cl\{x \in \mathbb{R}^d : \nu(x) > 0\}$ *is compact, where cl denotes the closure of a set.*

Then we call $F(\mathbb{R}^d)$ the space of fuzzy numbers. Obviously $\mathbb{R}^d \subset F(\mathbb{R}^d)$.
For $0 < \alpha \leq 1$, denote $[\nu]^\alpha = \{x \in \mathbb{R}^d; \nu(x) \geq \alpha\}$ and $[\nu]^0 = cl\{x \in \mathbb{R}^d; \nu(x) > 0\}$. By the definition of fuzzy numbers, it follows that for any $\alpha \in [0, 1]$, $[\nu]^\alpha$ is a bounded closed interval. The notation $[\nu]^\alpha = [\nu_l^\alpha, \nu_r^\alpha]$ denotes the α-level set of ν. We refer ν_l, ν_r as the left and right branches on ν respectively.

For $a, b \in F(\mathbb{R}^d), \eta \in \mathbb{R}^d$, the sum of two fuzzy numbers and the multiplication between a real and a fuzzy number are defined respectively by

$$[a + b]^\alpha = [a]^\alpha + [b]^\alpha = \{x + y : x \in [a]^\alpha, y \in [b]^\alpha\} \text{ and}$$
$$[\eta a]^\alpha = \eta[a]^\alpha = \{\eta x : x \in [a]^\alpha\}$$

for all $\alpha \in [0,1]$, where $[a]^\alpha + [b]^\alpha$ is the usual addition of two intervals of $\mathbb{R}^d$ and $\eta[a]^\alpha$ is the usual product of a number and a subset of $\mathbb{R}^d$.

Definition 2..4 *[17] The Hausdorff distance between fuzzy numbers is given by, $d_\infty : F(\mathbb{R}^d) \times F(\mathbb{R}^d) \to \mathbb{R}^+ \cup \{0\}$ as $d_\infty(x,y) := \sup_{\alpha\in(0,1]} \max\{d_H(|x_l^\alpha - y_l^\alpha|, |x_r^\alpha - y_r^\alpha)|\} = \sup_{\alpha\in(0,1]} \max\{d_H([x]^\alpha, [y]^\alpha)\}$. The metric space $(F(\mathbb{R}^d), d_\infty)$ is complete and the following properties for the metric d_∞ hold*

(i) $d_\infty(x+z, y+z) = d_\infty(x,y)$ *for all* $x, y, z \in F(\mathbb{R}^d)$

(ii) $d_\infty(x+z, y+w) = d_\infty(x,y) + d_\infty(z,w)$ *for all* $x, y, z, w \in F(\mathbb{R}^d)$

(iii) $d_\infty(kx, ky) = |k| d_\infty(x,y)$ *for all* $x, y \in F(\mathbb{R}^d)$.

Definition 2..5 *[18] Let $x, y \in F(\mathbb{R}^d)$. If there exists $z \in F(\mathbb{R}^d)$ such that $x = y + z$, then z is called as H-difference of x and y, and it is denoted by $x \ominus y$*

Definition 2..6 *[19] The generalised Hukuhara difference (gH-difference) of two fuzzy number $x, y \in F(\mathbb{R}^d)$ is $z \in F(\mathbb{R}^d)$, which is defined as follows*

$$x \ominus_{gH} y = z \Leftrightarrow (i)\ x = y + z \text{ or } (ii)\ y = x + (-1)z$$

Definition 2..7 *[20] The generalised Hukuhara derivative of a two fuzzy-valued function $f : (a,b) \to F(\mathbb{R}^d)$ at x_0 is defined as*

$$(f)'_{gH}(x_0) = \lim_{h \longrightarrow 0} \frac{f(x_0 + h) \ominus_{gH} f(x_0)}{h}$$

If $f'_{gH}(x_0) \in F(\mathbb{R}^d)$, is said to be f is gH-differentiable at x_0. Also, f is [(i)-gH]-differentiable at x_0 if (i) $(f)'_{gH}(x_0; a) = [(f_l)'(x_0; a), (f_r)'(x_0; a)]$ and f is [(ii)-gH]-differentiable at x_0, if (ii) $(f)'_{gH}(x_0; a) = [(f_r)'_{gH}(x_0; a), (f_l)'_{gH}(x_0; a)]$, where $0 \le a \le 1$.

Definition 2..8 *[7] A fuzzy set-valued random variable is a function $\hat{X} : \Omega \to F(\mathbb{R}^d)$, such that its level set*

$$\hat{X}_\alpha(\omega) = \{x \in \mathbb{R}^d : X(\omega)(x) \ge \alpha\}$$

is a set-valued random variable for every $\alpha \in (0,1]$.

Definition 2..9 *[7] A fuzzy set-valued random variable $\hat{X}$ is called integrably bounded if the real-valued random variable $\|\hat{X}_{0+}\|_K$ is integrable, where*

$$\hat{X}_{0+}(\omega) = cl\{x \in \mathbb{R}^d : \bar{X}(\omega)(x) > 0\}.$$

It is easy to know that $\|\hat{X}_{0+}\|_K = \sup_{\alpha \in (0,1]} \|\hat{X}_\alpha\|_K$.

Let $L^1[\Omega, \mathcal{F}, P; F_k(\mathbb{R}^d)]$ be the set of all integrably bounded fuzzy set-valued random variables.

The expectation of $\|\hat{X}\|$, denoted by $E[\hat{X}]$ is a fuzzy set in $F(\mathbb{R}^d)$ such that for every $\alpha \in [0,1]$, $E([\hat{X}])_\alpha = clE[\hat{X}_\alpha]$, where the closure is taken in $\mathbb{R}^d$.

Definition 2..10 *[7] $\hat{F} = \{\hat{F}(t) : t \in J\}$ is called a fuzzy set-valued stochastic process if $\hat{F} : J \times \omega \to F(\mathbb{R}^d)$ is such that for any fixed $t \in J$, $F(t,\cdot)$ is a fuzzy set-valued random variable. For any $\alpha \in (0,1]$, $\hat{F}_\alpha = \{\hat{F}_\alpha(t) : t \in J\}$ is a set-valued stochastic process called $\alpha -$ level set process.*

Definition 2..11 *[7] A fuzzy set-valued stochastic process $\hat{F} = \{\hat{F}(t) : t \in J\}$ is called adapted with respect to the filtration $\{\mathcal{F}_t : t \in J\}$, if $\hat{F}(t)$ is measurable with respect to $\mathcal{F}_t$, for each $t \in J$, and denoted by $\{\hat{F}(t), \mathcal{F}_t : t \in J\}$.*

Definition 2..12 *[7] A set-valued stochastic process $F = \{F(t) : t \in J\}$ is called progressively measurable, if for any $A \in \mathcal{B}(\mathbb{R}^d)$ and any $t \in J$, $\{(s,\omega) \in [0,t] \times \Omega : F(s,\omega) \cap A \neq \emptyset \in \mathcal{B}([0,t]) \times \mathcal{F}_t$. F is called L^1-bounded, if the real stochastic process $\{\|F(t)\|_K, \mathcal{F}_t : t \in J\} \in L^1(\mathbb{R})$.*

Definition 2..13 *[7] A $\mathbb{R}^d$-valued progressively process $\{f(t), \mathcal{F}_t : t \in J\} \in L^1(\mathbb{R}^d)$ is called an L^1-selection of $F = \{F(t), \mathcal{F}_t : t \in J\}$ if $f(t,\omega) \in F(t,\omega)$ for a.e. $(t,\omega) \in J \times \Omega$.*

Let $S^1(\{F(\cdot)\})$ or $S^1(F)$ denote the family of all L^1-selections of $F = \{F(t), \mathcal{F}_t : t \in J\}$, that is

$$S^1(F) = \Big\{\{f(t) : t \in J\} \in L^1(\mathbb{R}^d) : f(t,\omega) \in F(t,\omega), \text{ for a.e. } \quad (t,\omega) \in J \times \Omega\Big\}$$

Let $L^1(\mathcal{K}(\mathbb{R}^d))$ denote the set of all L^1-bounded progressively measurable $\mathcal{K}(\mathbb{R}^d)$-valued stochastic processes.

Definition 2..14 *[7] A fuzzy set-valued stochastic process $\hat{F} = \{\hat{F}(t) : t \in J\}$ is called progressively measurable, if for any $\alpha \in (0, 1]$, $\hat{F}_\alpha = \{\hat{F}_\alpha(t); t \in J\}$ is a progressively measurable set-valued stochastic process.*

Definition 2..15 *[7] A fuzzy set-valued stochastic process $\hat{F} = \{\hat{F}; t \in J\}$ is called integrably bounded, if $\hat{F}_{0+} = \{\hat{F}_{0+}(t); t \in J\}$ is L^1-bounded.*
Let $L^1[\Omega, \mathcal{F}, \nu; F(\mathbb{R}^d)]$ be the set of all integrably bounded progressively measurable fuzzy set-valued stochastic processes.

Definition 2..16 *[16] A fuzzy set-valued stochastic process $\hat{F} = \{\mathcal{F}_t, t \in J\} \in L^1(F_K(\mathbb{R}^d))$. For any $\alpha \in (0, 1], t \in J, \omega \in \Omega$, define*

$$(A)\int_0^t \hat{F}_\alpha(s,\omega)ds := \Big\{\int_0^t f_\alpha(s,\omega)ds : f_\alpha \in S^1(\hat{F}\alpha)\Big\}$$

where $\int_0^t f_\alpha(s,\omega)ds$ is the Lebesgue integral $(A)\int_0^t \hat{F}_\alpha(s,\omega)ds$ is called the Aumann type Lebesgue integral of the fuzzy set-valued stochastic process $\hat{F}$ with respect to time t. For any $0 \le u < t < T$,

$$(A)\int_0^t \hat{F}_\alpha(s,\omega)ds := (A)\int_0^t I_{[u,t]}(s)\hat{F}_\alpha(s,\omega)ds.$$

3. Fuzzy Fractional Calculus

We represent $C(J', F(\mathbb{R}^d))$ be a space of all fuzzy-valued functions, continuous on $J' := (a, b]$

$$C(J', F(\mathbb{R}^d)) := \{f/f : J' \to F(\mathbb{R}^d),\ f \text{ is continuous}\}$$

Definition 3..1 *[21] Let $f : [a, b] \to F(\mathbb{R}^d)$, for $0 < \beta \le 1$, the fuzzy Riemann-Liouville integral of fuzzy valued f is defined as*

$$(I^\beta_{a+}f)(t) \quad = \quad \frac{1}{\Gamma(\beta)}\int_a^t (t-s)^{\beta-1}f(s)ds, \quad t > a.$$

For all $0 \leq r \leq 1$, since $f(x,r) = [f_l(x,r), f_r(x,r)]$, the fuzzy Riemann-Liouville integral of fuzzy-valued function f can be calculated based on its left and right end function respectively.

Definition 3..2 *[21] Let $f_{gH}^{m} \in C(J', F(\mathbb{R}^d)) \cap L^1(J', F(\mathbb{R}^d))$. The fuzzy gH-fractional Caputo differentiability ($^C[gH]$-differentiability) of fuzzy valued function*

$$\begin{aligned}({}_{gH}^{C}D_{a+}^{\beta} f)(t) &= \frac{1}{\Gamma(1-\beta)} \int_a^t \frac{f'(s)}{(t-s)^{\beta}} ds \\ &= I_{a+}^{1-\beta}(D_{gH}f)(t),\end{aligned}$$

where $0 < \beta \leq 1$, then we say if f is $[(i) - gH]$-differentiable at x_0 then f is $^C[(i) - gH]$- differentiable at x_0 and f is $[(ii) - gH]$-differentiable at x_0 then f is $^C[(ii) - gH]$- differentiable at x_0.

Definition 3..3 *[22] A two-parameter function of the Mittag-Leffler type is defined by the series expansion*

$$E_{\alpha,\beta}(z) = \sum_{k=0}^{\infty} \frac{z^k}{\Gamma(k\alpha + \beta)}, \; z \in \mathbb{C}.$$

The Laplace integral of Mittag-Leffler function is

$$\int_0^{\infty} e^{-st} t^{\beta-1} E_{\alpha,\beta}(\pm At^{\alpha}) dt = \frac{s^{\alpha-\beta}}{(s^{\alpha}I \mp A)}.$$

That is

$$\mathcal{L}\{t^{\beta-1} E_{\alpha,\beta}(\pm At^{\alpha})\}(s) = \frac{s^{\alpha-\beta}}{(s^{\alpha}I \mp A)}.$$

4. Controllability of Fuzzy Fractional Stochastic Differential System

Consider the following nonlinear fuzzy Caputo fractional stochastic differential system (FCFSDS) is represented as

$$
\begin{aligned}
{}_{gH}^{C}D^{q}x(t) &= Ax(t) + Bu(t) + f(t, x(t)) + \int_0^t g(s, x(s))dw(s), \quad t \in J := [0, T], \\
x(0) &= x_0, \qquad (1)
\end{aligned}
$$

where $0 < q < 1$, A and B are fuzzy numbers, $u : J \rightarrow F(\mathbb{R}^d)$ is the control input, the nonlinear functions $f : J \times \Omega \times F(\mathbb{R}^d) \rightarrow F(\mathbb{R}^d)$, $g : J \times \Omega \times F(\mathbb{R}^d) \rightarrow \mathbb{R}^d \times \mathbb{R}^m$ are continuous, initial value $x_0 : \Omega \rightarrow F(\mathbb{R}^d)$ is a fuzzy random variable and $w(t) = \{w_1(t), w_2(t), \cdots, w_n(t)\}$ be an m-dimensional Wiener process defined on the probability space. Let $C(J, F(\mathbb{R}^d))$ be the Banach space of continuous with norm

$$
\|d_\infty(x, y)\| := \sup_{\alpha \in (0,1]} \max \|\{d_H([x]^\alpha, [y]^\alpha)\}\| < \infty.
$$

Solution Representation:

Consider the linear fuzzy fractional stochastic differential system is represented as

$$
\begin{aligned}
{}^{C}D^{q}x(t) &= Ax(t) + Bu(t) + f(t) + \int_0^t g(s)dw(s), \quad t \in J := [0, T], \\
x(0) &= x_0, \qquad (2)
\end{aligned}
$$

where $0 < q < 1$, A and B are fuzzy numbers, initial value $x_0 : \Omega \rightarrow F(\mathbb{R}^d)$, $u \in J \rightarrow F(\mathbb{R}^d)$ is the control input, the nonlinear functions $f : J \times \Omega \rightarrow F(\mathbb{R}^d)$ are continuous, $g : J \times \Omega \rightarrow \mathbb{R}^d \times \mathbb{R}^m$.

Taking the Laplace transform on (2) with respect to t on both sides, we get

$$
\begin{aligned}
s^q\hat{x}(s) - s^{q-1}x(0) &= A\hat{x}(s) + B\hat{u}(s) + \hat{f}(s) + \hat{g}(s) \\
(s^qI - A)\hat{x}(s) &= s^{q-1}x(0) + B\hat{u}(s) + \hat{f}(s) + \hat{g}(s)
\end{aligned}
$$

where I is an identity matrix.

$$
\hat{x}(s) = \frac{s^{q-1}}{(s^qI - A)}x(0) + \frac{1}{(s^\alpha I - A)}B\hat{u}(s) + \frac{1}{(s^\alpha I - A)}\hat{f}(s) + \frac{1}{(s^\alpha I - A)}\hat{g}(s).
$$

Applying inverse Laplace transform on both sides, we have

$$\begin{aligned} L^{-1}\{\hat{x}(s)\} &= L^{-1}\Big\{(s^{q-1})(s^qI-A)^{-1}\Big\}x(0) + L^{-1}\Big\{(s^qI-A)^{-1}\Big\} * BL^{-1}\Big\{\hat{u}(s)\Big\} \\ & L^{-1}\Big\{(s^qI-A)^{-1}\Big\} * L^{-1}\Big\{\hat{f}(s)\Big\} + L^{-1}\Big\{(s^qI-A)^{-1}\Big\} * L^{-1}\Big\{\hat{g}(s)\Big\}. \end{aligned}$$

Finally, substituting fuzzy Laplace transformation of Mittag-Leffler function, we get the solution of system (2) as

$$[x(t)]^\alpha = \Big[E_q(At^q)x_0 + \int_0^t (t-s)^{q-1}E_{q,q}(A(t-s)^q)\Big[Bu(s) + f(s) + \int_0^s g(\theta)dw(\theta)\Big]ds\Big]^\alpha$$

Similarly, one can write the solution for FCFSDS (1), if $x(t)$ is ${}^C[(i)-gH]$-differentiable and ${}^C[(ii)-gH]$-differentiable as,

$$\begin{aligned} [x(t)]^\alpha &= \Big[E_q(At^q)x_0 + \int_0^t (t-s)^{q-1}E_{q,q}(A(t-s)^q)Bu(s)ds + \int_0^t (t-s)^{q-1}E_{q,q}(A(t-s)^q) \\ &\times f(s,x(s)ds) + \int_0^t (t-s)^{q-1}E_{q,q}(A(t-s)^q)\Big(\int_0^s g(\theta,x(\theta))dw(\theta)\Big)ds\Big]^\alpha, \end{aligned} \quad (3)$$

or

$$\begin{aligned} [x(t)]^\alpha &= E_q(At^q)x_0 \ominus (-1)\Big[\int_0^t (t-s)^{q-1}E_{q,q}(A(t-s)^q)Bu(s)ds + \int_0^t (t-s)^{q-1}E_{q,q}(A(t-s)^q) \\ &\times f(s,x(s)ds) + \int_0^t (t-s)^{q-1}E_{q,q}(A(t-s)^q)\Big(\int_0^s g(\theta,x(\theta))dw(\theta)\Big)ds\Big] \end{aligned} \quad (4)$$

Definition 4..1 *The system (1) is said to be α-level controllable on J, if for every $x_0 \in F(\mathbb{R}^d)$ there exists a control $u : J \to F(\mathbb{R}^d)$ such that the fuzzy solution $x(t)$ of (1) satisfies $x(T) = x_T$ in time T. (i.e.,) $[x(T)]^\alpha = [(x_T)_l^\alpha, (x_T)_r^\alpha] = [(x_T)^\alpha]$.*

For convenience, define $a = \sup_{t\in J}\|E_{q,q}(A(t-s)^q\|^2$.
Let us assume the following hypotheses :

(H1): Let f and g satisfy the Lipschitz conditions that is there exists a positive constants K, L such that

$$\begin{aligned} \mathbb{E}\|d_\infty(f(t,x(t)), f(t,y(t)))\|^2 &\le K\mathbb{E}\|x(t)-y(t)\|^2 \text{ and} \\ \mathbb{E}\,\|g(\theta,x(\theta)) - g(\theta,y(\theta))\|^2 &\le L\mathbb{E}\|x(t)-y(t)\|^2 \end{aligned}$$

for all $x(t), y(t) \in F(\mathbb{R}^d)$.

(H2): The fuzzy mapping $W : E(\mathbb{R}^d) \to F(\mathbb{R}^d)$

$$W^{\alpha}(u) = \begin{cases} \int_0^T (T-s)^{q-1}(E_{q,q}(A(T-s)^q)B)^{\alpha}u(s)ds, & u \subset \bar{\Gamma}_u \\ 0 & otherwise \end{cases}$$

where $u \in E(\mathbb{R}^d)$ is a family of all nonempty convex subsets of $\mathbb{R}^d$ and $\bar{\Gamma}_u = \{[u]^{\alpha} \in \mathcal{K}(\mathbb{R}^d)$ for every $\alpha \in [0,1]\}$ such that each component $[u]^{\alpha}$ is a support u. Then there exists $W_i^{\alpha}(i = l, r)$ such that

$$\begin{aligned} W_l^{\alpha}(u_l) &= \int_0^T (T-s)^{q-1}(E_{q,q}(A(T-s)^q)B)_l^{\alpha}u_l(s)ds, \quad u_l(s) \in [u_l^{\alpha}(s), u_l(s)] \\ W_l^{\alpha}(u_r) &= \int_0^T (T-s)^{q-1}(E_{q,q}(A(T-s)^q)B)_r^{\alpha}u_r(s)ds, \quad u_r(s) \in [u_l(s), u_r^{\alpha}(s)] \end{aligned}$$

where W_l^{α}, W_r^{α} are bijective mappings.
(H3): $8\frac{T^{2q}}{q^2}a(K + LT) < 1$.

Theorem 4..2 *Suppose that the hypotheses (H1)- (H3) are satisfied. Then the system (1) is α-level controllable on J.*

Proof: Using the bijective mappings in (H2), introduce the $\alpha-$ level set of nonlinear control $u(s)$ defined by

$$\begin{aligned} [u(s)]^{\alpha} &= [u_l^{\alpha}(s), u_r^{\alpha}(s)] \\ &= \Big[(W_l^{\alpha})^{-1}\Big\{(x_T)_l^{\alpha} - (E_q(AT^q)x_0)_l^{\alpha} - \int_0^T (T-s)^{q-1}(E_{q,q}(A(t-s)^q)f(s,x(s)ds))_l^{\alpha} \\ &\quad - \int_0^T (T-s)^{q-1}\Big(E_{q,q}(A(T-s)^q)\Big(\int_0^s g(\theta,x(\theta))dw(\theta)\Big)ds\Big)_l^{\alpha}\Big|\mathcal{F}_s\Big\}, \\ &\quad (W_r^{\alpha})^{-1}\Big\{(x_T)_l^{\alpha} - (E_q(AT^q)x_0)_r^{\alpha} - \int_0^T (T-s)^{q-1}(E_{q,q}(A(T-s)^q)f(s,x(s)ds))_r^{\alpha} \\ &\quad - \int_0^T (T-s)^{q-1}\Big(E_{q,q}(A(T-s)^q)\Big(\int_0^s g(\theta,x(\theta))dw(\theta)\Big)ds\Big)_r^{\alpha}\Big|\mathcal{F}_s\Big\}\Big]. \end{aligned}$$

Then substituting this expression into the equation (3) as

$$
\begin{aligned}
[x(T)]^{\alpha} &= \Big[(E_q(AT^q)x_0)_l^{\alpha} + \int_0^T (T-s)^{q-1}(E_{q,q}(A(T-s)^q)B)_l^{\alpha} \\
&\times (W_l^{\alpha})^{-1}\Big\{(x_T)_l^{\alpha} - (E_q(AT^q)x_0)_l^{\alpha} - \int_0^T (T-s)^{q-1}(E_{q,q}(A(T-s)^q)f(s,x(s)ds))_l^{\alpha} \\
&- \int_0^T (T-s)^{q-1}\Big(E_{q,q}(A(T-s)^q)\Big(\int_0^s g(\theta,x(\theta))dw(\theta)\Big)ds\Big)_l^{\alpha}\Big|\mathcal{F}_s\Big\}(s)ds \\
&+ \int_0^T (T-s)^{q-1}(E_{q,q}(A(T-s)^q)f(s,x(s)ds))_l^{\alpha} + \int_0^T (T-s)^{q-1}\Big(E_{q,q}(A(T-s)^q) \\
&\times \Big(\int_0^s g(\theta,x(\theta))dw(\theta)\Big)ds\Big)_l^{\alpha}, (E_q(AT^q)x_0)_r^{\alpha} + \int_0^T (T-s)^{q-1}(E_{q,q}(A(T-s)^q)B)_r^{\alpha} \\
&\times (W_r^{\alpha})^{-1}\Big\{(x_T)_r^{\alpha} - (E_q(AT^q)x_0)_r^{\alpha} - \int_0^T (T-s)^{q-1}(E_{q,q}(A(T-s)^q)f(s,x(s)ds)_r^{\alpha} \\
&- \int_0^T (T-s)^{q-1}E_{q,q}(A(T-s)^q)\Big(\int_0^s g_r^{\alpha}(\theta,x(\theta))dw(\theta)\Big)ds\Big)_r^{\alpha}\Big|\mathcal{F}_s\Big\}(s)ds \\
&+ \int_0^T (T-s)^{q-1}(E_{q,q}A(T-s)^q)f(s,x(s)ds))_r^{\alpha} + \int_0^T (T-s)^{q-1}\Big(E_{q,q}(A(T-s)^q) \\
&\times \Big(\int_0^s g(\theta,x(\theta))dw(\theta)\Big)ds\Big)_r^{\alpha}\Big] \\
&= [(x_T)_l^{\alpha}, (x_T)_r^{\alpha}] \\
&= [(x_T)^{\alpha}].
\end{aligned}
$$

Similarly one can prove (4).

Let $\mathcal{B} = C(J, F(\mathbb{R}^d)$. Then one can easily check that the operators Φ_1, Φ_2 are continuous functions from $C(J, F(\mathbb{R}^d))$ into itself.

To Prove: Φ_1, Φ_2 are contraction mappings on $\mathcal{B}$.

Consider

$$
\begin{aligned}
(\Phi_1 x)(t) &= E_q(At^q)x_0 + \int_0^t (t-s)^{q-1}E_{q,q}(A(t-s)^q)Bu(s)ds \\
&+ \int_0^t (t-s)^{q-1}E_{q,q}(A(t-s)^q)f(s,x(s)ds) \\
&+ \int_0^t (t-s)^{q-1}E_{q,q}(A(t-s)^q)\Big(\int_0^s g(\theta,x(\theta))dw(\theta)\Big)ds \\
(\Phi_2 x)(t) &= E_q(At^q)x_0 \ominus (-1)\Big[\int_0^t (t-s)^{q-1}E_{q,q}(A(t-s)^q)Bu(s)ds \\
&+ \int_0^t (t-s)^{q-1}E_{q,q}(A(t-s)^q)f(s,x(s)ds) \\
&+ \int_0^t (t-s)^{q-1}E_{q,q}(A(t-s)^q)\Big(\int_0^s g(\theta,x(\theta))dw(\theta)\Big)ds\Big]
\end{aligned}
$$

For any $x, y \in \mathcal{B}$,

$$
\begin{aligned}
d_H\Big([(\Phi_1 x)(t)]^\alpha, [(\Phi_1 y)(t)]^\alpha\Big) &= d_H\Big(\Big[E_q(At^q)x_0 + \int_0^t (t-s)^{q-1}E_{q,q}(A(t-s)^q)BW^{-1}\Big\{x_T \\
&\quad - E_q(AT^q)x_0 - \int_0^T (T-s)^{q-1}E_{q,q}(A(T-s)^q)f(s,x(s))ds \\
&\quad - \int_0^T (T-s)^{q-1}E_{q,q}(A(T-s)^q)\Big(\int_0^s g(\theta,x(\theta))dw(\theta)\Big)ds\Big|\mathcal{F}_s\Big\}(s)ds \\
&\quad + \int_0^t (t-s)^{q-1}E_{q,q}(A(t-s)^q)f(s,x(s))ds \\
&\quad + \int_0^t (t-s)^{q-1}E_{q,q}(A(t-s)^q)\Big(\int_0^s g(\theta,x(\theta))dw(\theta)\Big)ds\Big]^\alpha\Big), \\
&\quad d_H\Big(\Big[E_q(At^q)x_0 + \int_0^t (t-s)^{q-1}E_{q,q}(A(t-s)^q)BW^{-1}\Big\{x_T \\
&\quad - E_q(AT^q)x_0 - \int_0^T (T-s)^{q-1}E_{q,q}(A(T-s)^q)f(s,y(s))ds \\
&\quad - \int_0^T (T-s)^{q-1}E_{q,q}(A(T-s)^q)\Big(\int_0^s g(\theta,y(\theta))dw(\theta)\Big)ds\Big|\mathcal{F}_s\Big\}(s)ds \\
&\quad + \int_0^t (t-s)^{q-1}E_{q,q}(A(t-s)^q)f(s,y(s))ds \\
&\quad + \int_0^t (t-s)^{q-1}E_{q,q}(A(t-s)^q)\Big(\int_0^s g(\theta,y(\theta))dw(\theta)\Big)ds\Big]^\alpha\Big) \\
&= d_H\Big(\Big[E_q(At^q)x_0\Big]^\alpha + \Big[\int_0^t (t-s)^{q-1}E_{q,q}(A(t-s)^q)BW^{-1}\Big\{x_T \\
&\quad - E_q(AT^q)x_0 - \int_0^T (T-s)^{q-1}E_{q,q}(A(T-s)^q)f(s,x(s))ds \\
&\quad - \int_0^T (T-s)^{q-1}E_{q,q}(A(T-s)^q)\Big(\int_0^s g(\theta,x(\theta))dw(\theta)\Big)ds\Big|\mathcal{F}_s\Big\}(s)ds\Big]^\alpha \\
&\quad + \Big[\int_0^t (t-s)^{q-1}E_{q,q}(A(t-s)^q)f(s,x(s)ds)\Big]^\alpha \\
&\quad + \Big[\int_0^t (t-s)^{q-1}E_{q,q}(A(t-s)^q)\Big(\int_0^s g(\theta,x(\theta))dw(\theta)\Big)ds\Big]^\alpha\Big), \\
&\quad d_H\Big(\Big[E_q(At^q)x_0\Big]^\alpha + \Big[\int_0^t (t-s)^{q-1}E_{q,q}(A(t-s)^q)BW^{-1}\Big\{x_T \\
&\quad - E_q(AT^q)x_0 - \int_0^T (T-s)^{q-1}E_{q,q}(A(T-s)^q)f(s,y(s))ds \\
&\quad - \int_0^T (T-s)^{q-1}E_{q,q}(A(T-s)^q) \times \Big(\int_0^s g(\theta,y(\theta))dw(\theta)\Big)ds\Big|\mathcal{F}_s\Big\}(s)ds\Big]^\alpha \\
&\quad + \Big[\int_0^t (t-s)^{q-1}E_{q,q}(A(t-s)^q)f(s,y(s))ds\Big]^\alpha \\
&\quad + \Big[\int_0^t (t-s)^{q-1}E_{q,q}(A(t-s)^q)\Big(\int_0^s g(\theta,y(\theta))dw(\theta)\Big)ds\Big]^\alpha\Big)
\end{aligned}
$$

$$
\begin{aligned}
\leq\ & d_H\Big(\Big[\int_0^t (t-s)^{q-1}E_{q,q}(A(t-s)^q)BW^{-1}\Big\{\int_0^T (T-s)^{q-1}E_{q,q}(A(T-s)^q)f(s,x(s))ds\Big\}\Big]^\alpha,\\
& \Big[\int_0^t (t-s)^{q-1}E_{q,q}(A(t-s)^q)BW^{-1}\Big\{\int_0^T (T-s)^{q-1}E_{q,q}(A(T-s)^q)f(s,y(s))ds\Big\}\Big]^\alpha\Big)\\
& +d_H\Big(\Big[\int_0^t (t-s)^{q-1}E_{q,q}(A(t-s)^q)BW^{-1}\Big\{\int_0^T (T-s)^{q-1}E_{q,q}(A(T-s)^q)\\
& \times\Big(\int_0^s g(\theta,x(\theta))dw(\theta)\Big)ds\Big\}\Big]^\alpha, \Big[\int_0^t (t-s)^{q-1}E_{q,q}(A(t-s)^q)BW^{-1}\\
& \times\Big\{\int_0^T (T-s)^{q-1}E_{q,q}(A(T-s)^q)\Big(\int_0^s g(\theta,y(\theta))dw(\theta)\Big)ds\Big\}\Big]^\alpha\Big)\\
& +d_H\Big(\Big[\int_0^t (t-s)^{q-1}E_{q,q}(A(t-s)^q)f(s,x(s))ds\Big]^\alpha,\\
& \Big[\int_0^t (t-s)^{q-1}E_{q,q}(A(t-s)^q)f(s,y(s))ds\Big]^\alpha\Big)\\
& +d_H\Big(\Big[\int_0^t (t-s)^{q-1}E_{q,q}(A(t-s)^q)\Big(\int_0^s g(\theta,x(\theta))dw(\theta)\Big)ds\Big]^\alpha,\\
& \Big[\int_0^t (t-s)^{q-1}E_{q,q}(A(t-s)^q)\Big(\int_0^s g(\theta,y(\theta))dw(\theta)\Big)ds\Big]^\alpha\Big)\\
\leq\ & d_H\Big(\Big[WW^{-1}\Big\{\int_0^T (T-s)^{q-1}E_{q,q}(A(T-s)^q)f(s,x(s))ds\Big\}\Big]^\alpha, \Big[WW^{-1}\\
& \Big\{\int_0^T (T-s)^{q-1}E_{q,q}(A(T-s)^q)f(s,y(s))ds\Big\}\Big]^\alpha\Big)\\
& +d_H\Big(\Big[WW^{-1}\Big\{\int_0^T (T-s)^{q-1}E_{q,q}(A(T-s)^q)\Big(\int_0^s g(\theta,x(\theta))dw(\theta)\Big)ds\Big\}\Big]^\alpha,\\
& \Big[WW^{-1}\Big\{\int_0^T (T-s)^{q-1}E_{q,q}(A(T-s)^q)\Big(\int_0^s g(\theta,y(\theta))dw(\theta)\Big)ds\Big\}\Big]^\alpha\Big)\\
& +d_H\Big(\Big[\int_0^t (t-s)^{q-1}E_{q,q}(A(t-s)^q)f(s,x(s))ds\Big]^\alpha, \Big[\int_0^t (t-s)^{q-1}E_{q,q}(A(t-s)^q)f(s,y(s))ds\Big]^\alpha\Big)\\
& +d_H\Big(\Big[\int_0^t (t-s)^{q-1}E_{q,q}(A(t-s)^q)\Big(\int_0^s g(\theta,x(\theta))dw(\theta)\Big)ds\Big]^\alpha, \Big[\int_0^t (t-s)^{q-1}E_{q,q}(A(t-s)^q)\\
& \times\Big(\int_0^s g(\theta,y(\theta))dw(\theta)\Big)ds\Big]^\alpha\Big)\\
\leq\ & d_H\Big(\Big[\int_0^T (T-s)^{q-1}E_{q,q}(A(T-s)^q)f(s,x(s))ds\Big]^\alpha, \Big[\int_0^T (T-s)^{q-1}E_{q,q}(A(T-s)^q)f(s,y(s))ds\Big]^\alpha\Big)\\
& +d_H\Big(\Big[\int_0^T (T-s)^{q-1}E_{q,q}(A(T-s)^q)\Big(\int_0^s g(\theta,x(\theta))dw(\theta)\Big)ds\Big]^\alpha,\\
& \Big[\int_0^T (T-s)^{q-1}E_{q,q}(A(T-s)^q)\Big(\int_0^s g(\theta,y(\theta))dw(\theta)\Big)ds\Big]^\alpha\Big)\\
& +d_H\Big(\Big[\int_0^t (t-s)^{q-1}E_{q,q}(A(t-s)^q)f(s,x(s))ds\Big]^\alpha, \Big[\int_0^t (t-s)^{q-1}E_{q,q}(A(t-s)^q)f(s,y(s))ds\Big]^\alpha\Big)\\
& +d_H\Big(\Big[\int_0^t (t-s)^{q-1}E_{q,q}(A(t-s)^q)\Big(\int_0^s g(\theta,x(\theta))dw(\theta)\Big)ds\Big]^\alpha, \Big[\int_0^t (t-s)^{q-1}E_{q,q}(A(t-s)^q)\\
& \times\Big(\int_0^s g(\theta,y(t))dw(t)\Big)ds\Big]^\alpha\Big)\\
& \leq \int_0^T (T-s)^{q-1}E_{q,q}(A(T-s)^q)\Big(d_H\Big[\Big(f(s,x(s)),f(s,y(s))\Big)ds\Big]^\alpha\Big)\\
& +\int_0^T (T-s)^{q-1}E_{q,q}(A(T-s)^q)\Big(d_H\Big[\Big(\int_0^s g(\theta,x(\theta)),g(\theta,y(\theta))dw(\theta)\Big)ds\Big]^\alpha\Big)\\
& +\int_0^t (t-s)^{q-1}E_{q,q}(A(t-s)^q)\Big(d_H\Big[\Big(f(s,x(s)),f(s,y(s))\Big)ds\Big]^\alpha\Big)\\
& +\int_0^t (t-s)^{q-1}E_{q,q}(A(t-s)^q)\Big(d_H\Big[\Big(\int_0^s g(\theta,x(\theta)),g(\theta,y(\theta))dw(\theta)\Big)ds\Big]^\alpha\Big)
\end{aligned}
$$

$$\leq \int_0^T (T-s)^{q-1} E_{q,q}(A(T-s)^q)\Big(d_\infty\Big[\Big(f(s,x(s)), f(s,y(s))\Big)ds\Big]^\alpha\Big)$$
$$+\int_0^T (T-s)^{q-1} E_{q,q}(A(T-s)^q)\Big(d_\infty\Big[\Big(\int_0^s g(\theta,x(\theta)), g(\theta,y(\theta))dw(\theta)\Big)ds\Big]^\alpha\Big)$$
$$+\int_0^t (t-s)^{q-1} E_{q,q}(A(t-s)^q)\Big(d_\infty\Big[f(s,x(s)), f(s),y(s))\Big)ds\Big]^\alpha\Big)$$
$$+\int_0^t (t-s)^{q-1} E_{q,q}(A(t-s)^q)\Big(d_\infty\Big[\Big(\int_0^s g(\theta,x(\theta)), g(\theta,y(\theta))dw(\theta)\Big)ds\Big]^\alpha\Big)$$

Therefore

$$\begin{aligned}
\mathbb{E}\Big\|d_\infty\Big((\Phi x)(t),(\Phi y)(t)\Big)\Big\|^2 \leq\ & \mathbb{E}\Big\|\Big[\int_0^T (T-s)^{q-1} E_{q,q}(A(T-s)^q)\Big(d_\infty\Big[\Big(f(s,x(s)), f(s,y(s))\Big)ds\Big]\Big)\\
&+\int_0^T (T-s)^{q-1} E_{q,q}(A(T-s)^q)\\
&\times\Big(d_\infty\Big[\Big(\int_0^s g(\theta,x(\theta)), g(\theta,y(\theta))dw(\theta)\Big)ds\Big]\Big)\\
&+\int_0^t (t-s)^{q-1} E_{q,q}(A(t-s)^q)\Big(d_\infty\Big[\Big(f(s,x(s)), f(s,y(s))\Big)ds\Big]\Big)\\
&+\int_0^t (t-s)^{q-1} E_{q,q}(A(t-s)^q)\\
&\times\Big(d_\infty\Big[\Big(\int_0^s g(\theta,x(\theta)), g(\theta,y(\theta))dw(\theta)\Big)ds\Big]\Big)\Big]\Big\|^2\\
\leq\ & 4\Big[\mathbb{E}\Big\|\int_0^T (T-s)^{q-1} E_{q,q}(A(T-s)^q)\Big(d_\infty\Big[\Big(f(s,x(s)), f(s,y(s))\Big)ds\Big]\Big)\Big\|^2\\
&+\mathbb{E}\Big\|\int_0^T (T-s)^{q-1} E_{q,q}(A(T-s)^q)\\
&\times\Big(d_\infty\Big[\Big(\int_0^s g(\theta,x(\theta)), g(\theta,y(\theta))dw(\theta)\Big)ds\Big]\Big)\Big\|^2\\
&+\mathbb{E}\Big\|\Big[\int_0^t (t-s)^{q-1} E_{q,q}(A(t-s)^q)\Big(d_\infty\Big[\Big(f(s,x(s)), f(s,y(s))\Big)ds\Big]\Big)\Big\|^2\\
&+\mathbb{E}\Big\|\int_0^t (t-s)^{q-1} E_{q,q}(A(t-s)^q)\Big(d_\infty\Big[\Big(\int_0^s g(\theta,x(\theta)), g(\theta,y(\theta))dw(\theta)\Big)ds\Big]\Big)\Big\|^2\Big]\\
\leq\ & 4\Big[\frac{T^{2q}}{q^2}a^2K\mathbb{E}\|x(t)-y(t)\|^2 + \frac{T^{2q+1}}{q^2}a^2L\mathbb{E}\|x(t)-y(t)\|^2\\
&+\frac{T^{2q}}{q^2}a^2K\mathbb{E}\|x(t)-y(t)\|^2 + \frac{T^{2q+1}}{q^2}a^2L\mathbb{E}\|x(t)-y(t)\|^2\Big]\\
\leq\ & 4\Big[2\frac{T^{2q}}{q^2}a^2K\mathbb{E}\|x(t)-y(t)\|^2 + 2\frac{T^{2q+1}}{q^2}a^2L\mathbb{E}\|x(t)-y(t)\|^2\\
\leq\ & 8\frac{T^{2q}}{q^2}a^2[K+LT]\mathbb{E}\|x(t)-y(t)\|^2
\end{aligned}$$

By hypotheses (H3), $8\frac{T^{2q}}{q^2}a^2(K+LT) < 1$.

Hence Φ_1 is a contraction mapping. By the Banach fixed point theorem, Φ_1 has a unique fixed point $x \in C(J : F(\mathbb{R}^d))$ which is the solution of the system (3). Hence the system (3) is α-level controllable on J.

Similarly, one can prove that Φ_2 is a contraction. By the Banach fixed point theorem, (4) has a unique fixed point $x \in C(J : F(\mathbb{R}^d))$ which is the solution of the system (4). Hence the system (4) is α-level controllable on J. Therefore fuzzy Caputo fractional stochastic system (1) has two unique solutions. Hence the system (1) is α-level controllable on J.

Conclusion

In this chapter, the controllability result of fuzzy fractional stochastic differential equations has been promoted. Based on fractional calculus, stochastic analysis approach, and Banach fixed point theorem, existence and uniqueness have been derived for the controllability of fuzzy Caputo fractional stochastic systems in finite dimensional space. Many practical problems could be modeled into linear fuzzy fractional stochastic differential equations. Hence suitable controllability of fuzzy mapping may be constructed to establish the result and an appropriate nonlinear fuzzy stochastic model may be formulated and one can use the proposed technique to derive the controllability results.

References

[1] Zadeh, L. A. (1965). Fuzzy sets. *Information and Control*, 8: 338-353.

[2] Diamond, P. and Kloeden, P. E, *Metric space of fuzzy sets*, World Scientific, 1994.

[3] Balasubramaniam, P. and Muralisankar, S. (2004). Existence and uniqueness of fuzzy integrodifferential equation with nonlocal conditions. *International Journal of Computer and mathematics with applications*, 47: 1115-1122.

[4] Malinowski, M. and Mitchta, M. (2010). Stochastic set differential equations. *Nonlinear Analysis: Theory methods and Applications*, 72: 1256-1274.

[5] Malinowski, M. and Mitchta, M. (2010). Fuzzy stochastic integral equations. *Dynamical Systems and Applications*, 19: 473-494.

[6] Malinowski, M. and Mitchta, M. (2011). Stochastic fuzzy differential equations with an application. *Kybernetika*, 47: 1-19.

[7] Li, J. and Wang, J. (2011). Fuzzy set-valued lebesgue integral and fuzzy stochastic differential equation. *Eighth International Conference on Fuzzy Systems and Knowledge Discovery (FSKD), IEEE*, 152-156.

[8] Malinowski, M. (2015). Fuzzy and Set-valued stochastic differential equations with local Lipchitz condition. *IEEE transactions on Fuzzy sy-tems*, 23: 1891-1898.

[9] Li, J. and Li, S. (2009). Aumann type Lebesgue integral and representation theorem. *International Journal of Computational Intelligence Syste-ms*, 1: 83-90.

[10] Balasubramaniam P. (2002). Controllability for the nonlinear fuzzy neutralfunctional differential equations. *Far East Journal of Applied mathematics*, 9 : 31-48.

[11] Park, J. H., Park, J. S. and Kwun, Y. C. (2006). Controllability for the semilinear fuzzy integrodifferential equations with nonlocal conditions. *Fuzzy Systems and Knowledge Discovery*, 221-230.

[12] Lee, B. Y., Park, D .G., Choi, G. T. and Kwun, Y. C. (2006). Controllability of a nonlinear fuzzy control system with nonlocal initial conditions in n-dimensional fuzzy vector space E_n^N. *International Journal of Fuzzy Logic and Intelligent Systems*, 6: 15-20.

[13] Lee, B. Y., Youm, H. E. and Kim, J. S. (2014). Exact controllability for abstract fuzzy differential equations in credibility space. *Internatio-nal Journal of Fuzzy Logic and Intelligent Systems*, 14: 145-153.

[14] Agarwal, R. P, Lakshmikantham, V. and Nieto, J. J. (2010). On the concept of solution for fractional differentialequations with uncertainty. *No-nlinear Analysis*, 72: 59-62.

[15] Allahviranloo, T., Salahshour, S. and Abbasbandy, S. (2012). Explicit solutions of fractional differential equations with uncertainty. *Soft Computing Fusion of Foundations, Methodologies and Applications*, 6: 297-302.

[16] Li, J. and Wang, J. (2012) Fuzzy set-valued stochastic Lebesgue integral. *Fuzzy sets and Systems*, 200: 48-64.

[17] Rivaz, A., Fard, O. S. and Bidgoli, T. A. (2017). On the existence and uni-queness of solutions for fuzzy fractional differential equations.*Tbilisi Mathematical Journal*, 10 : 197-205.

[18] Bede, B., Rudas, I. J. and Bencsik A. L. (2007). First order linear fuzzy differential equations under generalized differentiability. *Informa-tion Sciences*, 177 : 1648-1662.

[19] Stefanini, L. (2010). A generalization of Hukuhara difference for interval and fuzzy arithmetic. *Fuzzy Sets and Systems*, 161: 1564-1584.

[20] Allahviranloo, T., Armand, A. and Gouyandeh, Z. (2014). Fuzzy fractional differential equations under generalized fuzzy Caputo derivative. *Journal of Intelligent and Fuzzy Systems*, 26: 1481- 1490.

[21] Salahshour, S., Ahmadian, A., Senu, N., Baleanu D. and Agarwal P. 2015. On analytical solutions of the fractional differential equation with uncertainty : Application to the Basset problem. *Entropy*, 17: 885-902.

[22] Abbasbandy, S., Allahviranloo, T., Balooch Shahryari, M .R. and Salhshour, S. (2012). Fuzzy local fractional differential equations. *International Journal of Industrial Mathematics*, 4 : 231-246.

In: An Essential Guide to Fuzzy Systems ISBN: 978-1-53616-128-1
Editor: Michael Gr. Voskoglou

Chapter 5

DEEP FUZZY NEURAL NETWORK: AN EVOLVING PERSPECTIVE

***Amir Samadi*[1] *and M. Foad Samadi*[2,*]**

[1]Department of Electrical Engineering, Center of Excellence on Soft Computing and Intelligent Information Processing, Ferdowsi University of Mashhad, Mashhad, Iran

[2]Department of Electrical and Computer, Engineering, University of Windsor, Windsor, ON, Canada

ABSTRACT

Neural networks and fuzzy systems have developed powerful systems together that encompass both capability of inference under uncertainty, and also computational capacity for estimation and classification. A multitude of techniques as a combination of these methodologies is proposed in the literature to address a number of different applications such as "Fuzzy Neuron", "Adaptive Neuro-Fuzzy Inference System", and "Fuzzy Restricted Boltzmann Machine". However, processing of big data posed new challenges to the field of uncertain systems in the last decade. As more

* Corresponding Author's Email: fsamadi@uwindsor.ca.

information is provided, uncertainty naturally increases, and therefore a hybrid deep fuzzy neural network is introduced as a powerful tool to address problems involving uncertainty and computation of big data. This chapter reviews the hybrids of neural networks and fuzzy logic and discusses how their combination can improve the quality of learning, decision making, inference, and computation. Fuzzy neural network structures such as "Extended Neo-Fuzzy Neuron", "Fuzzy Restricted Boltzmann Machine", and "Recurrent Fuzzy Neural Network" that, based on cascading properties, can render themselves to deep-fuzzy neural networks are introduced.

Keywords: fuzzy neural networks, deep learning, deep neural network, Big data

1. INTRODUCTION

Nowadays, we are exposed to a vast stream of information that can be used in many areas, including image/speech processing, system identification, estimation, and control. Fuzzy Logic (FL) and Artificial Neural Networks (ANN) are two well-known intelligent methodologies which have been employed to fulfill these tasks with various architectures and hybridization configurations. Their hybrid structure is capable of tackling nonlinearity, and uncertainty inherent in dynamic systems.

ANNs have been developed as a capable tool in the realm of information processing, which have long been a practical solution to control problems [1-3]. They are considered as one of the prominent methods of intelligent control in recent decades. ANN by training the weights between neuron layers, learn the functional relation among datasets and construct a model of the nonlinear system. On the other hand, FL makes a mathematical framework for approximate reasoning. The uncertainty surrounding the massive amount of data poses a challenge for identification and control of the system, and generally any other utilization of data. Therefore, to summarize, ANN can extract the essential features whereas FL handles uncertainty and express imprecision in terms of linguistic variable [4]. Recently, powerful systems have been emerged by compounding FL and

ANN that encompass both capability of inference under uncertainty, and also computational capacity for regression and classification (e.g., [4-6]). Ref. [7] classifies the neuro-fuzzy hybrid into three groups of sequential hybrid, auxiliary hybrid, and embedded hybrid. The structure of each class of Neuro-Fuzzy network is composed of two systems where one system represents Fuzzy system, and the other system represents neural network. They are explained as follows and their schematic is shown in Figure 1.

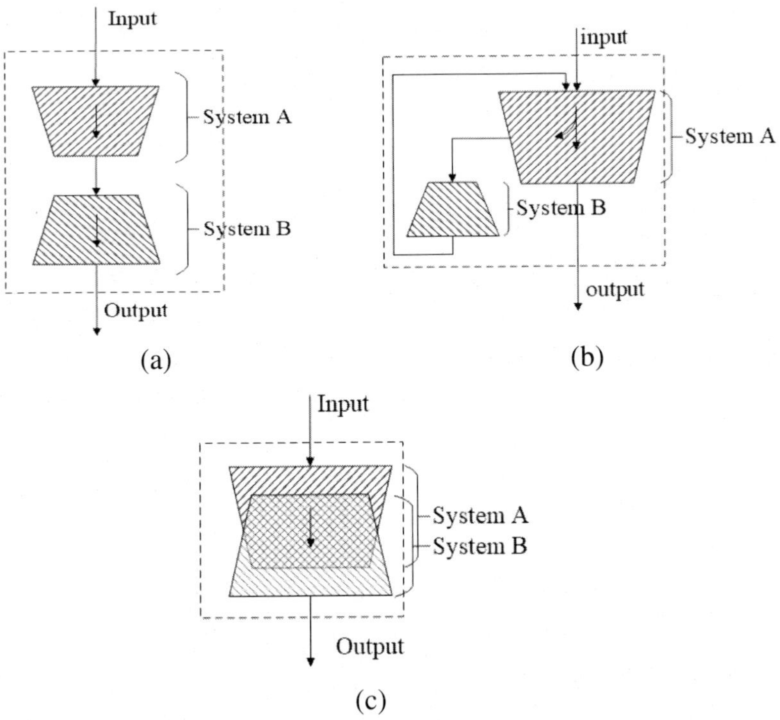

Figure 1. (a) Sequential hybrid. (b) Auxiliary hybrid. (c) Embedded Hybrid.

- Sequential hybrid: The two systems are used in pipelined fashion. For example, fuzzy system is employed as a data preprocessing system to obtain optimal parameters for the neural network.
- Auxiliary hybrid: One system calls the other system as a subroutine. For instance, a fuzzy system regulates the learning rate of the ANN during the training procedure.

- Embedded hybrid: the two systems are fused totally. For instance, consider an ANN which receives fuzzy inputs, processes it and extract fuzzy outputs as well.

A number of different techniques as a combination of these methodologies are proposed to address a number of different applications such as "Fuzzy Neuron" (see [8, 9]), "Adaptive Neuro-Fuzzy Inference System" (see [10]), "Fuzzy Restricted Boltzmann Machine" (see [11]), "Recurrent Fuzzy Neural Network" (see [12, 13]), and "Extended Neo-Fuzzy Neuron" (see [14]). The last three techniques that are more recent are reviewed here briefly.

Deep neural network is defined as an ANN with more than three layers (including input and output layers) of neurons where it can simulate and model a system with certain level of complexity of dynamics. Image classification, stock price prediction, self-driving cars, video synthesizing, text generation, and voice recognition are some applications of deep learning due to the level of underlying complexity. Generally, the complexity of a deep neural network is adjusted to complication level of the corresponding task based on cascading property of the network, which if more complexity needed, it is met by adding more hidden layers to network. This means addition of more weights and biases- which in turn improves the approximation ability of network to cope with complex functions. The network can also be trained such that each layer of network is trained on a separate set of features, where the combination of these extracted features leads to perform a complex task.

The rest of this chapter is formed as follows: The three prominent networks of recurrent fuzzy neural network, extended neo-fuzzy neuron, and. fuzzy restricted boltzmann machine are introduced in Section 2, 3, and 4, respectively. This will help the reader to get insight to main deep fuzzy neural networks introduced in literature. Reader will get familirized with main proposed structures, advantages of each network and their application. Section 5 summarizes the concluding remarks.

2. Recurrent Fuzzy Neural Network

Fuzzy neural network (FNN) is in general an ANN which has incorporated fuzzy input and/or fuzzy weights [15-18]. The employment of FNN is however limited in the temporal applications such as anticipating, recognition, identification, and dynamic control of temporal sequences (e.g., stock price predicting). For such problems, a FNN capable of coping with the underlying temporal dynamics is desired which has led to development of emerging recurrent fuzzy neural network (RFNN). RFNN is a system composed of FNN where tapped delays are introduced on feed-forward structure. Time delays help the structure to memorize information to solve dynamic problems [19-23].

RFNN system was proposed by Isik and Farrokhi [12] for the first time within which the dynamic structure of the rule base of FL not only depends on inputs but also the previous state of the network, thereby consequent part of the rule depends on the current and previous amount of antecedent. Therefore, RFNN has universal approximation property in addition to temporal information storage and the dynamic mapping ability which makes it a viable solution for temporal applications as mentioned above. C. H. Lee and C. C. Teng used RFNN for time series prediction, control and identificationof nonlinear systems [21]. J. S. wang and C. S. G. Lee employed it for controller design of an autonomous underwater vehicle [24]. G. D. Wu et al. used RFNN for speech recognition [25]. In [26], the rule base of RFNN is constructed by Takagi–Sugeno–Kang (TSK) and hence correspondent network is named as TSK-type recurrent fuzzy network (TRFN). Reference [27] extends TRFN to fuzzified TSK-type recurrent fuzzy network (FTRFN).

A regular RFNN has depicted in Figure 2 (type 1 refers to the first RFNN structure introduced), which contains the main five layers as follows:

- *Layer 1-Input layer*: The first layer similar to a normal neural network is the input layer which passes input values to the next layer. The difference is that inputs of this layer consist of linguistic variables.

- *Layer 2-Fuzzification Layer*: The membership degrees of Inputs are computed in the second layer, hence this layer is identified as fuzzification layer. Although any membership function type such as Triangular, Trapezoid, Gaussian, and Bell-shaped can be used here, Gaussian membership function due to differentiability has been employed more than other types.
- *Layer 3-Rule Layer*: The third layer is a representation of the antecedent part of fuzzy rules, where the firing strength of the rule is computed by a fuzzy operation like fuzzy AND operation. Therefore each node in the rule layer denoted by the fuzzy AND symbol $\prod$, and product of its inputs. The rule layer in addition to be operating on the last layer, receives the feedback of the output of next layer from the last sample time (Z^{-1} indicates the time delay).
- *Layer 4-Fuzzy Inference Layer*: In this layer, the consequent part of the fuzzy rule has been simulated by summation of weighted inputs and a bias which represents the constant term in the neural network. Weights in the inference layer are the neural network parameters.
- *Layer 5-Output Layer*: In the last layer, defuzzified output of the consequent part is computed. The most common defuzzification method is center of gravity whereas other methods can also be used. Gravity defuzzification can be shown as follows:

$$u = \frac{\sum_{i=1}^{m} \mu_{Bi}(w_i)w_i^*}{\sum_{i=1}^{m} \mu_{Bi}(w_i)} \tag{1}$$

where u represents the defuzzified output of the neuron.

The most widespread algorithms used to adjust membership functions parameters and defuzzification weights are gradient descent learning and back-propagation (BP) algorithms. Real-time recurrent learning (RTRL) [28] and back-propagation through time (BPTT) are also designed for training in recurrent networks [29]. Generally, all learning algorithms are classified under two types of structure learning algorithm and parameter learning algorithm. The structure learning algorithm not only minimizes the fuzzy rules generated, but also it minimizes the number of fuzzy sets for each input, whereas parameter learning algorithm focuses on connection weights and model parameters [30].

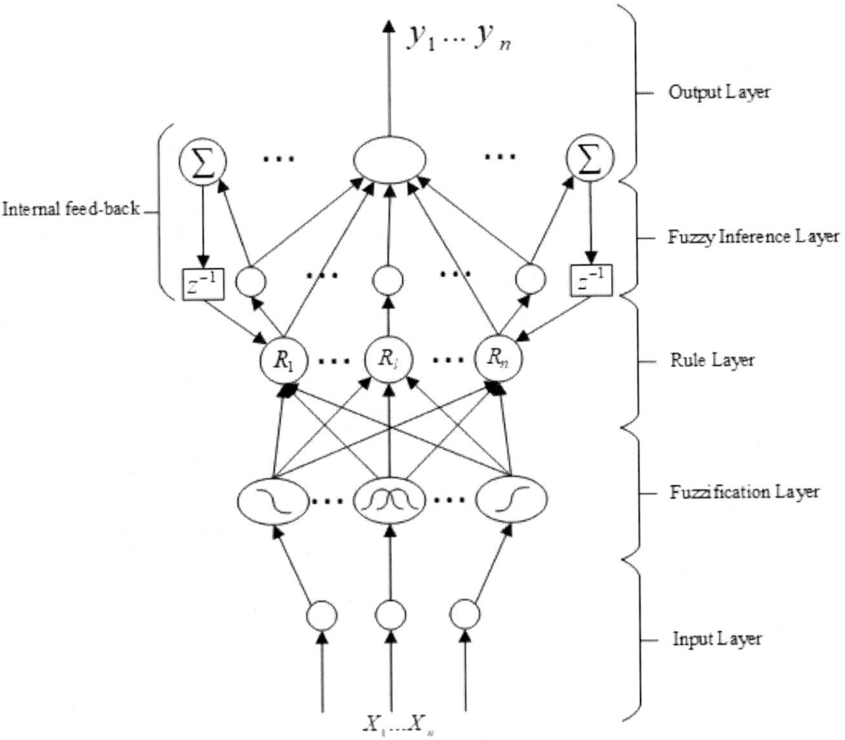

Figure 2. A recurrent fuzzy neural network type 1.

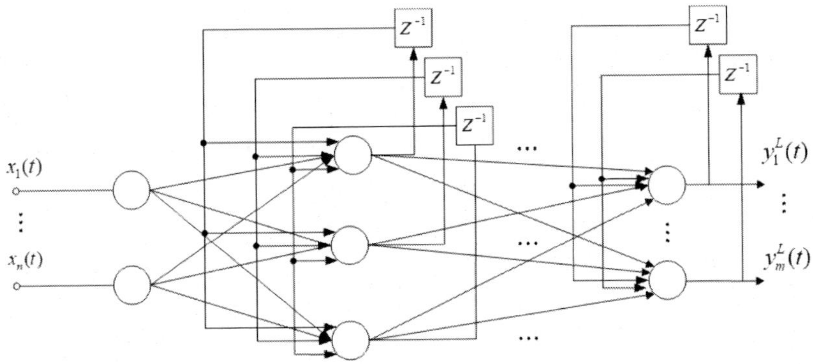

Figure 3. A recurrent fuzzy neural network type 2.

Although the fuzzy inference can be extended to multi-layers to fit to a certain level of complexity, Aliev et al., [31, 32] introduced a new RFNN identified as type 2 in this chapter. The new structure is more appropriate for use in problems with high level of complexity. This structure which is quite similar to recurrent multi-layer perceptron can add more hidden layers to increase the complexity level.

The time delay elements (Z^{-1}) represent memory cells, which keep values of neurons at the previous time step and feed them back to the corresponding inputs at the next time step. Suppose network layers are numbered consecutively from 0 to *L*. The first layer distributes the inputs signals to the network.

$$y_i^0(t) = x_i^0(t) \tag{2}$$

The output of hidden layer neurons (layers 1 to $L-1$) are computed as follows:

$$y_i^l(t) = \phi\left(\theta_i^l + \sum_j x_j^l(t)w_{ij}^l + \sum_j y_i^l(t-1)v_{ij}^l\right) \tag{3}$$

where $x_j^l(t)$ represents the *j*-th fuzzy input to the *i*-th neuron in layer *l* at the time step *t*, $y_i^l(t)$indicates the output signal of the *i*-th neuron at the time step *t*, w_{ij}^l shows the fuzzy weight between to neuron *i* from layer *l* and neuron *j* in layer *l-1*, θ_i^l represents the fuzzy bias of neuron *i*, $y_i^l(t-1)$ is the output neuron *j* at the time step (*t* - 1), v_{ij}^l indicates the recurrent weight between neuron *i* in layer *l* and neuron *j* in the same layer, and ϕ denotes the activation function. Linear activation function can be utilized in last layer for regression tasks and other activation functions such as softmax may be employed for classification task as follows.

$$\phi(x) = \frac{x}{x+|x|} \tag{4}$$

Each forward and recurrent weights, and biases are expressed as a fuzzy number in the forms such as triangular, gaussian, bell-shaped, and trapezoid [33]. For instance, a triangular fuzzy number for weights and biases formed as $w_{ij}^{l} = T(w_{Lij}^{l}, w_{Aij}^{l}, w_{Rij}^{l})$, $, v_{ij}^{l} = T(v_{Lij}^{l}, v_{Aij}^{l}, v_{Rij}^{l})$, and $\theta_{i}^{l} = T(\theta_{Li}^{l}, \theta_{Ai}^{l}, \theta_{Ri}^{l})$.

There are two main approaches to train RFNN. First one is a back-propagation algorithm (BP)-based method where derivative of the minimization function is required, and the second approach could be a heuristic minimization method such as genetic algorithms (GA) to minimize loss function and calculate the appropriate fuzzy weights and biases. Aliev et al. used the GA-based algorithm [31], which is effective in cases where it is difficult, or impossible to find derivatives.

3. Extended Neo-Fuzzy Neuron

Neuro-fuzzy system (or neo-fuzzy neuron (NFN)) is hybrid neuro-fuzzy-systems (HNFS) [34-36] system which is considered as a combination of Fuzzy Inference Systems (FISs) and ANNs [37-39]. NFN is organized with multiple inputs and one output where they are related by a non-linear mechanism (Figure 4). This structure ensures the universal approximation as discussed in [40] where input and target relationship has enough simplicity which can reduce the size of required training data and also decrease iteration requirement [41, 42]. Hence, NFN present a simple and efficient solution model. For instance Y. Bodyanskiy et al. [43] employed NFN for modeling and anticipating time series properties. Also, the same authros optimized and speeded up the NFN learning procedure, which allowed to employ the NFN in real-time data mining as discussed in [44]. K.T. Chaturvedi et al. [44] used NFN to model the economic transition. A. Soualhi et al., [45] proposed the NFN for predicting of bearing condition. References [46, 47] present some other applications of NFN.

NFN's structure inherited the zero-order Takagi-Sugeno-Kang (TSK) inference system. C. M. Salgado et al. in [48], proposed an NFN for classifying five different medical datasets by extending the order of TSK in the structure. Also Y. Zhang et al. discussed a hierarchical hybrid structure (deep TSK classifier) which has interpretable linguistic fuzzy rules [49].

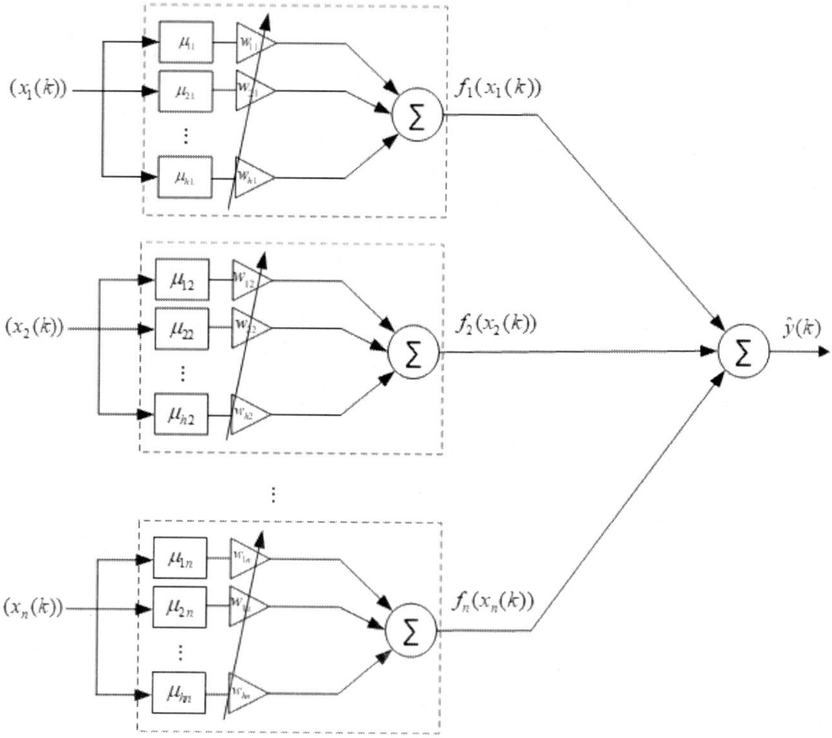

Figure 4. A Neo-Fuzzy Neuron (NFN) architecture.

The NFN as depicted in Figure 4 contains multiple inputs $x = (x_1, \dots, x_n) \in R^n$, and one output $\hat{y}$ which is resulted from the fuzzified outcomes of n-dimensional input *x*. Traingular membership function $\mu_{li}(x_i)$ is employed to perform the fuzzification as follows:

$$\mu_{li} = \begin{cases} \frac{x_i - c_{l-1,i}}{c_{l,i} - c_{l-1,i}}, & c_{l-1,i} \leq x_i < c_{l,i} \\ \frac{c_{l+1,i} - x_i}{c_{l+1,i} - c_{l,i}}, & c_{l,i} \leq x_i < c_{l+1,i} \\ 0, & otherwise \end{cases} \tag{5}$$

where $c_{l,i}$ indicates center of each triangular fuzzy number. Then the fuzzified inputs are multiplied by synaptic weights w_{li} which are trained by datasets for each nonlinear synapse or mapping between inputs, neurons and outputs.

$$f_i(x_i) = \sum_{l=1}^{h} w_{li}\mu_{li}(x_i) \tag{6}$$

Finally, the output is computed as follows

$$\hat{y} = \sum_{l=1}^{n} f_i(x_i) \tag{7}$$

The NFN that satisfied zero-order Takagi-Sugeno inference is quite similar to Wang-Mendel [50, 51] neuro-fuzzy system. A zero-order Takagi-Sugeno rule expressed as follows which is in accordance with NFN's output

IF x_i is x_{li} THEN AN OUTPUT is w_{li}, $l = 1,2, \dots, h$

Bodyanskiy [14, 52-54] introduces neo-fuzzy neuron (ENFN) which is an extenstion of NFN and simulates p-order Takagi-Sugeno, where instead of using w_{li}, p-order Takagi-Sugeno is employed as follows:

IF x_i is x_{li} THEN AN OUTPUT is $w_{li}^0 + w_{li}^1 x_i^1 + w_{li}^2 x_i^2 + \cdots + w_{li}^p x_i^p$

Therefore, the summation of synapse will be formed as follows:

$$f_i(x_i) = \sum_{l=1}^{h} \mu_{li}(x_i)(w_{li}^0 + w_{li}^1 x_i^1 + w_{li}^2 x_i^2 + \cdots + w_{li}^p x_i^p) \tag{8}$$

Figure 5 shows the architecture of an extended non-linear synapse (ENS), which the $\mu_{li}(x_i)$ represents a membership function.

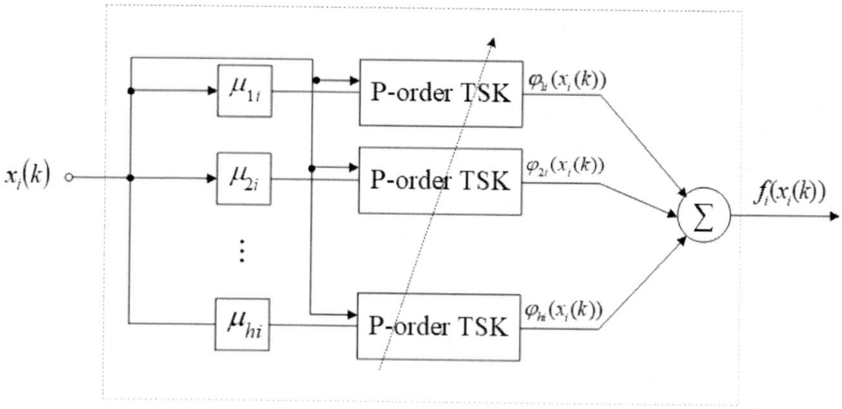

Figure 5. An extended non-linear synapse.

$$\hat{y} = \sum_{i=1}^{n} f_i(x_i) \tag{9}$$

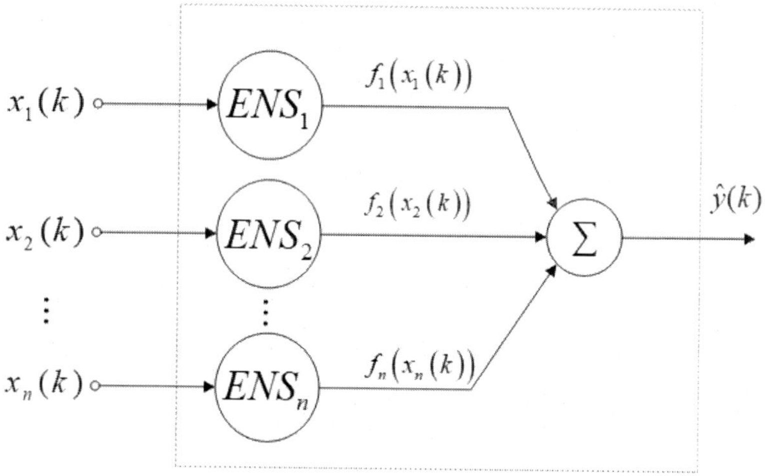

Figure 6. An extended neo-fuzzy neuron.

Now by assembling *n* ENS module for *n* inputs as shown in Figure 6, output $\hat{y}$ is computed by summation of ENSs.

The training procedure exploits gradient descend algorithm as follows:

$$E(k) = \frac{1}{2}\left(\hat{y}(k) - T(k)\right)^2 = \frac{1}{2}\left(\sum_{l=1}^{n} f_i(x_i) - T(k)\right)^2 = \tag{10}$$

$$\frac{1}{2}\left(\sum_{i=1}^{n}\sum_{l=1}^{h} w_{li}\, \mu_{li}(x_i) - T(k)\right)^2$$

$$e(k) = \hat{y}(k) - T(k) \tag{11}$$

$$w_{li}(k) = w_{li}(k-1) - \eta e(k)\mu_{li}(x_i) \tag{12}$$

where T(k) denotes the target signal, E(k) indicates the model squared error, e(k) is learning error, η marks learning rate parameter, and ŷ(k) stands for neuron output. To speed up the training procedure, the algorithm can be modified as follows (see [55]).

$$w(k) = w(k-1) - \left(r(k)\right)^{-1} e(k)\, \mu(x) \tag{13}$$

$$r(k) = \alpha r(k-1) - \|\, \mu(x)\|^2 \tag{14}$$

The architecture of multidimensional extended neo-fuzzy neuron has been depicted in Figure 7 which proposed by Y. Bodyanskiy et al. [56, 57]. It is possible to add more layers or neurons to make network suitable for various applications such as estimation and prediction, classification, feature learning, etc. Therefore, it has a cascading property that enables the structure to be extended to adapt itself to a certain level of complexity. Hence, this network is categorized as deep neural network. A multilayer perceptron (MLP) or softmax function is utilized at the end layer, after the inference layer, to determine the final outputs of network.

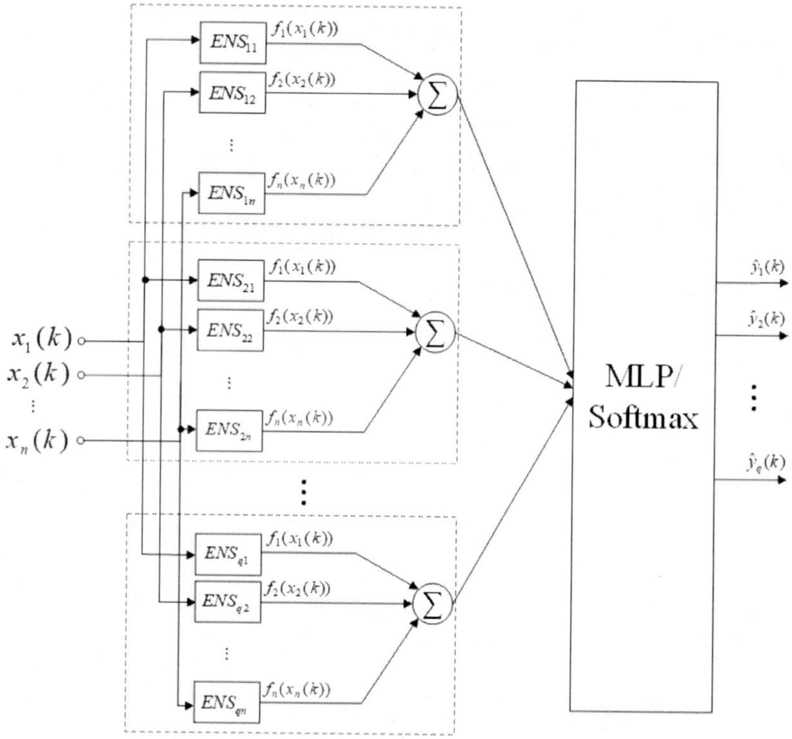

Figure 7. Multidimensional extended neo-fuzzy neuron.

4. Fuzzy Restricted Boltzmann Machine

Boltzmann's Machine (BM) is a kind of ANNs that was invented by Hinton and his colleagues in 1985 [58]. It can be used to learn the probability distribution of a given set of samples. Restricted Boltzmann Machine (RBM), in contrast, has been constructed by imposing some restrictions on the BM architecture to simplify the learning procedure of the network. It is worthwhile mentioning that, this model is used as the constructive blocks in one of the Deep Neural Network (DNN) architectures called the Deep Belief Network (DBN) [59]. The learning process in RBM can be performed as supervised, semi-supervised, and unsupervised manner with various applications[60] such as reconstruction, estimation [61], dimensions reduction[62], classification [63], collaborative filtering [64], feature learning [65], and information modeling [66]. RBM and DBN are discussed in details in [67, 68]. In RBMs, the parameters that describe the relationship between units in the visible and hidden layers are limited to a crisp number. Therefore, this structure leads to limitation in information representation since variables inherently have uncertain behavior. Furthermore, RBM does not manifest enough robustness when the data contain some noise. To address these issues and decrease the distortion that is originated by the linear relationship between visible and hidden units, the Fuzzy Restricted Boltzmann Machines (FRBM) and its respective learning algorithm is presented [11, 69-71]. In the proposed network, the model parameters are considered as fuzzy numbers (Figure 8). Thus, it is shown that FRBMs have better performance in data representation and reveals more robustness property when datasets are contaminated by noise.

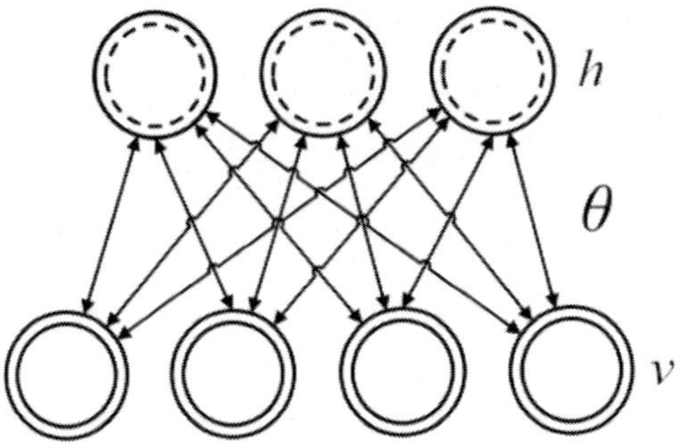

Figure 8. Schematic of fuzzy restricted boltzmann machine.

RBM is an energy-based probabilistic network in which the energy function defines the probability distribution and the probability is defined as:

$$P(v,h,\theta) = \frac{e^{-E(v,h,\theta)}}{Z} \tag{15}$$

$$Z = \sum_v \sum_h e^{-E(v,h,\theta)} \tag{16}$$

where $\mathrm{E}(\mathrm{v},\mathrm{h},\theta)$ is the energy function, θ denotes the parameters of the model and Z represents the summation over all states which is identfied as

Partition FunctionVectors v and h represent the unit vector of visible units and hidden units, respectively. The energy function for RBMs is defined as:

$$E(v,h) = -\sum_{i=1}^{n} b_i v_i - \sum_{j=1}^{m} c_j h_j - \sum_{i=1}^{n}\sum_{j=1}^{m} \omega_{ij} h_j v_i \tag{17}$$

where $w=(\omega_{ij})_{n\times m}$ and ω_{ij} represents the weight connecting the ith visible unit and j^{th} hidden unit, while $b = (b_i)_{1\times n}$ and $c = (c_j)_{1\times m}$ are bias terms associated with the visible units and hidden units, respectively. Now, if all weights and biases are considered as fuzzy numbers, the energy function will be changed into:

$$E(v,h) = -\sum_{i=1}^{n} \bar{b}_i v_i - \sum_{j=1}^{m} \bar{c}_j h_j - \sum_{i=1}^{n}\sum_{j=1}^{m} \bar{\omega}_{ij} h_j v_i \tag{18}$$

where $\bar{\omega}$, $\bar{b}$, and $\bar{c}$ denote conection weights, biases of visible and hidden unit respectively. References [11] and [69] prove that the probability in the visible unit is obtained by summation over hidden units in (5)

$$P(v) = \frac{1}{2}\left(\frac{\sum_h e^{-E^L(v;h)}}{Z} + \frac{\sum_h e^{-E^R(v;h)}}{Z}\right) \tag{19}$$

where *L* and *R* denote left an right side of model parameters, respectively. For instance consider $\bar{\omega}$ as a symmetric triangular fuzzy number, it has to expressed as $\bar{\omega} = (\omega^L, \omega^R)$ (Figure 9) similar to $\bar{b}$, and $\bar{c}$.

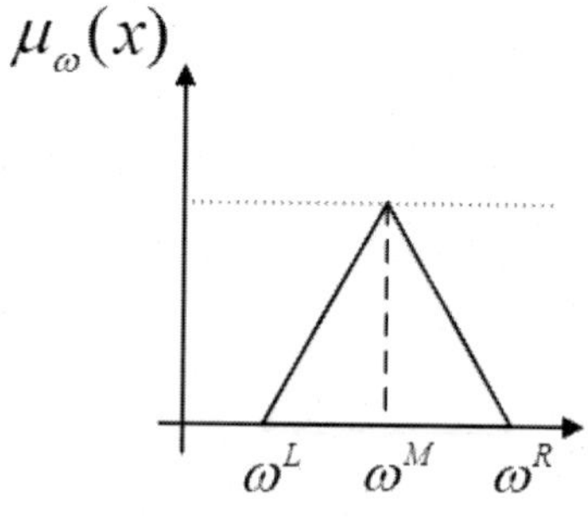

Figure 9. Traingular fuzzy number.

Derivation of the log-probability of the training data with respect to weights and biases can be obtained as follows:

$$\frac{\partial \log P(v^n)}{\partial \omega^L{}_{ij}} = \langle v_i h_j \rangle_{data} - \langle v_i h_j \rangle^L{}_{model} \tag{20}$$

$$\frac{\partial \log P(v^n)}{\partial \omega^R{}_{ij}} = \langle v_i h_j \rangle_{data} - \langle v_i h_j \rangle^R{}_{model} \tag{21}$$

$$\frac{\partial \log P(v^n)}{\partial b^L{}_i} = \langle v_i \rangle_{data} - \langle v_i \rangle^L{}_{model} \tag{22}$$

$$\frac{\partial \log P(v^n)}{\partial b^R{}_i} = \langle v_i \rangle_{data} - \langle v_i \rangle^R{}_{model} \tag{23}$$

$$\frac{\partial \log P(v^n)}{\partial c^L{}_j} = \langle h_j \rangle_{data} - \langle h_j \rangle^L{}_{model} \tag{24}$$

$$\frac{\partial \, log \, P(v^n)}{\partial c^R{}_j} = \langle h_j \rangle_{data} - \langle h_j \rangle^R{}_{model} \tag{25}$$

where $\langle \mathrm{v_i h_j} \rangle$ is the mathematical expectation over data or model, Also $\langle \mathrm{v_i h_j} \rangle^{\mathrm{L}}$, and $\langle \mathrm{v_i h_j} \rangle^{\mathrm{R}}$ represent left, and right corresponding mathematical expectation. According to (20-25), the learning rules for the weights of the networks can be written as:

$$\Delta\omega_{ij}^L = \alpha\big(\langle v_i^l h_j^l \rangle_{data} - \langle v_i^l h_j^l \rangle_{model}\big) \tag{26}$$

$$\Delta\omega_{ij}^R = \alpha\big(\langle v_i^l h_j^l \rangle_{data} - \langle v_i^l h_j^l \rangle_{model}\big) \tag{27}$$

$$\Delta b_i^L = \alpha\big(\langle v_i \rangle_{data} - \langle v_i \rangle^L{}_{model}\big) \tag{28}$$

$$\Delta b_i^R = \alpha\big(\langle v_i \rangle_{data} - \langle v_i \rangle^R{}_{model}\big) \tag{29}$$

$$\Delta c_j^L = \alpha\left(\langle h_j \rangle_{data} - \langle h_j \rangle^L{}_{model}\right) \tag{30}$$

$$\Delta c_j^R = \alpha\left(\langle h_j\rangle_{data} - \langle h_j\rangle^R{}_{model}\right) \tag{31}$$

where α is the learning rate.

The training algorithm of RBM is an approximate Maximum-Likelihood (ML) learning algorithm which is recognized as Contrastive Divergence method. It starts with a random initialization in a layer and then performs Gibbs sampling. Gibbs sampling is exploited iteratively until the network reaches the equilibrium state. Afterward, the weights of the RBM are obtained by taking the gradient of log-likelihood of the training data which yields the direction for weight update as below:

$$\frac{\partial \, log \, p(v^0)}{\partial \omega_{ij}} = \langle v_i^0 h_j^0\rangle - \langle v_i^\infty h_j^\infty\rangle \tag{32}$$

$$\frac{\partial \log P(v^n)}{\partial \omega^R{}_{ij}} = \langle v_i^0 h_j^0\rangle_{data} - \langle v_i^\infty h_j^\infty\rangle^R{}_{model} \tag{33}$$

$$\frac{\partial \log P(v^n)}{\partial b^L{}_i} = \langle v_i^0\rangle_{data} - \langle v_i^\infty\rangle^L{}_{model} \tag{34}$$

$$\frac{\partial \log P(v^n)}{\partial b^R{}_i} = \langle v_i^0\rangle_{data} - \langle v_i^\infty\rangle^R{}_{model} \tag{35}$$

$$\frac{\partial \, log \, P(v^n)}{\partial c^L{}_j} = \langle h_j^0\rangle_{data} - \langle h_j^\infty\rangle^L{}_{model} \tag{36}$$

$$\frac{\partial \log P(v^n)}{\partial c^R{}_j} = \langle h_j^0\rangle_{data} - \langle h_j^\infty\rangle^R{}_{model} \tag{37}$$

where $\langle \mathrm{v_i^0 h_j^0}\rangle$ is the correlation between visible and hidden units in the initial state and $\langle \mathrm{v_i^\infty h_j^\infty}\rangle$ is the correlation between visible and hidden units in the equilibrium state. As the length of chain increases, Gibbs sampling is performed for a larger number of iterations, the generated sample's contribution to the likelihood reduces and the samples can be used to estimate negative gradient. Theoretically, this contribution tends to zero

[72]. However, the chain is never long enough to estimate the exact value of gradient [73].

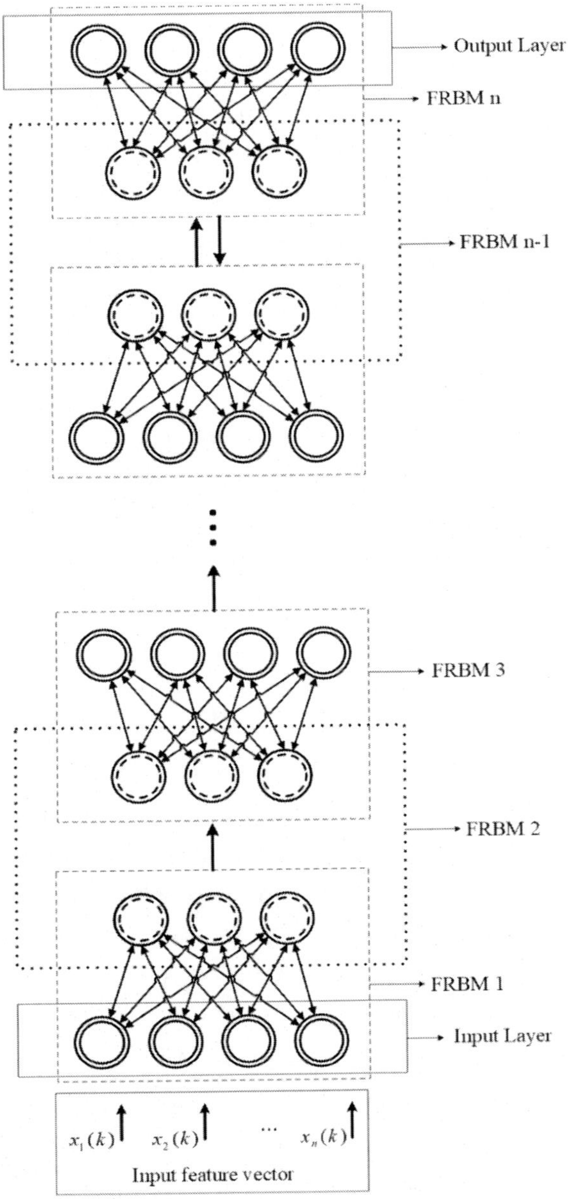

Figure 10. A deep belief network.

RBMs are employed to establish a deep neural network named Deep Belief Network (DBN) [59], structured by stacking and training in a greedy layer-by-layer method. A DBN that constructed by FRBM instead of RBM is depicted in Figure 10. There are two algorithms for training DBN which are greedy layerwise [74], and wake-sleep algorithms [59]. In the greedy layerwise approach each FRBM is first trained independently in the so-called pre-training stage and then network is considered as MLP and trained by BP and SGD in the fine-tuning stage. Depending on the complexity of task, the number of layers of DBN can be increased or reduced, hence by cascading property, this fuzzy-based network is considered as a deep fuzzy neural network.

Conclusion and Discussion

Emerging big data and vast information requires our analysis, identification, and modeling methodologies to be able to cope with an ever-more complexity level. Deep networks are developed as a viable solution to address emerging complex systems. In this realm, cascading property of the network play the main role in adaptability of the network to extend and fit its complexity according to the associated dynamics of the problem. On the other hand, as more information is provided, uncertainty is increased, therefore adding fuzziness capability to network is a proper approach to face with the uncertainty and computation of big data, hence hybrid deep fuuzy neural networks introduced. This chapter is served as an introduction to the state-of-the-art methodology of deep fuzzy neural networks. A summary of each network is reviewed, and their application is dicussed. This area is a developing area and has drawn a great deal of intereset in researchers of different fields due to the versatility of the method. The interested reader is encourged to consult with provided references to learn more about each presented network.

REFERENCES

[1] Chris Tseng H. and B. Almogahed, "Modular neural networks with applications to pattern profiling problems," *Neurocomputing,* Article vol. 72, no. 10-12, pp. 2093-2100, 2009.

[2] Gupta M. and J. Qi, "Fusion of fuzzy logic and neural networks with applications to decision and control problems," In *American Control Conference, 1991*, 1991: IEEE, pp. 30-31.

[3] Barto A. G., R. S. Sutton, and C. W. Anderson, "Neuronlike adaptive elements that can solve difficult learning control problems," *IEEE transactions on systems, man, and cybernetics,* no. 5, pp. 834-846, 1983.

[4] Wang L.-X., *"A course in fuzzy systems and control prentice hall,"* Facsimile edition, 1997.

[5] Chen Y.-C. and C.-C. Teng, "A model reference control structure using a fuzzy neural network," *Fuzzy sets and Systems,* vol. 73, no. 3, pp. 291-312, 1995.

[6] Kiguchi K. and T. Fukuda, "Intelligent position/force controller for industrial robot manipulators-application of fuzzy neural networks," *IEEE Transactions on Industrial Electronics,* vol. 44, no. 6, pp. 753-761, 1997.

[7] Alavala C. R., *Fuzzy logic and neural networks: basic concepts & application*. New Age International, 2008.

[8] Buckley J. J. and Y. Hayashi, "Numerical relationships between neural networks, continuous functions, and fuzzy systems," *Fuzzy Sets and Systems,* Review vol. 60, no. 1, pp. 1-8, 1993.

[9] Buckley J. J. and Y. Hayashi, "Hybrid neural nets can be fuzzy controllers and fuzzy expert systems," *Fuzzy Sets and Systems,* vol. 60, no. 2, pp. 135-142, 1993.

[10] Jang J.-S., "ANFIS: adaptive-network-based fuzzy inference system," *IEEE transactions on systems, man, and cybernetics,* vol. 23, no. 3, pp. 665-685, 1993.

[11] Chen C. L. P., C. Y. Zhang, L. Chen, and M. Gan, "Fuzzy Restricted Boltzmann Machine for the Enhancement of Deep Learning," *IEEE*

Transactions on Fuzzy Systems, Article vol. 23, no. 6, pp. 2163-2173, 2015, Art no. 7047917.

[12] Isik C. and M. Farrokhi, "Recurrent neuro-fuzzy systems," In *1997 Annual Meeting of the North American Fuzzy Information Processing Society-NAFIPS (Cat. No. 97TH8297)*, 1997: IEEE, pp. 362-366.

[13] Subathra B. and T. K. Radhakrishnan, "Recurrent neuro fuzzy and fuzzy neural hybrid networks: A review," *Instrumentation Science and Technology,* Review vol. 40, no. 1, pp. 29-50, 2012.

[14] Bodyanskiy Y. V., O. K. Tyshchenko, and D. S. Kopaliani, "A multidimensional cascade neuro-fuzzy system with neuron pool optimization in each cascade," *arXiv preprint arXiv:1610.06485,* 2016.

[15] O'Hagan M., "A fuzzy neuron based upon maximum entropy ordered weighted averaging," In *International Conference on Information Processing and Management of Uncertainty in Knowledge-Based Systems*, 1990: Springer, pp. 598-609.

[16] Yamakawa T., "Pattern recognition hardware system employing a fuzzy neuron," In *Proceedings of the International Conf of Fuzzy Logic and Neural Networks, Lizuka Japan*, 1990, pp. 943-948.

[17] Requena I. and M. Delgado, "R-FN: A model of fuzzy neuron," in *Proc. 2nd Int. Conf. on Fuzzy Logic & Neural Networks, Iizuka, Japan*, 1992, pp. 793-796.

[18] Buckley J. J. and Y. Hayashi, "Fuzzy neural networks: A survey," *Fuzzy Sets and Systems,* Article vol. 66, no. 1, pp. 1-13, 1994.

[19] Lin J.-H. and C. Isik, "A maximum entropy radial basis function network based neuro-fuzzy controller," in *Proceedings of IEEE 5th International Fuzzy Systems*, 1996, vol. 1: IEEE, pp. 156-161.

[20] Ku C.-C. and K. Y. Lee, "Diagonal recurrent neural networks for dynamic systems control," *IEEE transactions on neural networks,* vol. 6, no. 1, pp. 144-156, 1995.

[21] Lee C.-H. and C.-C. Teng, "Identification and control of dynamic systems using recurrent fuzzy neural networks," *IEEE Transactions on fuzzy systems,* vol. 8, no. 4, pp. 349-366, 2000.

[22] Lin C.-J. and C.-C. Chin, "Recurrent wavelet-based neuro fuzzy networks for dynamic system identification," *Mathematical and computer modelling,* vol. 41, no. 2-3, pp. 226-239, 2005.

[23] Lin F.-J., R.-J. Wai, and C.-M. Hong, "Hybrid supervisory control using recurrent fuzzy neural network for tracking periodic inputs," *IEEE Transactions on Neural Networks,* vol. 12, no. 1, pp. 68-90, 2001.

[24] Wang J.-S. and C. G. Lee, "Self-adaptive recurrent neuro-fuzzy control of an autonomous underwater vehicle," *IEEE transactions on robotics and automation,* vol. 19, no. 2, pp. 283-295, 2003.

[25] Wu G.-D. and Z.-W. Zhu, "An enhanced discriminability recurrent fuzzy neural network for temporal classification problems," *Fuzzy Sets and Systems,* vol. 237, pp. 47-62, 2014.

[26] Juang C.-F., "A TSK-type recurrent fuzzy network for dynamic systems processing by neural network and genetic algorithms," *IEEE Transactions on Fuzzy Systems,* vol. 10, no. 2, pp. 155-170, 2002.

[27] Juang C.-F. and K.-C. Ku, "A recurrent fuzzy network for fuzzy temporal sequence processing and gesture recognition," *IEEE Transactions on Systems, Man, and Cybernetics, Part B (Cybernetics),* vol. 35, no. 4, pp. 646-658, 2005.

[28] Williams R. J. and D. Zipser, "A learning algorithm for continually running fully recurrent neural networks," *Neural computation,* vol. 1, no. 2, pp. 270-280, 1989.

[29] Williams R. J. and J. Peng, "An efficient gradient-based algorithm for on-line training of recurrent network trajectories," *Neural computation,* vol. 2, no. 4, pp. 490-501, 1990.

[30] Lin R.-C., W.-D. Weng, and C.-T. Hsueh, "Design of an SCRFNN-based nonlinear channel equaliser," *IEE Proceedings-Communications,* vol. 152, no. 6, pp. 771-779, 2005.

[31] Aliev R. A., B. Guirimov, B. Fazlollahi, and R. R. Aliev, "Evolutionary algorithm-based learning of fuzzy neural networks. Part 2: Recurrent fuzzy neural networks," *Fuzzy Sets and Systems,* vol. 160, no. 17, pp. 2553-2566, 2009.

[32] Aliev R. A., B. Fazlollahi, R. R. Aliev, and B. Guirimov, "Linguistic time series forecasting using fuzzy recurrent neural network," *Soft Computing,* vol. 12, no. 2, pp. 183-190, 2008.

[33] Dubois D. and H. Prade, *Fundamentals of fuzzy sets*. Springer Science & Business Media, 2012.

[34] Rutkowski L., *Computational intelligence: Methods and techniques* (Computational Intelligence: Methods and Techniques). 2008, pp. 1-514.

[35] Kruse R., C. Borgelt, C. Braune, S. Mostaghim, and M. Steinbrecher, *Computational intelligence: A methodological introduction*. Springer, 2016.

[36] Bodyanskiy Y., "Computational intelligence techniques for data analysis," In *Lecture Notes in Informatics (LNI), Proceedings - Series of the Gesellschaft fur Informatik (GI)*, 2005, vol. P-72, pp. 15-36.

[37] Miki T. and T. Yamakawa, "Analog implementation of neo-fuzzy neuron and its on-board learning," *Computational Intelligence and Applications,* pp. 144-149, 1999.

[38] Uchino E. and T. Yamakawa, "Soft computing based signal prediction, restoration, and filtering," in *Intelligent Hybrid Systems*: Springer, 1997, pp. 331-351.

[39] Yamakawa T., "A neo fuzzy neuron and its applications to system identification and prediction of the system behavior," In *Proc. of the 2nd Int. Conf. on Fuzzy Logic & Neural Networks*, 1992, pp. 477-483.

[40] Kolodyazhniy V., Y. Bodyanskiy, and P. Otto, "Universal approximator employing neo-fuzzy neurons," In *Computational Intelligence, Theory and Applications*: Springer, 2005, pp. 631-640.

[41] Myklebust O., "Zero defect manufacturing: a product and plant oriented lifecycle approach," *Procedia CIRP,* vol. 12, pp. 246-251, 2013.

[42] Lee J., E. Lapira, B. Bagheri, and H.-a. Kao, "Recent advances and trends in predictive manufacturing systems in big data environment," *Manufacturing Letters,* vol. 1, no. 1, pp. 38-41, 2013.

[43] Bodyanskiy Y., I. Pliss, and O. Vynokurova, "Flexible neo-fuzzy neuron and neuro-fuzzy network for monitoring time series

properties," *Information Technology and Management Science,* vol. 16, no. 1, pp. 47-52, 2013.

[44] Chaturvedi K. T., M. Pandit, and L. Srivastava, "Modified neo-fuzzy neuron-based approach for economic and environmental optimal power dispatch," *Applied Soft Computing,* vol. 8, no. 4, pp. 1428-1438, 2008.

[45] Soualhi A., G. Clerc, H. Razik, and F. Rivas, "Long-term prediction of bearing condition by the neo-fuzzy neuron," In *2013 9th IEEE International Symposium on Diagnostics for Electric Machines, Power Electronics and Drives (SDEMPED)*, 2013: IEEE, pp. 586-591.

[46] Zurita D., M. Delgado, J. A. Carino, J. A. Ortega, and G. Clerc, "Industrial time series modelling by means of the neo-fuzzy neuron," *IEEE access,* vol. 4, pp. 6151-6160, 2016.

[47] Pandit M., L. Srivastava, and V. Singh, "On-line voltage security assessment using modified neo fuzzy neuron based classifier," in *2006 IEEE International Conference on Industrial Technology*, 2006: IEEE, pp. 899-904.

[48] Salgado C. M., J. L. Viegas, C. S. Azevedo, M. C. Ferreira, S. M. Vieira, and J. M. Sousa, "Takagi–Sugeno fuzzy modeling using mixed fuzzy clustering," *IEEE Transactions on Fuzzy Systems,* vol. 25, no. 6, pp. 1417-1429, 2017.

[49] Zhang Y., H. Ishibuchi, and S. Wang, "Deep Takagi–Sugeno–Kang fuzzy classifier with shared linguistic fuzzy rules," *IEEE Transactions on Fuzzy Systems,* vol. 26, no. 3, pp. 1535-1549, 2018.

[50] Wang L. X. and H. Ying, "Adaptive fuzzy systems and control: design and stability analysis," *Journal of Intelligent and Fuzzy Systems-Applications in Engineering and Technology,* vol. 3, no. 2, p. 187, 1995.

[51] Wang L.-X. and J. M. Mendel, "Fuzzy basis functions, universal approximation, and orthogonal least-squares learning," *IEEE transactions on Neural Networks,* vol. 3, no. 5, pp. 807-814, 1992.

[52] Bodyanskiy Y. V. and N. Kulishova, "Extended neo-fuzzy neuron in the task of images filtering," *Radioelectronics, Computer Science, Control,* Vol. 1, pp.112-119, 2014.

[53] Bodyanskiy Y. V., O. K. Tyshchenko, and D. S. Kopaliani, "An extended neo-fuzzy neuron and its adaptive learning algorithm," *arXiv preprint arXiv:*1610.06483, 2016.

[54] Bodyanskiy Y., S. Popov, and T. Rybalchenko. *Multilayer neuro-fuzzy network for short term electric load forecasting, Lecture Notes in Computer Science (including subseries Lecture Notes in Artificial Intelligence and Lecture Notes in Bioinformatics)*, vol. 5010 LNCS, pp. 339-348, 2008.

[55] Bodyanskiy Y., I. Kokshenev, and V. Kolodyazhniy, "An adaptive learning algorithm for a neo fuzzy neuron," in *EUSFLAT Conf.*, 2003: Citeseer, pp. 375-379.

[56] Hu Z., Y. V. Bodyanskiy, N. Y. Kulishova, and O. K. Tyshchenko, "A multidimensional extended neo-fuzzy neuron for facial expression recognition," *International Journal of Intelligent Systems and Applications,* vol. 9, no. 9, p. 29, 2017.

[57] Bodyanskiy Y., N. Kulishova, and O. Chala, "The Extended Multidimensional Neo-Fuzzy System and Its Fast Learning in Pattern Recognition Tasks," *Data,* vol. 3, no. 4, p. 63, 2018.

[58] Group P. R., "*Parallel distributed processing: Explorations in the microstructure of cognition:* Vol. 1," Foundations. Cambridge, MA: MIT Press, 1986.

[59] Hinton G. E., S. Osindero, and Y.-W. Teh, "A fast learning algorithm for deep belief nets," *Neural computation,* vol. 18, no. 7, pp. 1527-1554, 2006.

[60] Salakhutdinov R., J. B. Tenenbaum, and A. Torralba, "Learning with hierarchical-deep models," *IEEE transactions on pattern analysis and machine intelligence,* vol. 35, no. 8, pp. 1958-1971, 2013.

[61] Kuremoto T., S. Kimura, K. Kobayashi, and M. Obayashi, "Time series forecasting using a deep belief network with restricted Boltzmann machines," *Neurocomputing,* vol. 137, pp. 47-56, 2014.

[62] Hinton G. E. and R. R. Salakhutdinov, "Reducing the dimensionality of data with neural networks," *science,* vol. 313, no. 5786, pp. 504-507, 2006.

[63] Ciregan D., U. Meier, and J. Schmidhuber, "Multi-column deep neural networks for image classification," in *Computer vision and pattern recognition (CVPR), 2012 IEEE conference on*, 2012: IEEE, pp. 3642-3649.

[64] Vincent P., H. Larochelle, I. Lajoie, Y. Bengio, and P.-A. Manzagol, "Stacked denoising autoencoders: Learning useful representations in a deep network with a local denoising criterion," *Journal of Machine Learning Research,* vol. 11, no. Dec, pp. 3371-3408, 2010.

[65] Bengio Y., A. Courville, and P. Vincent, "Representation learning: A review and new perspectives," *IEEE transactions on pattern analysis and machine intelligence,* vol. 35, no. 8, pp. 1798-1828, 2013.

[66] Hinton G. E. and R. R. Salakhutdinov, "Replicated softmax: an undirected topic model," in *Advances in neural information processing systems*, 2009, pp. 1607-1614.

[67] Bengio Y., "Learning deep architectures for AI," *Foundations and trends® in Machine Learning,* vol. 2, no. 1, pp. 1-127, 2009.

[68] Hinton G. E., "A practical guide to training restricted Boltzmann machines," In *Neural networks: Tricks of the trade*: Springer, 2012, pp. 599-619.

[69] Feng S. and C. L. P. Chen, "A Fuzzy Restricted Boltzmann Machine: Novel Learning Algorithms Based on the Crisp Possibilistic Mean Value of Fuzzy Numbers," *IEEE Transactions on Fuzzy Systems,* Article vol. 26, no. 1, pp. 117-130, 2018, Art no. 7782438.

[70] Shuang F. and C. L. P. Chen, "Fuzzy restricted boltzmann machine and deep belief network: A comparison on image reconstruction," In *2017 IEEE International Conference on Systems, Man, and Cybernetics, SMC 2017*, 2017, vol. 2017-January, pp. 1828-1833.

[71] Samui S., I. Chakrabarti, and S. K. Ghosh. *Improving the Performance of Deep Learning Based Speech Enhancement System Using Fuzzy Restricted Boltzmann Machine, Lecture Notes in Computer Science* (including subseries Lecture Notes in Artificial Intelligence and

Lecture Notes in Bioinformatics), vol. 10597 LNCS, pp. 534-542, 2017.

[72] Bengio Y. and O. Delalleau, "Justifying and generalizing contrastive divergence," *Neural computation,* vol. 21, no. 6, pp. 1601-1621, 2009.

[73] Sutskever I. and T. Tieleman, "On the convergence properties of contrastive divergence," In *Proceedings of the thirteenth international conference on artificial intelligence and statistics*, 2010, pp. 789-795.

[74] Salakhutdinov R., "Learning deep generative models," *Annual Review of Statistics and Its Application,* vol. 2, pp. 361-385, 2015.

In: An Essential Guide to Fuzzy Systems ISBN: 978-1-53616-128-1
Editor: Michael Gr. Voskoglou

Chapter 6

FUZZY POSSIBILITY THEORY: AN INTEGRATION OF VOLTAGE SAG TYPE DETECTION AND ITS IMPACT ON EQUIPMENT BEHAVIOR

Rituparna Mitra**, *Arup Kumar Goswami* and *Prashant Kumar Tiwari

Electrical Engineering Department, National Institute of Technology, Silchar, Assam, India

ABSTRACT

Magnitude and Phase angle of system voltage determine types of voltage sag. Equipment sensitivity essence decides voltage sag type. Fuzzy probability theory has been applied as a basic tool for determination of voltage sag type. Fuzzy possibility theory is an effective tool for making distinction and surfacing of equipment behavior and adjoining sensitivity for voltage sag type condition. Alternatively, voltage sag type detection is possible with multiple metering device placement in

* Corresponding Author's Email: rituparnamitra1990@gmail.com.

every load points with the help of enormous historical data. In this sphere data collection procedure is time consuming and uncertain. This data uncertainty can be dealt with fuzzy probability theory for accurate and realistic determination of voltage sag type. Voltage sag type detection and equipment behavior under sag condition are uncertain. Integration of fuzzy possibility theory has been maneuvered to deal with the situation correctly and effectively. The outcome of this work is appropriate selection of voltage sag mitigation strategy and protection of the equipment of the system from tripping and malfunctioning. Magnitude and Phase angle of system voltage determine types of voltage sag. Equipment sensitivity essence decides voltage sag type. Fuzzy probability theory has been applied as a basic tool for determination of voltage sag type. Fuzzy possibility theory is an effective tool for making distinction and surfacing of equipment behavior and adjoining sensitivity for voltage sag type condition. Alternatively, voltage sag type detection is possible with multiple metering device placement in every load points with the help of enormous historical data. In this sphere data collection procedure is time consuming and uncertain. This data uncertainty can be dealt with fuzzy probability theory for accurate and realistic determination of voltage sag type. Voltage sag type detection and equipment behavior under sag condition are uncertain. Integration of fuzzy possibility theory has been maneuvered to deal with the situation correctly and effectively. The outcome of this work is appropriate selection of voltage sag mitigation strategy and protection of the equipment of the system from tripping and malfunctioning.

Keywords: voltage sag, fuzzy probability, fuzzy possibility theory, equipment trip

INTRODUCTION

Voltage sag is a definite unavoidable phenomenon for a power distribution system. Voltage sag is an important aspect to maintain good power quality in a system. Maintenance of power quality is important for smooth operation of equipment connected in the system. Power quality of a system can be maintained by maintaining voltage profile of it. When the system voltage drops by 10 - 90% of its nominal value for few cycles to one minute, the situation is called as voltage sag. Voltage sag may occur in a system due to heavy load energization, online starting of induction motors, single line to ground faults, double line fault and triple line symmetrical faults, etc. Occurrence of voltage sag has an impact on equipment and processes of the electrical system. The equipment/process may trip or mal-operate in voltage sag condition. Different types of voltage sag can be determined by sag magnitude and angles. An equipment reacts differently for different types of voltage sag. The equipment reaction can decide the process control scheme.

Previously voltage sag type has been detected using different methods. In 2015, voltage sag and swell characteristics have been dealt with three phase voltage ellipse parameter. Based on instantaneous magnitude of three individual phases, rotating elliptical three phase voltage vector have been drawn to identify the type of voltage sag/swell. An algorithm has been adopted with minor and major axis and inclination angle of that ellipse to classify and characterize voltage sag/swell (Mollah Rezaul Alam 2015). Three phase three angle algorithm has been introduced to characterize voltage sag from measured three phase voltages. The algorithm also compared with symmetrical component and six phase algorithm for validation (M. Madrigal 2007). Remaining complex voltage has been used to characterize balanced and unbalanced voltage sag in power system (Lidong Zhang 2000). Canadian Electrical Association surfaced involving 22 utilities to determine the frequency of voltage sags for industrial and commercial customers (M. B. Don O. Koval 1997) (R. A. Don O. Koval 1998). Using characteristics voltage and zero sequence current, mathematically two indices named phase-to-neutral and phase-to-phase voltage have been proposed to characterize voltage sag in (Padmanabh Thakur 2013).

Impact of voltage sag is not only applicable to industrial consumers. Commercial and domestic customers are also affected by voltage sag by several means. Both types of customers face monetary loss due to production loss, reputation loss, data loss, loss in wages and raw materials, etc. Different equipment reacts differently in voltage sag condition. The severity of voltage sag and equipment behavior in voltage sag had been reported in (J. Wang 2005). Stochastic probabilistic approach had been adopted to assess voltage tolerance curve of sensitive equipment in an industrial plant and to determine number of equipment trips due to voltage sag (C. P. Gupta 2006). Monte Carlo simulation has been used for choosing fault type, location and resistance and probabilistically predicted chances of equipment trip in voltage sag (André dos Santos 2016). Recurrent voltage sag in low voltage loads results automatic reclosing of power distribution systems. Voltage sag contour data has been used to draw power acceptability curve of loads connected to low voltage in 3-D CBEMA format (Jong-Fil Moon 2007). Directly fed induction motors are mostly sensitive to voltage sag. In unsymmetrical voltage sag condition energy lost by induction motors had been simulated in MATLAB using speed-torque characteristics (Vijaya Huchche 2018) (J. Pedra 2007) (A. Rola´n 2012). Impact of induction motor due to unbalanced voltage sag caused by single line-to-ground and line-to-line faults is reported in (G¨ulali Yal¸cinkaya 1998). Dynamic behavior of doubly fed induction generator in symmetrical voltage sag condition has been analyzed with fault ride through capability (Alejandro Rol´an 2011). Impact of induction motors on voltage sag characteristics are observed through transient simulation in PSCAD/EMTDC (Jovica V. Milanovic´ 2008). During voltage sag induction motor reduces its speed. After fault clearing the induction motor reaccelerates and again creates voltage sag in the system (Bollen, The Influence of Motor Reacceleration on Voltage Sags 1995). For deep and long voltage sag in low voltage bus, impact of induction motors on voltage sag characteristics are explained in (Jovia V. Milanovic´ 2008). Sensitivity of AC adjustable speed drives in voltage sag condition has been tested in different conditions for rectangular single, two and three phased voltage sags and for nonrectangular balanced three phase voltage sag (K. S. S. Ž. Djokic´ 2005). Based on IEC 61000-3-7, probabilistic distribution has been used for system disturbance and unified fuzzy inference engine used to represent severity and equipment sustainability in voltage sag condition simultaneously (Chan-Nan Lu 2007). A cement plant faces Kiln outage and other equipment failures due to voltage sag had been reported (Finch 2016). Doubly fed induction generator is used in wind turbine. Deadbeat controller scheme used to estimate positive sequence of stator voltage and current vector in voltage sag condition (C. M. R. Osorio

2017). Fuzzy random assessment method with alpha cut used as a tool to determine failure probability of sensitive equipment under voltage sag condition (Xian-Yong Xiao 2010). Practically behavior of a standard three phase squirrel cage induction motor rated 5.5 kW, 1500 rpm, 380 V in voltage sag and short interruption condition has been reported (Juan C. Gomez 2002).

Personal computers (PC) are very much sensitive to voltage sags. Malfunction criterion of PCs for rectangular voltage sag has been reported in (J. D. S. Ž. Djokic′ 2005). Any equipment has specific immunity time to withstand voltage sag. As the time crosses the equipment tends to malfunction or trip. A case study on thermal power plant had been done to determine the immunity time of boiler coal feeding system (Li Wang 2017). AC coil contactors are equally sensitive to voltage sags. The condition has been tested with ideal and non-ideal voltage, two stage voltage sags and starting of induction motors. Generic sensitivity curve have been drawn for contactors (Saˇsa ˇ Z. Djokic′ 2004). Reaction of adjustable speed drives in voltage sag condition is observed under dc-link variant voltage and it's related to source impedance, dc-link inductance and output load (Jos′e Luis Dur′an-G′omez 1999). Impression of symmetrical voltage sag on three-phase three-leg transformers are explained experimentally and also modelled in PSpice (Luis Guasch 2004).

In this chapter, the overall impact of voltage sag on equipment as well as in industrial production process are determined. Four basic types of voltage sag occurrence due to balanced and unbalanced fault types are considered to observe characteristics of several equipment in the system. The occurred voltage sags are categorized in sag types A, B, C and D with voltage sag magnitude, phase angle jumps and point in wave. This categorization has been done with fuzzy probability theory. Individual sag type has different impact on equipment behavior in voltage sag condition. Whether the equipment will operate or trip due to voltage sag depends on sag type. For individual sag type and for different operating criterion in voltage sag condition, equipment's behavior is determined by fuzzy possibility theory. Voltage tolerance limits and voltage sag type are the inputs of fuzzy possibility theory. Determination of equipment trip due to different types of voltage sag is the goal of this approach.

VOLTAGE SAG DETECTION USING FUZZY PROBABILITY THEORY

Previously a few researchers have dealt with the fact to characterize voltage sag occurrence. Voltage sag have characterized in four types based on voltage sag magnitude, angular deflection and phase angle jump generally. In this paper, four types of voltage sag named as Type A, Type B, Type C, Type D has been taken into account (Bollen, Understanding Power Quality Problems 2000). Basic equations and phasor diagrams of these voltage sag type are represented in Table 1 and Figure 1 respectively.

Unbalanced faults are the main causes of voltage sag. Most occurred fault is single line to ground fault. Voltage sag type also depends on the fault type and also on the load connection. The dependency among these three factors are summarized in Table 2.

The voltage sag type is determined in the following way:

For sag type B

1. Pre-fault voltage is measured with respect to time.
2. Zero sequence voltage is determined.
3. Remaining voltage determined by subtracting zero sequence voltage from pre-fault voltage.
4. Determine the rms values of remaining voltage.
5. Three voltage difference amongst the phases are calculated.
6. Lowest magnitude of three phase sag midst all rms voltages.

For sag type C and D

1. RMS voltage of three phases are determined.

2. RMS values of three voltage difference are determined.
3. The lowest value of the rms voltage difference determines sag type C.
4. Smallest magnitude of rms phase difference voltage value determines sag type D.

Table 1. Equations of four types of sag

Type A	Type B
$V_a = V$ $V_b = -\frac{1}{2}V - \frac{1}{2}jV\sqrt{3}$ $V_c = -\frac{1}{2}V + \frac{1}{2}jV\sqrt{3}$	$V_a = V$ $V_b = -\frac{1}{2} - \frac{1}{2}j\sqrt{3}$ $V_c = -\frac{1}{2} + \frac{1}{2}j\sqrt{3}$
Type C	**Type D**
$V_a = 1$ $V_b = -\frac{1}{2} - \frac{1}{2}jV\sqrt{3}$ $V_c = -\frac{1}{2} + \frac{1}{2}jV\sqrt{3}$	$V_a = V$ $V_b = -\frac{1}{2}V - \frac{1}{2}j\sqrt{3}$ $V_c = -\frac{1}{2}V + \frac{1}{2}j\sqrt{3}$

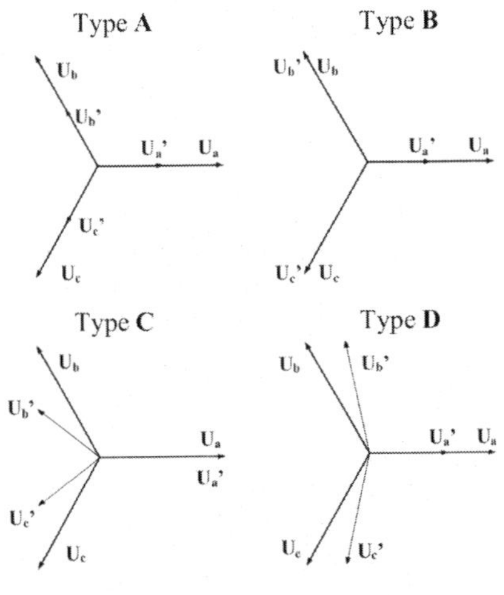

Figure 1. Phasor diagram of four types of sag.

Table 2. Voltage Sag Type based on Fault and Load type

	Three Phase Fault	Phase-to-Phase Fault	Single Phase Fault
Star Connected Load	Type A	Type C	Type B
Delta Connected Load	Type A	Type D	Type C

Voltage sag type has been determined here using fuzzy probability theory (Buckley 2006). . Let A and B are crisp subset of X. $X = \{x_1, x_2, ..., x_n\}$. Fuzzy probability distribution $\overline{P}(\{x_i\}) = \overline{a}_1, 0 < \overline{a}_i < 1, 1 \le i \le n$. If $A = \{x_1, ..., x_k\}, 1 \le k \le n$, then

$$\overline{P}(A)[\alpha] = \{\sum_{i=1}^{k} a_i \mid S\} \text{ for } 0 < \alpha < 1 \tag{1}$$

$$S = \{(x_1, ..., x_n) \mid x_i \ge 0 \forall i, \sum_{i=1}^{n} x_i = 1\} \tag{2}$$

$$Dom[\alpha] = (\prod_{i=1}^{n} \overline{a}_i[\alpha]) \bigcap S \tag{3}$$

where $\overline{P}(A)[\alpha]$ is alpha cut of $\overline{P}(A)$.

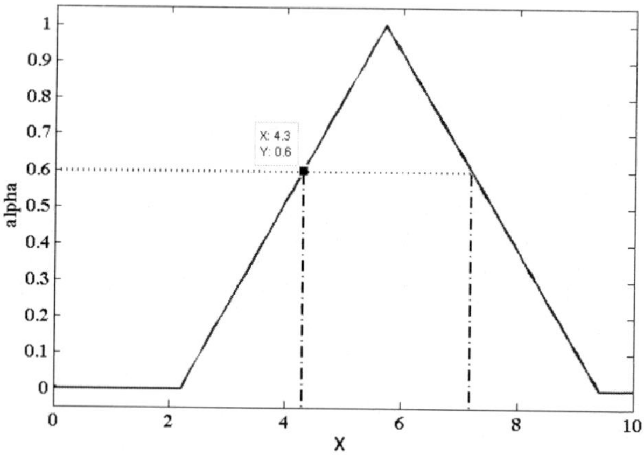

Figure 2. Fuzzy probability for voltage sag type detection.

Table 3. Alpha Cuts

α	$\overline{P_{A'}}[\alpha]$
0	[2.2 9.4]
0.2	[2.9 8.6]
0.4	[3.6 7.9]
0.6	[4.3 7.2]
0.8	[5 6.4]
1	5.7

The procedure for determination of voltage sag type is explained below with an example:

We have to detect voltage sag type among four options A, B, C and D. For a specific sag duration (60ms), the total occurrence of voltage sag due to unbalanced fault is 780. We would like to determine the chance of occurrence of sag type A. The crisp value available to us: p(A) = 0.35, p(B) = 0.25, p(C) = 0.22 and p(D) = 0.18. As these are the crisp probability, we are transforming them into fuzzy numbers.

Let $\dfrac{\overline{p_A} = (0.3/0.35/0.4), \overline{p_B} = (0.2/0.25/0.3),}{\overline{p_C} = (0.17/0.22/0.27), \overline{p_D} = (0.13/0.18/0.23)}$.

Let $\overline{P_{A'}}$ is the fuzzy probability of not occurrence of voltage sag type A.

The equation of fuzzy probability for alpha cut is:

$$\overline{P}(A')[\alpha] = \{p_B + p_C + p_D \mid S\} \tag{4}$$

For all α, $p_A \in \overline{p_A}[\alpha], p_B \in \overline{p_B}[\alpha], p_C \in \overline{p_C}[\alpha], p_D \in \overline{p_D}[\alpha], p_A + p_B + p_C + p_D = 1$.

We know, $p(A) = 1 - p(A')$, so the chance of occurrence of voltage sag type A is presented as triangular fuzzy number = (2.2/5.7/9.4). After defuzzification using centroid method, the chance of voltage sag type A occurrence in a distribution system is 30%.

EQUIPMENT TRIP CALCULATION USING FUZZY POSSIBILITY THEORY

Usually equipment in distribution system follows rectangular equipment tolerance curve as shown in Figure 3.

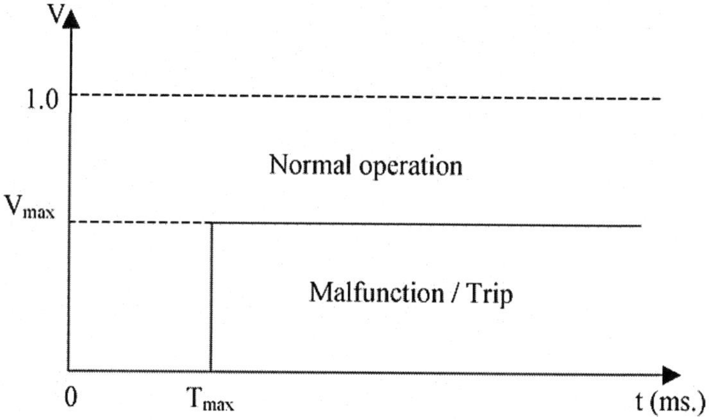

Figure 3. Usual voltage tolerance curve.

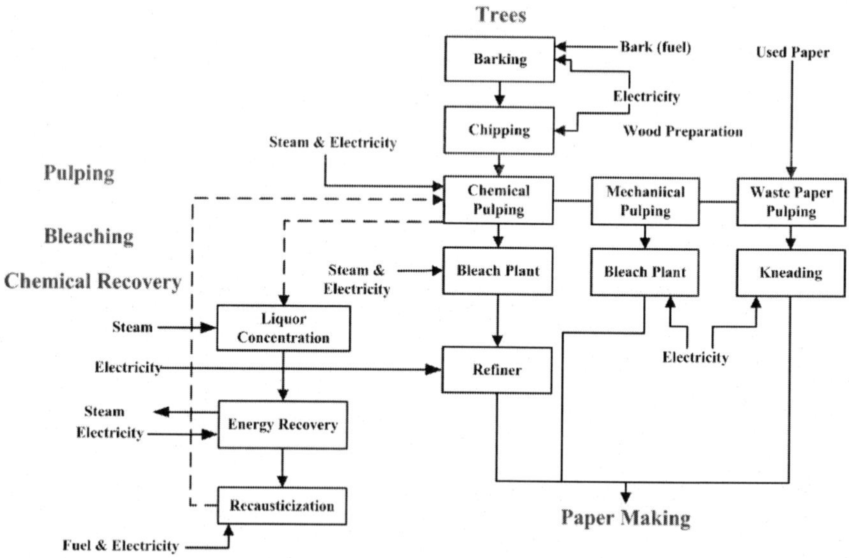

Figure 4. Outline of a paper industry.

In this paper the proposed approach is validated with a paper industry located at 37 bus Barak Valley Distribution System in bus number 12. The outline of paper production in the paper industry is shown in Figure 4.

The essential electrical equipment required in paper industry are: AC Contactor, DC Generator, Motor Starter, Power Supply, Transformer, Transmitter, and Variable Frequency Drive. The behavior of these electrical machines under voltage sag is descripted briefly below.

- *AC Contactor*: A typical AC contactor has 3 parts named as control coil, magnetic circuit and mechanical spring. The control coil of contactor is induction type. During normal operation voltage leads by 90 degree from current. Disengagement of the parts in contactor happens based on voltage sag magnitude and duration. After 20 cycles of voltage sag, the contactors outage happens. Due to contactor trip, the voltage can drop from 100% to 0% of its nominal value for three phases. During voltage sag, this equipment doesn't follow sinusoidal wave. Point in wave of voltage sag recovers has a deep impact on contactor's performance. Phase shift and decreased voltage after voltage sag also changes the operation of contactors (A. E. Turner October 1996). Characteristic curve for contactors is shown in Figure 5.
- *Transformers*: There is significant impact of voltage sag on transformer inrush current and on power transformers. Voltage sag starts and ends at high voltage peak. So the chance of occurrence of shallow sag is more than severe sag. In this condition, magnetic flux of transformer is low, resulting transformer saturation has been not observed frequently. Due to voltage sag, magnetic flux becomes flat while transformer saturates (Bronzeado 2006) (Amin Shokri 2014).
- *Adjustable Speed Drives (ASD)*: In paper mills special machines like different adjustable drives are connected with DC motors for production of pulp from raw material, conversion of pulp into paper and final packaging, etc. Wide speed and torque control from the drive system required for production. Generally, ASDs

have three parts, named as rectifier, DC Bus and inverter. The response of ASD's in voltage sag condition depends on hardware, controller and on response time of controllers. During voltage sag, DC bus voltage drops. If voltage sag condition sustains for shorter time, the capacitor will be able to supply energy to the bus. But if voltage sag persists, DC voltage will drop to a lower level and eventually the inverter will trip (G. Alonso Orcajo 2005). For three phase or balanced fault, energy supply is absent due to reverse biased diode rectifier. As the energy will be served from DC capacitor, the bus voltage starts to decay with load torque, mechanical speed, inverter and motor efficiency. During voltage sag type C and D occurrence for phase-to-phase fault, three phase rectifier is single phased. As output terminals are two in this case, rms input current of ASD increases.

Fuzzy Possibility Theory

Possibility theory is actually an evidence theory. The search space or body of evidence contains nested focal elements. The terms related to possibility theory are necessity and possibility. Differently can be expressed as belief and plausibility. In possibility theory the data is normally distributed. In 1978 Prof. Zadeh introduced the term "possibility theory". In possibility distribution, π is a set from state of affairs S is mapped to unit interval [0, 1]. The function π holds the detailed information about individual agent and helps to distinguish plausibility of the agent surprisingly when its least expected.

$\pi(s) = 0$; when state s is impossible and rejected
$\pi(s) = 1$; when state s in possible, plausible and unsurprising

Since possibility theory is able to handle uncertainty and insufficient data condition, two significant boundaries can be explained.

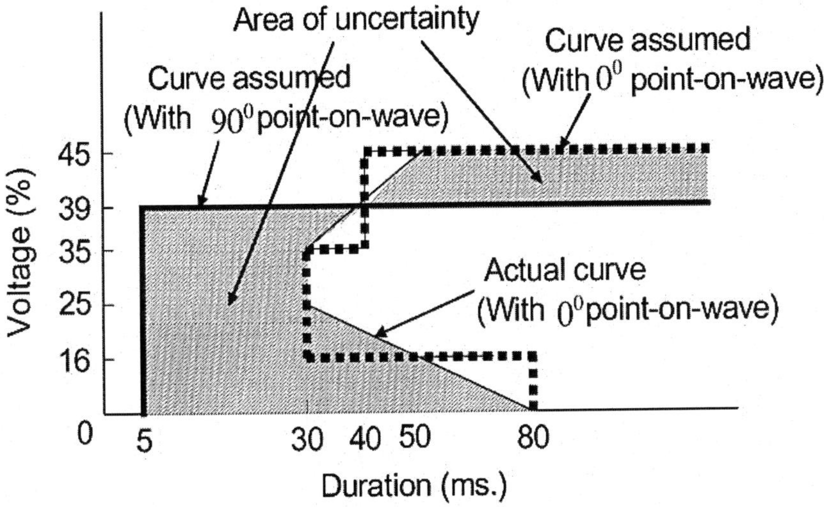

Figure 5. Voltage tolerance curve of contactors.

- Complete Knowledge: for some states s0, π(s0) = 1 and for other s, π(s) = 0, only s0 is possible.
- Complete Ignorance: π(s) = 1, for s∈S (all states are possible).

If A is a subset of states, any queries regarding the state can be obtained from degree of possibility and necessity (if the possibility state lies between [0, 1]).

$$\prod(A) = \sup_{s\in A}\pi(s);\;\; N(A) = \inf_{s\notin A} 1-\pi(s) \tag{5}$$

The possibility degree $\prod(A)$ evaluates the consistency of state A with knowledge π. And $N(A)$ quantify the certainty of A based on knowledge. The possibility-necessity duality is: $N(A)=1-\prod(A^c)$, $\prod(S) = N(S) = 1 \;\; and \;\; \prod(\varphi) = N(\phi) = 0$. Basic properties of possibility theory are:

- Maxitivity: $\prod(A\bigcup B) = \max(\prod(A), \prod(B))$
- Necessity measure:

$$N(A \bigcap B) = \min(N(A), N(B)) \quad (6)$$

Estimation of actual possibility with evidence for all sets of A is defined as guaranteed possibility. The supporting mathematical expression is:

$$\Delta(A) = \inf_{s \in A} \pi(s)\,. \quad (7)$$

Interval valued fuzzy (IVF) sets are used here to map A from the universe S to the closed interval [0, 1]. IVF is actually a special case of type 2 fuzzy system. The chance of effectiveness to deal with uncertainty is maximized with fuzzy possibility approach (Zadeh 1978) (D. Dubois 2000).

RESULT AND DISCUSSION

The proposed approach is validated with a real time Barak Valley Distribution system, in Assam, India. The single line diagram of the system is shown in Figure 6. The in-feed transmission line of this system is 220kV. The distribution system is interconnected with 132 and 33 kV transmission lines. Five 132/33 kV and one 220/132 kV substation is located in Barak Valley. 132 kV industrial customer (Hindusthan Paper Corporation limited) is feed from bus 12. The approach has been tested and justified for bus 12. Total 158 numbers of voltage sag occurred in the system. Voltage sag distribution based on magnitude and time for bus 12 is shown in Figure 7. The source impedance and feeder data of bus 12 is mentioned in Table 4 (calculated using 100 MVA as base value).

Using fuzzy probability theory voltage sag type has been determined. 158 numbers of voltage sag has been occurred in bus 12. The procedure for type detection is already explained with an example in the previous section. Likewise, for other types of voltage sag occurrence is also determined and mentioned in Table 5. From the table the chances of occurrence of each type of voltage sag type can be determined. Probability of individual voltage sag type occurrence are: a) Type A- 35%, b) Type B- 25%, c) Type C- 22%, d) Type D- 18%. The probability of occurrence of specific voltage sag type is required for determination of number of trips due to respective voltage sag type. Impact of voltage sag type on equipment's operating condition can be ascertained from this.

Among all the electrical equipment required in paper industry, three of them has been chosen for consideration. Equipment mostly sensitive to voltage sag are chosen for further analysis. The most sensitive equipment among them is adjustable speed drives. ASDs application in paper industry is very wide. Winders and coilers require DC drive for constant torque for low speed or standstill motor condition. Large diameter paper reels with high-inertia load connected with DC motor requires high-bandwidth control of DC drives. Voltage tolerance range of selective equipment is mentioned in Table 6.

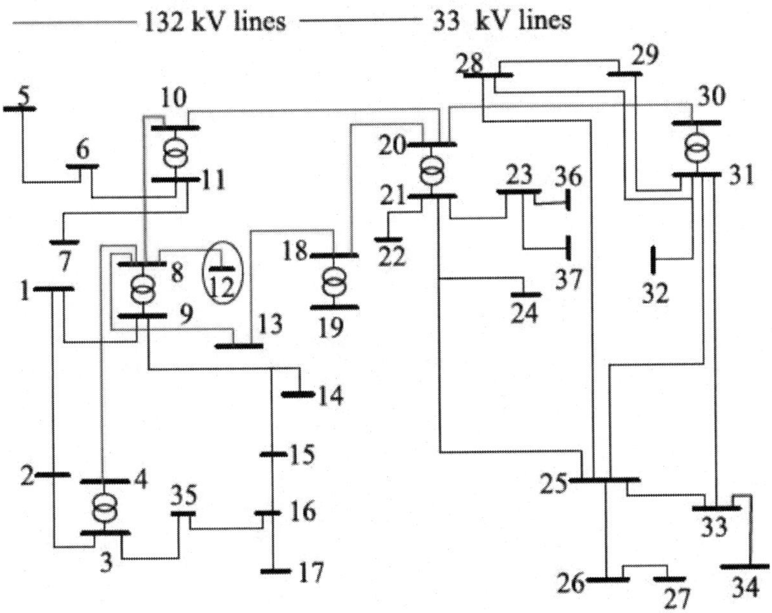

Figure 6. Barak Valley distribution system.

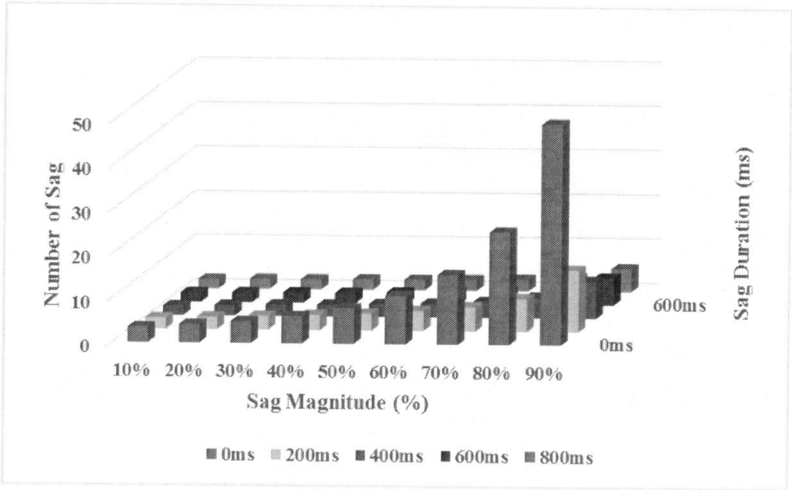

Figure 7. Voltage sag characteristics.

Table 4. Source and Feeder Impedance of 132 kV Bus

132 kV bus	Zero Sequence	Positive and Negative Sequence
Source Impedance	0.046 + j2.7%	0.09 + j2.8%
Feeder	0.23 + j0.65% per km	0.1 + j0.25% per km

Table 5. Type wise Number of Voltage Sag occurrence

Sag Type	A	B	C	D	Total
No. of Sag	55	40	35	28	158

Table 6. Voltage Tolerance of Different Equipment

	Voltage Tolerance		
	Minimum	Maximum	Average
AC Contactor	30 ms, 60%	10 ms, 75%	20 ms, 65%
Motor Starter	80ms, 40%	20 ms, 60%	50ms, 50%
Variable Frequency Drive	80ms, 60%	30 ms, 80%	50ms, 75%

ASD's behave in three ways during voltage sag condition. In case I the speed of motor decreases with voltage sag and recovers speed after voltage sag clearance. In case II motor speed reduces to zero and restarts automatically and accelerates back to normal speed. In case III motor speed reduced to zero and requires manual restart. For lower rated motors, there is no significant difference in voltage tolerance curve for full loading or half loading condition. For voltage sag type A, ASD can operate normally, no trip. For type B, temporarily the drive operates outside operating range but automatically recovers. For type C, the drive safely shuts down and for sag type D, the drive shuts down and requires manual starting. Voltage sag tolerance and tripping of ASD due to sag events is shown in Figure 8. Number of sag witnessed by ASD due to different types of sag condition is mentioned in Table 7. Different types of motor behavior due to voltage sag is depicted in Figure 9. General voltage tolerance curve of ASDs is shown in Figure 10.

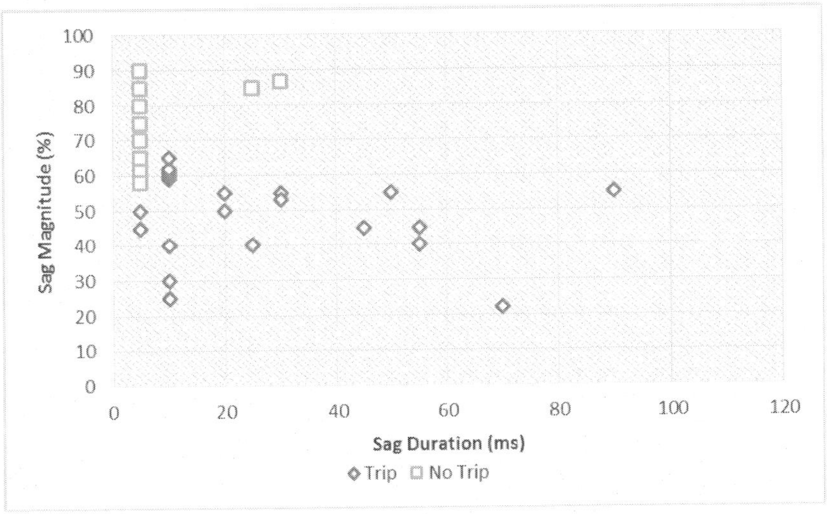

Figure 8. Impact of voltage sag on ASD.

Table 7. Performance of ASD in Voltage Sag

Sag	**I**	**II**	**III**
0%, 30 ms	15	-	-
50%, 100 ms	5	7	3
70%, 1000 ms	2	10	3

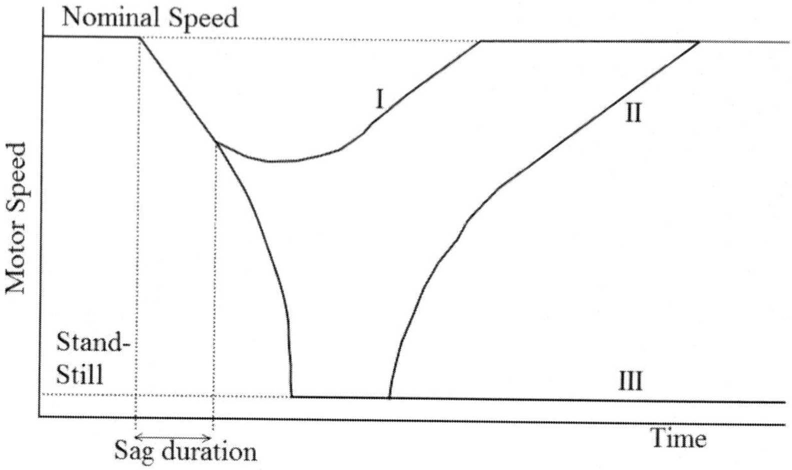

Figure 9. Motor Behavior of ASD in voltage sag.

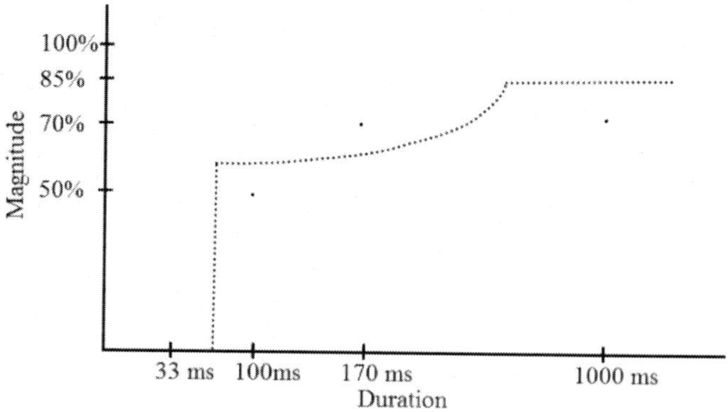

Figure 10. Voltage tolerance curve of ASD.

The behavior of contactors in voltage sag condition doesn't depend on their size. Electrical contactors are designed for a specific voltage level. If voltage drops below a certain level, the contactors will disconnect. Contactors may disconnect even if system voltage is not below the minimum steady state hold in voltage. Contactors have random pattern during voltage sag. Voltage tolerance curve of contactors are shown in Figure 11. For voltage sag type C and D, the contacts definitely de-energizes.

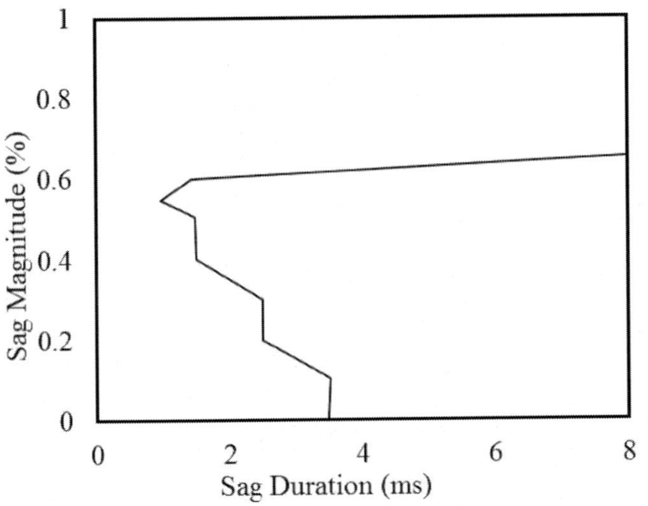

Figure 11. Voltage Tolerance curve of Contactors.

There is not very prominent impact of voltage sag in transformers tripping. As transformers are the electromagnetic device and transfers voltage from one terminal to another, also during voltage sag condition transformers transfers voltage with lesser intensity of voltage sag. In practical application, transformers can be used as voltage sag lessening equipment. The impact of voltage sag gets reduced by transformers because of high inrush current flow. Therefore, due to voltage sag no trip occurs for transformers.

CONCLUSION

Voltage sag is one of the crucial issue for power quality maintenance. Due to voltage sag the system performance reduces and the equipment connected to the system malfunctions and ultimately trips. In this paper fuzzy probability and possibility theory have been used to distinguish voltage sag type and equipment behavior under different voltage sag type accordingly. From this approach, it's observed that amongst four voltage sag type, type C and D are the most dangerous. Moderate to ultra-sensitive equipment trips for sag type C and D occurrence. Paper industry is one of the most sensitive equipped industry susceptible to voltage sag. Most functions of this industry is controlled by adjustable speed drives and they are very much sensitive towards voltage sag. For type A and B sag, the ASD operates normally or regains its normal operating condition within short time. But for voltage sag C and D, it trips and requires manual restart. The scenario for contactors is also the same. They are non-responsive for voltage sag type A and B, but for type C and D it de-energizes. Transformers are non-receptive towards voltage sag. They lowers down the voltage sag during operation. Transformer operation is generally affected by phase angle jump. That induces inrush current in the circuit and the terminal of the transformer voltage comes down with shallow voltage sag. The outcome of this approach is to select judiciously voltage sag mitigation device for sensitive equipment. Mitigation device must have high sustainability during voltage sag for prevention of equipment trip and will reduce monetary loss by limiting downtime cost and production loss.

REFERENCES

Alejandro, Rolán., Felipe, Córcoles. & Joaquín, Pedra. (2011). "Doubly Fed Induction Generator Subject to Symmetrical Voltage Sags." *IEEE Transactions on Energy Conversion*, *26* (4), 1219-1229.

Amin Shokri, S., Asghar, Gholamian. & Mohammad, Yazdani-Asrami. (2014). "Effect of Voltage Sag on Transformers Inrush Current." *International Journal of Mechatronics, Electrical and Computer Technology (Austrian E-Journals of Universal Scientific Organization)*, *4* (12), 811-839.

André, dos Santos. & Maria, Teresa Correia de Barros. (2016). "Predicting Equipment Outages Due to Voltage Sags." *IEEE Transactions on Power Delivery*, *31* (4), 1683-1691.

Alonso Orcajo, G. J. M., Cano, R. C. H., Rojas, G. M. G., Melero, M. F. & Cabanas, F. Pedrayes. (2005). "Voltage sags in industrial systems." *Renewable Energy & Power Quality Journal (RE&PQJ)*, *1* (3), 296-301.

Bollen, Math H. J. (1995). "The Influence of Motor Reacceleration on Voltage Sags." *IEEE Transactions on Industrial Applications*, *31* (4), 667-674.

—. (2000). Understanding Power Quality Problems. New York: John Wiley& Sons.

Bronzeado, Herivelto. & Oliveira, Jose. (2006). "Effect of Voltage Dips on Power Transformers." *12th International Conference on Harmonics and Quality of Power - ICHQP*.

Buckley, James J. (2006). Fuzzy Probability and Statistics. Netherlands: Springer.

Chan-Nan, Lu. & Cheng-Chieh, Shen. (2007). "Estimation of Sensitive Equipment Disruptions Due to Voltage Sags." *IEEE Transactions on Power Delivery*, *22* (2), 1132-1137.

Dubois, D., Kerre, E., Mesiar, R. & Prade, H. (2000). "Fuzzy interval analysis." In Fundamentals of Fuzzy Sets, by H. Prade D. Dubois, 483-581. Kluwer, Boston, Mass: Springer.

Don, O. Koval. & Brent Hughes, M. (1997). "Canadian National Power Quality Survey: Frequency of Industrial and Commercial Voltage Sags." *IEEE Transactions on Industrial Applications*, *33* (3), 622-627.

Don, O. Koval., Romela, A. Bocancea., Kai, Yao. & Brent Hughes, M. (1998). "Canadian National Power Quality Survey: Frequency and Duration of Voltage Sags and Surges At Industrial Sites." *IEEE Transactions on Industry Applications*, *34* (5), 904-910.

Djokic, S. Ž., Desmet, J., Vanalme, G., Milanovic´, J. V. & Stockman, K. (2005). "Sensitivity of Personal Computers to Sensitivity of Personal Computers to." *IEEE Transactions on Power Delivery*, *20* (1), 375-383.

Djokic, S. Ž., Stockman, K., Milanovic´, J. V., Desmet, J. J. M. & Belmans, R. (2005). "Sensitivity of AC Adjustable Speed Drives to Voltage Sags and Short Interruptions." *IEEE Transactions on Power Delivery*, *20* (1), 494-505.

Finch, Alan C. (2016). "A Cement Plant's Experience in Investigating Power Sags Leads to a Reduction in Kiln Outages by Utilizing Power Hardening Methods." *IEEE Transactions on Industry Applications*, *52* (5), 4435-4441.

Gupta, C. P. & Member, Jovica V. Milanovic. (2006). "Probabilistic Assessment of Equipment Trips due to Voltage Sags." *IEEE Transactions on Power Delivery*, *21* (2), 711-718.

Gülali, Yal cinkaya., Math, H. J. Bollen. & Peter, A. Crossley. (1998). "Characterization of Voltage Sags in Industrial Distribution Systems." *IEEE Transactions on Industrial Applications*, *34* (4), 682-688.

Jong-Fil, Moon., Sang-Yun, Yun. & Jae-Chul, Kim. (2007). "Quantitative Evaluation of the Impact of Repetitive Voltage Sags on Low-Voltage Loads." *IEEE Transactions on Power Delivery*, *22* (4), 2395 - 2400.

Josü, Luis Dur´an-G´omez., Prasad, N. Enjeti. & Byeong, Ok Woo. (1999). "Effect of Voltage Sags on Adjustable-Speed Drives: A Critical Evaluation and an Approach to Improve Performance." *IEEE Transactions on Industry Applications*, *35* (6), 1440-1449.

Jovia, V. Milanovic´., Sarat, C. Vegunta. & Myo, T. Aung. (2008). "The Influence of Induction Motors on Voltage Sag Propagation—Part II: Accounting for the Change in Sag Performance at LV Buses." *IEEE Transactions on Power Delivery*, *23* (2), 1072-1078.

Jovica, V. Milanovic´., Myo, T. Aung. & Sarat, C. Vegunta. (2008). "The Influence of Induction Motors on Voltage Sag Propagation—Part I: Accounting for the Change in Sag Characteristics." *IEEE Transactions on Power Delivery*, *23* (2), 1063 - 1071.

Juan, C. Gomez., Medhat, M. Morcos., Claudio, A. Reineri. & Gabriel, N. Campetelli. (2002). "Behavior of Induction Motor Due to Voltage Sags and Short Interruptions." *IEEE Transactions on Power Delivery*, *17* (2), 434-440.

Li, Wang., Xuezhong, Zhu., Huaying, Zhang., Yuming, Zhao., Zhengguo, Zhu., Mingtian, Fan. & Zixuan, Zheng. (2017). "Research of process immunity time of boiler coal feeding system in thermal power plant and solution to voltage sags." *IET (24th International Conference & Exhibition on Electricity Distribution (CIRED))*, (1), 841-845.

Lidong, Zhang. & Math, H. J. Bollen. (2000). "Characteristic of Voltage Dips (Sags) in Power Systems." *IEEE Transactions on Power Delivery*, *15* (2), 827-832.

Luis, Guasch., Felipe, Córcoles., Joaquín, Pedra. & Luis, Sáinz. (2004). "Effects of Symmetrical Voltage Sags on Three-Phase Three-Legged Transformers." *IEEE Transactions on Power Delivery*, *19* (2), 875-883.

Madrigal, M. & Member, B. H. Rocha. (2007). "A Contribution for Characterizing Measured Three-Phase Unbalanced Voltage Sags Algorithm." *IEEE Transactions on Power Delivery*, *22* (3), 1885-1890.

Mollah, Rezaul Alam., Muttaqi, K. M. & Abdesselam, Bouzerdoum. (2015). "Characterizing Voltage Sags and Swells Using Characterizing Voltage Sags and Swells Using." *IEEE Transactions on Industry Applications, 51* (4), 2780 - 2790.

Osorio, C. M. R., Chaves, J. S. S., Murari, A. L. L. F. & Sguarezi Filho, A. J. (2017). "Comparative Analysis of the Doubly Fed Induction Generator (DFIG) Under Balanced Voltage Sag." *IEEE Latin America Transactions, 15* (5), 869-876.

Pedra, J., Co′ rcoles, F. & Sainz, L. (2007). "Effects of unsymmetrical voltage sags on squirrel-cage induction motors." *IET Gener. Transm. Distrib., 1* (5), 769–775.

Padmanabh, Thakur., Asheesh, K. Singh. & Ramesh, C. Bansal. (2013). "Novel way for classification and type detection of voltage sag." *IET Generation, Transmission & Distribution, 7* (4), 398–404.

Rolan, A., Co′ rcoles, F. & Pedra, J. (2012). "Behaviour of the doubly fed induction generator exposed to unsymmetrical voltage sags." *IET Electric Power Applications, 6* (8), 561–574.

Sasa, Z. Djokic., Jovica, V. Milanovic. & Daniel, S. Kirschen. (2004). "Sensitivity of AC Coil Contactors to Voltage Sags, Short Interruptions, and Undervoltage Transients." *IEEE Transactions on Power Delivery, 19* (3), 1299-1307.

Turner, A. E. & Collins, E. R. (October 1996). "The Performance of Ac Contactors during Voltage Sags." *7th Int. Conf. on Harmonics and Quality of Power*. Los Vegas, NV., 589-595.

Vijaya, Huchche., Nita, Patne. & Anjali, Junghare. (2018). "Computation of Energy Loss in an Induction Motor During Unsymmetrical Voltage Sags—A Graphical Method." *IEEE Transactions on Industrial Informatics, 14* (5), 2023-2030.

Wang, J., Chen, S. & Lie, T. T. (2005). "System Voltage Sag Performance Estimation." *IEEE Transactions on Power Delivery, 20* (0), 1738-1747.

Xian-Yong, Xiao., Chao, Ma., Hong-Geng, Yang., Wu, Chen. & Hua-Qiang, Li. (2010). "Failure Probability Analysis of Sensitive Equipment Due to Voltage Sags Using Fuzzy-Random Assessment Method." *IEEE Transactions on Power Delivery, 25* (4), 2970-2975.

Zadeh, L. A. (1978). "Fuzzy Sets as a Basis For a Theory of Possibility." *Fuzzy Sets and Systems, 1*, 3-28.

Rolan, A., Co´ rcoles, F. & Pedra, J. (2012). "Behaviour of the doubly fed induction generator exposed to unsymmetrical voltage sags." *IET Electric Power Applications*, *6* (8), 561–574.

Sasa, Z. Djokic., Jovica, V. Milanovic. & Daniel, S. Kirschen. (2004). "Sensitivity of AC Coil Contactors to Voltage Sags, Short Interruptions, and Undervoltage Transients." *IEEE Transactions on Power Delivery*, *19* (3), 1299-1307.

Turner, A. E. & Collins, E. R. (October 1996). "The Performance of Ac Contactors during Voltage Sags." *7th Int. Conf. on Harmonics and Quality of Power*. Los Vegas, NV., 589-595.

Vijaya, Huchche., Nita, Patne. & Anjali, Junghare. (2018). "Computation of Energy Loss in an Induction Motor During Unsymmetrical Voltage Sags—A Graphical Method." *IEEE Transactions on Industrial Informatics*, *14* (5), 2023-2030.

Wang, J., Chen, S. & Lie, T. T. (2005). "System Voltage Sag Performance Estimation." *IEEE Transactions on Power Delivery*, *20* (0), 1738-1747.

Xian-Yong, Xiao., Chao, Ma., Hong-Geng, Yang., Wu, Chen. & Hua-Qiang, Li. (2010). "Failure Probability Analysis of Sensitive Equipment Due to Voltage Sags Using Fuzzy-Random Assessment Method." *IEEE Transactions on Power Delivery*, *25* (4), 2970-2975.

Zadeh, L. A. (1978). "Fuzzy Sets as a Basis For a Theory of Possibility." *Fuzzy Sets and Systems*, *1*, 3-28.

In: An Essential Guide to Fuzzy Systems
Editor: Michael Gr. Voskoglou
ISBN: 978-1-53616-128-1

Chapter 7

FEATURE SELECTION USING A BHATTACHARYYA DISTANCE

Iuliana- Florentina Iatan*
Department of Mathematics and Computer Science,
Technical University of Civil Engineering, Bucharest, Romania

ABSTRACT

We know that some of the classification algorithms are more efficient in the spaces with low dimension and fail to be practical when applied in the spaces of higher dimension; the transformations of the original n-dimensional feature space into a space with a smaller dimension are therefore strongly recommended. On this purpose, in this chapter we present an algorithm, based on the Bhattacharyya distance method. The criterion of the feature selection presented in this chapter is a very special one in that it is directly related to the probability of the Bayes error.

Keywords: feature selection, Bhattacharra distance, Chernoff bounds, Bayes error

---

* Corresponding Author's Email: iuliafi@yahoo.com.

INTRODUCTION

The four best known approaches for pattern recognition are:

1. statistical classification
2. syntactic or structural matching
3. recognition by intelligence computational techniques
4. recognition by expert systems based on some languages that are specific to artificial intelligence.

The aim of this chapter is to highlight the importance of using the statistical methods in the field of pattern recognition, underpinning to the development of the neural methods of pattern recognition.

The statistic approach has been more intensively studied and used in practice than the traditional approaches of pattern recognition. However, the theory of artificial neural network techniques has been getting significant importance.

Interestingly, the neural networks are good for pattern recognition, but they do not explain the mechanism of decisions.

The fuzzy logic which raises with the imprecise information can be useful for explaining decisions, but can not automatically collect the fuzzy rules on which the decisions are made.

The neuro-fuzzy systems combine the characteristics of the fuzzy systems (which can process the fuzzy information using the fuzzy algorithms) with those of the neural networks (which have the ability to learn and have a parallel structure which can rapidly process the information) to form a neural network that can learn from the environment.

The most of all classification or pattern recognition methods are to decide to which class an unknown pattern belongs.

Pattern Recognition refers to the classification or description of objects or patterns. The patterns themselves can range from characters in an image of printed text to biological waveforms, such as EKG or EEG waveforms. Pattern recognition can use some mathematical methods of statistical decision theory and this approach is named *statistical pattern recognition.*

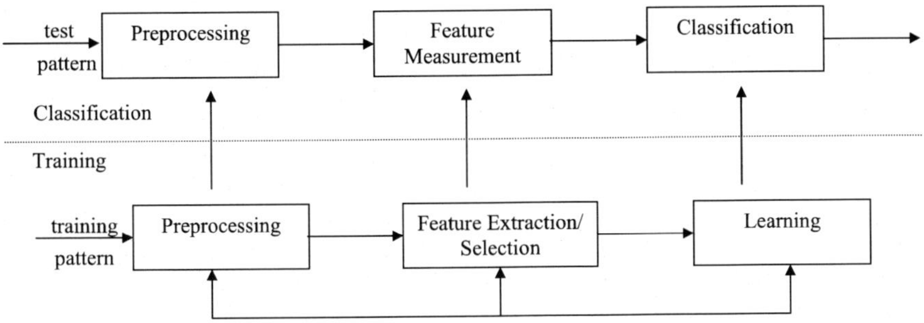

Figure 1. The statistical pattern recognition system.

In the statistical recognition, a pattern is represented by n- dimensional components and the decision on classification is based on a similarity measure, which is expressed as a measure of a distance or a discriminant function.

Statistical methods are important for pattern recognition owing to the concrete systems of pattern recognition. The statistical pattern recognition system is shown [7] in Figure 1 and it operates in two modes: training (learning) and classification (testing).

The role of the preprocessing module is to remove noise from the pattern of interest.

In the training mode the feature selection module finds the appropriate features for representing the input patterns and the classifier is trained to partition the feature space.

In the classification mode, the trained classifier assigns the input pattern to one of the pattern classes under consideration based on the measured features.

In statistical pattern recognition, the features are assumed to have a probability density function conditioned on the pattern class.

Figure 2 shows [7] that the statistical methods solve many problems from in field of pattern recognition.

Various strategies are used in the statistical pattern recognition which depend on the information available on the conditional probability densities corresponding to the classes, i.e., on the information on the parameters which appear in the analytic expression of the probability density function corresponding to the class ω_j, denoted $p(X \mid \omega_j)$, $j = \overline{1, M}$, M being the number of the classes.

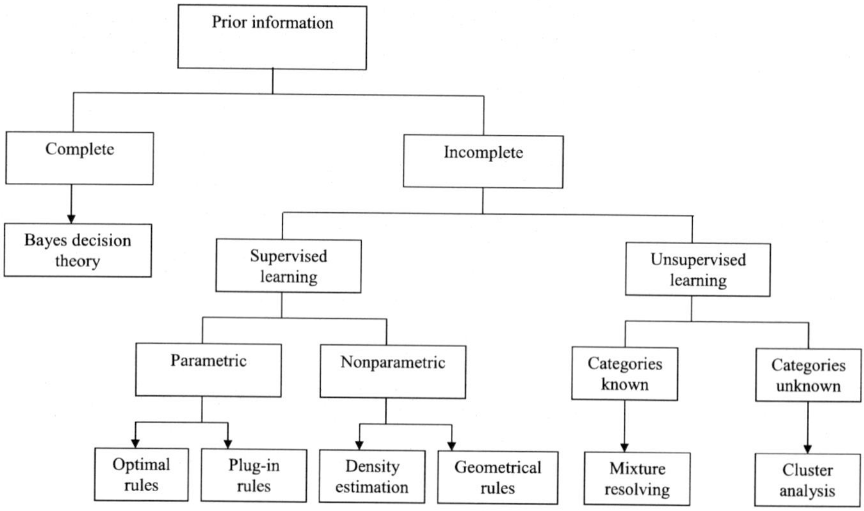

Figure 2. Various approaches in statistical pattern recognition.

If all of the class-conditional densities are completely specified, the optimal Bayes decision rule can be used to design a classifier.

However, the class-conditional densities are usually unknown in practice and must be learned from the available training patterns.

Chernoff Bounds for the Bayesian Error

The probability of *error* ε for the Bayes test is [10]:

$$\varepsilon = P(\omega_1)\varepsilon_1 + P(\omega_2)\varepsilon_2; \tag{1}$$

here:

$$\varepsilon_1 = \int_t^{\infty} f(h \mid \omega_1)\mathrm{d}h, \tag{2}$$

$$\varepsilon_2 = \int_{-\infty}^{t} f(h \mid \omega_2)\mathrm{d}h, \tag{2}$$

and

$$\begin{aligned} h(X) &= -\ln l(X), \\ l(X) &= \frac{f(X \mid \omega_1)}{f(X \mid \omega_2)}, \end{aligned} \tag{3}$$

$$t = \ln \frac{P(\omega_1)}{P(\omega_2)}. \tag{4}$$

In the relation (3), $l(X)$ is the *likelihood ratio*, $f(X \mid \omega_i), i = \overline{1,2}$ being the conditional probability densities of the classes.

The term t means the *threshold value* of the likelihood ratio for a decision.

We shall denote

$$u(s) = -\ln \int_{-\infty}^{\infty} \mathrm{e}^{sh} f(h \mid \omega_1)\mathrm{d}h, \ s \geq 0. \tag{5}$$

We evaluate:

$$\varepsilon_1 = \int_t^{\infty} f(h \mid \omega_1)\mathrm{d}h = \int_t^{\infty} e^{-u(s)-sh} f(h \mid \omega_1)\mathrm{d}h \tag{6}$$

As from the fact that $s \geq 0$ it results:

$$e^{-sh} \le e^{-st}, \quad h \ge t,$$

from the relation (6) we achieve:

$$\varepsilon_1 \le \mathrm{e}^{-u(s)-st}. \tag{7}$$

The upper limit of ε_1 is obtained by minimizing the expression $-u(s)-st$ with respect to s. The optimal value of s can be computed from:

$$-\frac{\mathrm{d}u}{\mathrm{d}s} - t = 0,$$

hence

$$-\frac{\mathrm{d}u}{\mathrm{d}s} = t. \tag{8}$$

The upper limit of ε_1 from (7) with the optimum condition (8) is called *Chernoff limit*.

We can write:

$$f(X \mid \omega_1)^{-s} \cdot f(X \mid \omega_2)^{1+s} = \\ = f(X \mid \omega_1)^{1-s} \cdot f(X \mid \omega_2)^{s} \cdot \left(\frac{f(X \mid \omega_1)}{f(X \mid \omega_2)} \right)^{-1}. \tag{9}$$

From (3) it results

$$\mathrm{e}^{h(X)} = l(X)^{-1} = \left(\frac{f(X \mid \omega_1)}{f(X \mid \omega_2)} \right)^{-1}. \tag{10}$$

From the relations (9) and (10) we deduce:

$$f(X\mid\omega_1)^{-1}\cdot f(X\mid\omega_2)=\mathrm{e}^{h(X)}; \tag{11}$$

hence

$$f(h\mid\omega_2)=\mathrm{e}^{h}\cdot f(h\mid\omega_1). \tag{12}$$

On the basis of the relations (2) and (12) we shall evaluate:

$$\begin{aligned}\varepsilon_2 &= \int_{-\infty}^{t} f(h\mid\omega_2)\,\mathrm{d}h = \int_{-\infty}^{t} \mathrm{e}^{h}\cdot f(h\mid\omega_1)\,\mathrm{d}h = \\ &= \int_{-\infty}^{t} \mathrm{e}^{-u(s)+(1-s)h}\cdot f(h\mid\omega_1)\,\mathrm{d}h.\end{aligned} \tag{13}$$

But for $s\leq 1$ we have:

$$e^{(1-s)h}\leq e^{(1-s)t},\ h\leq t. \tag{14}$$

From (13) and (14) we have:

$$\varepsilon_2\leq \mathrm{e}^{-u(s)+(1-s)t}\int_{-\infty}^{t} f(h\mid\omega_1)\,\mathrm{d}h,$$

hence:

$$\varepsilon_2\leq \mathrm{e}^{-u(s)+(1-s)t},\ \text{for } h\leq t \text{ and } s\leq 1. \tag{15}$$

From (15) we can notice that the optimum S is expressed through the same condition (8).

The relation (7) can be written as:

$$\varepsilon_1 \leq e^{-u(s)}.\left(e^{-t}\right)^s.$$

and the relation (4) can be expressed in the form:

$$e^{-t} = \frac{P(\omega_2)}{P(\omega_1)}.$$

The relation (1) becomes:

$$\varepsilon = P(\omega_1)\varepsilon_1 + P(\omega_2)\varepsilon_2 \leq P(\omega_1)\cdot e^{-u(s)}.\left[\frac{P(\omega_2)}{P(\omega_1)}\right]^s + P(\omega_2)\cdot e^{-u(s)}.\left[\frac{P(\omega_1)}{P(\omega_2)}\right]^{1-s};$$

therefore:

$$\varepsilon \leq e^{-u(s)} P(\omega_1)^{1-s} P(\omega_2)^s. \tag{16}$$

CLASS SEPARABILITY CRITERIA

In order to formulate the class separability criteria, in the discriminant analysis (representing one of the multidimensional analysis techniques that are most commonly used in practice) it is necessary to define the *within-class scatter matrix of the classes* and the *between-class scatter matrix.* The term of discriminant analysis designates a family of some techniques used to classify the patterns that are characterized by a number of continuous or discrete variables.

A *within-class scatter matrix of a class* shows the dispersion of the class samples around the mean vector μ_i and it is expressed by the covariance matrix Σ_i.

The average internal dispersion matrix is defined through:

$$S_w = \sum_{i=1}^{2} P(\omega_i) E\left[(X - \mu_i)(X - \mu_i)' \mid \omega_i\right] = \sum_{i=1}^{2} P(\omega_i) \Sigma_i \cdot \tag{17}$$

The *within-class scatter matrix* can be defined in the following ways:

$$S_{b_1} = \sum_{i=1}^{2} P(\omega_i)(\mu_i - \mu_0)(\mu_i - \mu_0)' , \tag{18}$$

$$S_{b_2} = (\mu_1 - \mu_2)(\mu_1 - \mu_2)' , \tag{19}$$

$$S_{b_3} = E\left[\left(X^{(1)} - X^{(2)}\right)\left(X^{(1)} - X^{(2)}\right)'\right] =$$
$$= E\left(X^{(1)} X^{(1)'}\right) + E\left(X^{(2)} X^{(2)'}\right) - 2E\left(X^{(1)} X^{(2)'}\right);$$

therefore:

$$S_{b_3} = \Sigma_1 + \Sigma_2 + (\mu_1 - \mu_2)(\mu_1 - \mu_2)', \tag{20}$$

where:

- μ_0 means the mean vector of the mixture, namely:
  $$\mu_0 = E(X) = P(\omega_1)\mu_1 + P(\omega_2)\mu_2 , \tag{21}$$
- $X^{(i)}$ represents the samples belonging to the class $\omega_i, i = \overline{1,2},$ that are assumed to be independent.

The class separability criteria will be deduced from these matrices. The main aim is to achieve:

- a between-class scatter dispersion which is as high as possible,
- a within-class scatter dispersion which is as small as possible.

There are several possibilities for selecting these criteria [4, 10]:

1. $J_1 = \mathrm{tr}\left(S_2^{-1} S_1\right)$,
2. $J_2 = \ln\left|S_2^{-1} S_1\right| = \ln \frac{|S_1|}{|S_2|}$,
3. $J_3 = \mathrm{tr}(S_1) - \alpha[\mathrm{tr}(S_2) - c]$,

(22)

$J_4 = \mathrm{tr}(S_1)/\mathrm{tr}(S_2)$.

Remark. Usually, one uses:

- S_{bi} ($i=1$ or $i = 2$ or $i = 3$) instead of S_1,
- S_w or $S_w + S_{b_i}$ ($i=1$ or $i=2$ or $i=3$) instead of S_2.

Feature Selection to Maximize the J_1 Criterion

We want to select m ($m<n$) features contained in the vector $Y = (y_1, \ldots, y_m)'$ by applying the transformation matrix A, with the dimension $m \times n$ to the vector $X = (x_1, \ldots, x_n)'$, such that:

$$Y = AX.$$

The dispersion matrix of the type S_1 and S_2 corresponding to the random vector Y are represented by:

$$S_{1m} = AS_1 A'$$

and respectively

$$S_{2m} = AS_2A'.$$

We shall denote:

- $\lambda_i, \Phi_i, i = \overline{1,n}$ the eigenvalues and respectively the eigenvectors of $S_2^{-1}S_1$,
- $\gamma_j, \Psi_j, j = \overline{1,m}$ the eigenvalues and respectively the eigenvectors of $S_{2m}^{-1}S_{1m}$,

We assume that the linear transformation of the simultaneous diagonalization of the matrices S_1, S_2 has been applied such that:

$$\begin{cases} AS_1A' = \Lambda = \mathrm{diag}(\lambda_1, \cdots, \lambda_n) \\ AS_2A' = \mathrm{I}_n. \end{cases}$$

It results [9]:

$$J_1(n) = \mathrm{tr}\left(S_2^{-1}S_1\right) = \sum_{i=1}^{n}\lambda_i, \tag{23}$$

$$J_1(m) = \mathrm{tr}\left(S_{2m}^{-1}S_{1m}\right) = \sum_{j=1}^{m}\gamma_j, \tag{24}$$

We shall prove that $J_1(m)$ is invariant for any transformation matrix B, of the type $m \times m$.

As:

$$\mathrm{tr}\left[(BS_{2m}B')^{-1}(BS_{1m}B')\right]=\mathrm{tr}\left(B'^{-1}S_{2m}^{-1}B^{-1}BS_{1m}B'\right)=\mathrm{tr}\left(B'^{-1}S_{2m}^{-1}S_{1m}B'\right)=$$
$$=\mathrm{tr}\left(S_{2m}^{-1}S_{1m}B'B'^{-1}\right)=\mathrm{tr}\left(S_{2m}^{-1}S_{1m}\right)=J_1(m),$$

the feature selection to maximize the J_1 criterion, for a given m assumes to find the optimal transformation A such that to be maximized the sum of the eigenvalues of the matrix $S_{2m}^{-1}S_{1m}$.

In order to determine the matrix A, the term $J_1(m)$ can be maximized in the following way. One supposes:

$$\begin{aligned}&\Delta J_1(m)=\\&=\mathrm{tr}\left\{[(A+\Delta A)S_2(A'+\Delta A')]^{-1}[(A+\Delta A)S_1(A'+\Delta A')]\right\}-\\&-\mathrm{tr}\left[(AS_2A')^{-1}(AS_1A')\right]=0,\end{aligned} \tag{25}$$

for every ΔA.

If one approximates $(\mathrm{I}_n+\Delta)^{-1}\cong \mathrm{I}_n-\Delta$ and one ignores the terms with the grade at least two in ΔA, from (25) one achieves:

$$\begin{aligned}&\Delta J_1(m)\cong\\&-\mathrm{tr}\left[(AS_2A')^{-1}(\Delta AS_2A'+AS_2\Delta A')(AS_2A')^{-1}(AS_1A')+\right.\\&\left.+(AS_2A')^{-1}(\Delta AS_1A'+AS_1\Delta A')\right]=\\&=-\mathrm{tr}\left\{(AS_2A')^{-1}\Delta AS_2\left[A'(AS_2A')^{-1}(AS_1A')-S_2^{-1}S_1A'\right]\right\}-\\&-\mathrm{tr}\left\{\left[(AS_1A')(AS_2A')^{-1}A-AS_1S_2^{-1}\right]S_2\Delta A'(AS_2A')^{-1}\right\}\end{aligned} \tag{26}$$

Taking into account that AS_1A', AS_2A', S_1, S_2 are some symmetric matrices and (26) can be written in the form:

$$\Delta J_1(m)=-\mathrm{tr}[\cdot]-\mathrm{tr}[\cdot]'$$

we shall have [4]:

$$\Delta J_1(m) \cong -2 \cdot \operatorname{tr}\left\{(AS_2A')^{-1}\Delta AS_2\left[A'(AS_2A')^{-1}(AS_1A') - S_2^{-1}S_1A'\right]\right\} = 0. \tag{27}$$

As $\Delta J_1(m) = 0$ for every ΔA, from (27) one deduces:

$$A'(AS_2A')^{-1}(AS_1A') = S_2^{-1}S_1A'. \tag{28}$$

We can write:

$$(AS_2A')^{-1}(AS_1A') = S_{2m}^{-1}S_{1m} = \Psi \mathrm{M} \Psi', \tag{29}$$

where M and Ψ represents the diagonal matrix corresponding to the eigenvalues, respectively the matrix corresponding to the eigenvectors of the matrix $S_{2m}^{-1}S_{1m}$.

The relation (28) becomes:

$$A'\Psi \mathrm{M} \Psi' = S_2^{-1}S_1A'; \tag{30}$$

from here one achieves:

$$(A'\Psi)\mathrm{M} = S_2^{-1}S_1(A'\Psi), \tag{31}$$

respectively:

$$\gamma_i(A'\Psi)_i = S_2^{-1}S_1(A'\Psi)_i, \ i = \overline{1,m}, \tag{32}$$

$(A'\Psi)_i$ being the column vector of the rank i from the matrix $A'\Psi$.

From (32) we can notice that γ_i, $(A'\Psi)_i$, $i=\overline{1,m}$ are the eigenvalues respectively the eigenvectors corresponding to the matrix $S_2^{-1}S_1$.

As linear transformations do not affect the value of the criterion function $J_1(m)$ it results that the optimal transformation A is associated to the m dimensional space, where the eigenvalues corresponding to the matrix $S_{2m}^{-1}S_{1m}$ are $\gamma_i = \lambda_i$, $i=\overline{1,m}$, with $\lambda_1 > \ldots > \lambda_n$.

Hence:

$$A' = [\Phi_1, \Phi_2, \ldots, \Phi_m],$$

where $\Phi_1, \Phi_2, \ldots, \Phi_m$ are the eigenvectors corresponding to the first eigenvalues of the matrix $S_2^{-1}S_1$.

The maximum value of $J_1(m)$ for the selection of the m features corresponding to the first m eigenvalues from the above ordering is:

$$J_1(m) = \mathrm{tr}\,\mathrm{M} = \sum_{i=1}^{m} \lambda_i \,. \tag{33}$$

Example 1. One suposses $S_1 = S_{b_1}$ and $S_2 = S_w$. As the rank of the matrix S_{b_1} is equal to 2 it results that $S_w^{-1}S_{b_1}$ has only two nonzero eigenvalues.

Therefore:

$$J_1 = \lambda_1 + \lambda_2 = \mathrm{tr}\left(S_w^{-1}S_{b_1}\right) =$$
$$= \sum_{i=1}^{2} P(\omega_i)(\mu_i - \mu_0)' \left[P(\omega_1)\Sigma_1 + P(\omega_2)\Sigma_2\right]^{-1} (\mu_i - \mu_0),$$
$$\lambda_3 = \ldots = \lambda_n = 0.$$

Example 2. One suposses $S_1 = S_{b_2}$ and $S_2 = S_w$. As the rank of the matrix S_{b_2} is equal to 1 it results that $S_w^{-1} S_{b_1}$ has a single nonzero eigenvalue.

We shall have:

$$J_1 = \lambda_1 = \operatorname{tr}\left(S_w^{-1} S_{b_2}\right) = (\mu_1 - \mu_2)' \left[P(\omega_1)\Sigma_1 + P(\omega_2)\Sigma_2\right]^{-1} (\mu_1 - \mu_2)$$
$$= \sum_{i=1}^{2} P(\omega_i)(\mu_i - \mu_0)' \left[P(\omega_1)\Sigma_1 + P(\omega_2)\Sigma_2\right]^{-1} (\mu_i - \mu_0),$$

$$\lambda_2 = \ldots = \lambda_n = 0.$$

Bhattacharyya Distance Criterion

From the relations (5) and (11) it results that the relation (16) is written:

$$\varepsilon \le P(\omega_1)^{1-s} \cdot P(\omega_2)^s \int_{-\infty}^{\infty} \left[f(X \mid \omega_1)^{-1} \cdot f(X \mid \omega_2)\right]^s f(X \mid \omega_1) \mathrm{d}X =$$
$$= P(\omega_1)^{1-s} \cdot P(\omega_2)^s \int_{-\infty}^{\infty} f(X \mid \omega_1)^{1-s} \cdot f(X \mid \omega_2)^s \,\mathrm{d}X, 0 \le s \le 1.$$

Therefore:

$$\varepsilon \le P(\omega_1)^{1-s} \cdot P(\omega_2)^s \cdot \int_{-\infty}^{\infty} f(X \mid \omega_1)^{1-s} \cdot f(X \mid \omega_2)^s \,\mathrm{d}X, 0 \le s \le 1. \tag{34}$$

The right member of the relation (16) or $u(s)$ from (5) can be constituted as some class separability criteria.

If the class density probabilities have a normal distribution, one proves that $u(s)$ becomes [10]:

$$u(s)=\frac{1}{2}s(1-s)(\mu_1-\mu_2)'[(1-s)\Sigma_1+s\Sigma_2]^{-1}(\mu_1-\mu_2)+ \\ +\frac{1}{2}\ln\frac{|(1-s)\Sigma_1+s\Sigma_2|}{|\Sigma_1|^{1-s}|\Sigma_2|^{s}}. \tag{35}$$

Because of the difficulty of finding the optimum value of s one selects $s=0.5$, which contributes to an upper limit of ε, although it is not the smallest upper limit.

From the relation (16) one achieves:

$$\varepsilon\le\varepsilon_S=[P(\omega_1)P(\omega_2)]^{1/2}\,\mathrm{e}^{-u(1/2)}.$$

By using the relations (5) and (11) it results:

$$u\left(\frac{1}{2}\right)=-\ln\int_{-\infty}^{\infty}[f(X|\omega_1)\cdot f(X|\omega_2)]^{1/2} f(X|\omega_1)\,\mathrm{d}X,$$

namely:

$$u\left(\frac{1}{2}\right)=-\ln\int_{-\infty}^{\infty}[f(X|\omega_1)\cdot f(X|\omega_2)]^{1/2}$$

and it is called *Bhattacharyya distance*.

Therefore:

$$\varepsilon\le\varepsilon_S=[P(\omega_1)P(\omega_2)]^{1/2}\int_{-\infty}^{\infty}[f(X|\omega_1)\cdot f(X|\omega_2)]^{1/2}.$$

The feature selection using the Bhattacharyya distance is very difficult, involving the integration of the product of probability density functions.

In the case of the Gaussian distributions, the feature selection means to determine a transformation matrix A, having the dimension $m \times n$ (for a given value of m) which has to maximize the expression:

$$u_m(s) = \frac{1}{2} s(1-s)\left[A(\mu_1 - \mu_2)'\right] \cdot \left[(1-s)A\Sigma_1 A' + sA\Sigma_2 A'\right]^{-1} \cdot \left[A(\mu_1 - \mu_2)\right] + \\ + \frac{1}{2} \ln \frac{\left|(1-s)A\Sigma_1 A' + sA\Sigma_2 A'\right|}{\left|A\Sigma_1 A'\right|^{1-s}\left|A\Sigma_2 A'\right|^{s}},$$

where $\mu_1 - \mu_2$, Σ_1 and Σ_2 from (35) becomes $A(\mu_1 - \mu_2)$, $A\Sigma_1 A'$ and respectively $A\Sigma_2 A'$.

If the covariance matrices are equal, $\Sigma_1 = \Sigma_2 = \Sigma$ then:

$$u_m(s) = \frac{1}{2} s(1-s)\left[A(\mu_1 - \mu_2)'\right](A\Sigma A')^{-1}\left[A(\mu_1 - \mu_2)\right],$$

which can be written:

$$u_m(s) = \frac{1}{2} s(1-s)\operatorname{tr}\left\{(A\Sigma A')^{-1}\left[A(\mu_1 - \mu_2)(\mu_1 - \mu_2)' A'\right]\right\}. \tag{36}$$

One intends to maximize the Bhattacharyya distance, namely the minimization of the Chernoff bound corresponding to the probability of error.

The optimal transformation A is expressed [10] through a matrix consisting of the eigenvectors of the matrix $S_2^{-1}S_1$, where:

$$\begin{cases} S_1 = (\mu_1 - \mu_2)(\mu_1 - \mu_2)' \\ S_2 = \Sigma. \end{cases}$$

As the rank of the matrix S_1 is equal to 1, there is a single nonzero eigenvalue.

Hence:

$$\begin{cases} \lambda_1 = \mathrm{tr}\left(S_2^{-1} S_1\right) = \left(\mu_1 - \mu_2\right)' \Sigma^{-1} \left(\mu_1 - \mu_2\right) \\ \lambda_2 = \lambda_3 = \cdots = \lambda_n = 0. \end{cases}$$

The first eigenvector is [9]:

$$\phi_1 = \frac{\Sigma^{-1}\left(\mu_1 - \mu_2\right)}{\left[\left(\mu_1 - \mu_2\right)' \Sigma^{-1} \left(\mu_1 - \mu_2\right)\right]^{1/2}}. \tag{37}$$

Therefore a single feature carries all the information related to the separability of the classes, while all the other features are redundant.

APPLYING THE BHATTACHARYYA DISTANCE FOR FEATURE SELECTION

In this section we shall present the Matlab program which illustrates how the Bhattacharyya distance method is applied for the feature selection in the case of three vector classes.

Using this program we will prove that the selection of the respective features contributes to the maximization of the Bhattacharyya distance.

One generates in each class, in a random way, five vectors with a normal distribution, having the dimension 4. The matrix $Xi, i = \overline{1,3}$ contains the vectors from the class i as its columns.

```
X1 = randn(4,5);
X2 = randn(4,5);
X3 = randn(4,5);
```

The corresponding covariance matrices and matrices are determined for each of the three classes.

```
m1 = zeros(4,1); m2 = zeros(4,1); m3 = zeros(4,1);
sig1 = zeros(4); sig2 = zeros(4); sig3 = zeros(4);
n1 = 5; n2 = 5; n3 = 5;
for i = 1:5
    m1 = m1 + X1(:,i);
    m2 = m2 + X2(:,i);
    m3 = m3 + X3(:,i);
end
m1 = m1/n1;
m2 = m2/n2;
m3 = m3/n3;
for i = 1:5
    sig1 = sig1 + (m1-X1(:,i))*(m1-X1(:,i))';
    sig2 = sig2 + (m2-X2(:,i))*(m2-X2(:,i))';
    sig3 = sig3 + (m3-X3(:,i))*(m3-X3(:,i))';
end
sig1 = sig1/n1;
sig2 = sig2/n2;
sig3 = sig3/n3;
```

Since the vectors have a normal distribution in the three classes and the covariance matrices of the classes are different, the Bhattacharyya distance achieved from the relation (35) for $s = 0.5$ will be used as a *class separability criterion*.

We shall denote:

$$J1 = \frac{1}{2} s(1-s)(\mu_1 - \mu_2)' \left[(1-s)\Sigma_1 + s\Sigma_2\right]^{-1} (\mu_1 - \mu_2),$$

$$J2 = \frac{1}{2} \ln \frac{\left|(1-s)\Sigma_1 + s\Sigma_2\right|}{|\Sigma_1|^{1-s} |\Sigma_2|^{s}},$$

namely

$$u(s) = J1 + J2 . \tag{38}$$

Firstly, we shall maximize $J1$ and then $J2$ (the notations correspond to the criterion from (22)).

To maximize $J1$ we shall choose:

- as $S1$, the matrix

  $$S_{b_1} = \sum_{i=1}^{3} P(\omega_i)(\mu_i - \mu_0)(\mu_i - \mu_0)',$$

  where:

  $$\mu_0 = \sum_{i=1}^{3} P(\omega_i)\mu_i;$$

- as $S2$, the matrix

  $$S_w = \sum_{i=1}^{3} P(\omega_i)\Sigma_i.$$

Therefore:

```
m0 = (m1 + m2 + m3)/3;
S11 = ((m1-m0)*(m1-m0)' + (m2-m0)*(m2-m0)' + (m3-m0)*
    (m3-m0)')/3;
S21 = (sig1 + sig2 + sig3)/3;
```

We can see that to achieve the maximum value of $J1$, we have to select two features (corresponding to the two largest eigenvalues of matrix $S_2^{-1}S_1$).

By using the following Matlab sequence we shall determine the eigenvectors and the eigenvalues of the matrix $S_2^{-1}S_1$.

[v1,u1] = eigs(inv(S21)*S11);

We shall achieve:

v1 =

-0.34534621838309 -0.28413078750648 0.87951335607237 -0.20226449296838

0.36270037425068 -0.53049198226234 -0.20535477241175 0.97835691636272

0.26970231535141 0.68996930342483 0.42767612590256 -0.02700358470308

0.82246281925751 -0.40223166543395 -0.03713226724843 -0.03431654852475

u1 =

1.95000235553974	0	0	0
0	0.97564407442682	0	0
0	0	-0.00000000000000	0
0	0	0	0

The matrix $v1$ contains as its columns, the eigenvectors of the matrix $S_2^{-1}S_1$, and the diagonal matrix $u1$ has the elements represented by the eigenvalues of the matrix $S_2^{-1}S_1$.

The maximum value of $J1$ is:

$$J1 = \mathrm{tr}\left(S_2^{-1}S_1\right) = \sum_{i=1}^{2} \lambda_i .$$

We have:

J1 = trace(inv(S21)*S11);

or

```
J1 = sum(diag(u1));
```

namely

```
J1 =
     2.92564642996656.
```

To maximize $J2$ we shall choose:

- as $S1$, the matrix

  $$S_{b_1} + S_w = \sum_{i=1}^{3} P(\omega_i)(\mu_i - \mu_0)(\mu_i - \mu_0)' + \sum_{i=1}^{3} P(\omega_i)\Sigma_i,$$

  where:

  $$\mu_0 = \sum_{i=1}^{3} P(\omega_i)\mu_i;$$

- as $S2$, the matrix

  $$S_w = \sum_{i=1}^{3} P(\omega_i)\Sigma_i.$$

Therefore:

```
S12 = S11 + S21;
S22 = S21;
```

To achieve the mximum value of $J2$ we have to select two features (corresponding to the two largest eigenvalues of matrix $S_2^{-1}S_1$).

By using the following Matlab sequence we shall determine the eigenvectors and the eigenvalues of the matrix $S_2^{-1}S_1$.

```
[v2,u2] = eigs(inv(S22)*S12).
```

It results:

v2 =

0.34534621838309 -0.28413078750648 -0.15907697602510 -0.87951335607237
-0.36270037425068 -0.53049198226234 0.98656256649090 0.20535477241175
-0.26970231535141 0.68996930342483 -0.00460155526266 -0.42767612590256
-0.82246281925751 -0.40223166543395 -0.03698166825287 0.03713226724843

u2 =

2.95000235553974	0	0	0
0	1.97564407442682	0	0
0	0	1.00000000000000	0
0	0	0	1.00000000000000

The maximum value of $J2$ is:

$$J2 = \ln \frac{\det S1}{\det S2} = \sum_{i=1}^{2} \ln \lambda_i$$

and it will be achieved either from

J2 = sum(log(diag(u2)))

either from

J2 = log(det(S12)/det(S22)).
We have

J2 =
1.76270042766288.

If $J1$ is dominant (namely $J1 > J2$) in the relation (38), then:

1. one selectes the proper vectors corresponding to the first two columns in the matrix $v1$; in Matlab we have:
   vv1 = [v1(:,1) v1(:,2)];
2. the four-dimensional initial vectors are transformed (from the three classes) into the bidimensional vectors in the following way:
   Y1 = vv1'*X1;
   Y2 = vv1'*X2;
   Y3 = vv1'*X3.

In the case when $J2 > J1$ one apllies the the previous steps, but the eigenvectors corresponding to the first two columns from the matrix $v2$ are selected.

We can use the Bayes criterion to highlight the performances of the Bhattacharyya distance method in the *class separability*.

We have to compute the means and the covariace matrices of the classes in the new space (a bidimensional space):

```
mm1 = zeros(2,1);
mm2 = zeros(2,1);
mm3 = zeros(2,1);
sigg1 = zeros(2);
sigg2 = zeros(2);
sigg3 = zeros(2);
for i = 1:5
    mm1 = mm1 + Y1(:,i);
    mm2 = mm2 + Y2(:,i);
    mm1 = mm1 + Y3(:,i);
end
mm1 = mm1/n1;
mm2 = mm2/n2;
mm3 = mm3/n3;
for i = 1:5
```

```
    sigg1 = sigg1 + (mm1-Y1(:,i))*(mm1-Y1(:,i))';
    sigg2 = sigg2 + (mm2-Y2(:,i))*(mm1-Y2(:,i))';
    sigg3 = sigg3 + (mm3-Y3(:,i))*(mm3-Y3(:,i))';
end
sigg1 = sigg1/n1;
sigg2 = sigg2/n2;
sigg3 = sigg3/n3;
```

Then, one computes the values of the discriminant functions:

$$g_i(X) = -\frac{1}{2}(X-\mu_i)'\Sigma_i^{-1}(X-\mu_i) - \frac{1}{2}\ln|\Sigma_i| - \frac{n}{2}\ln 2\pi,\ i=\overline{1,3}$$

and one selects the class which corresponds to the maximum discriminant function.

The following Matlab sequence classifies the first vector from each of the three classes in the bidimensional space.

```
g11 = -0.5*(Y1(:,1)-mm1)'*sigg1^(-1)*(Y1(:,1)-mm1)-log(2*pi*
    (det(sigg1))^0.5);
g12 = -0.5*(Y1(:,1)-mm2)'*sigg2^(-1)*(Y1(:,1)-mm2)-log(2*pi*
    (det(sigg2))^0.5);
g13 = -0.5*(Y1(:,1)-mm3)'*sigg3^(-1)*(Y1(:,1)-mm3)-log(2*pi*
    (det(sigg3))^0.5);
g21 = -0.5*(Y2(:,1)-mm1)'*sigg1^(-1)*(Y2(:,1)-mm1)-log(2*pi*
    (det(sigg1))^0.5);
g22 = -0.5*(Y2(:,1)-mm2)'*sigg2^(-1)*(Y2(:,1)-mm2)-log(2*pi*
    (det(sigg2))^0.5);
g23 = -0.5*(Y2(:,1)-mm3)'*sigg3^(-1)*(Y2(:,1)-mm3)-log(2*pi*
    (det(sigg3))^0.5);
g31 = -0.5*(Y3(:,1)-mm1)'*sigg1^(-1)*(Y3(:,1)-mm1)-log(2*pi*
    (det(sigg1))^0.5);
g32 = -0.5*(Y3(:,1)-mm2)'*sigg2^(-1)*(Y3(:,1)-mm2)-log(2*pi*
    (det(sigg2))^0.5);
```

```
g33 = -0.5*(Y3(:,1)-mm3)'*sigg3^(-1)*(Y3(:,1)-mm3)-log(2*pi*
    (det(sigg3))^0.5);
```

one achieves:

```
g11 =
    -1.43432440030691
g12 =
    -7.84934262843320
g13 =
    -2.13868034822775
g21 =
    -19.93682127280234
g22 =
    -1.21758869083571
g23 =
    -4.26588269031842
g31 =
    -4.02169638326603
g32 =
    -7.64652058223834
g33 =
    -2.16563726299448
```

Remark. One can notice that all the three vectors are correctly classified.

FUTURE WORK

I am proposing a future development of another approach with a view to combine the two computational intelligent techniques: the statistical methods and the neuro-fuzzy systems in order to build a system with the ability to treat the cognitive uncertainties.

REFERENCES

[1] Ashcroft, M. (2016). *The Essentials of Data Analytics and Machine Learning, Persontyle*. Data Science Centre of Excellence.

[2] Choi, E. and Lee, C. (2003). Feature extraction based on the Bhattacharyya distance. *Pattern Recognition*, 36, pp. 1703-1709.

[3] Drotár P, Gazda J. and Smékal Z. (2015). An experimental comparison of feature selection methods on two-class biomedical datasets. *Computers in Biology and Medicine,* 66, pp. 1-10.

[4] Fukunaga, K. (1990) *Introduction to statistical pattern recognition,* Academic Press, New York.

[5] Hoque, N., Bhattacharyya, D. K. and Kalita, J. K. (2014). MIFS-ND: A mutual information-based feature selection method. *Expert Systems with Applications*, 41, pp. 6371–6385.

[6] Iatan, I. (2017). *Issues in the Use of Neural Networks in Information Retrieval,* Springer.

[7] Iatan, I. (2010). *Statistical Methods for Pattern Recognition,* Lambert Academic Publishing AG& Co. KG, Saarbrücken, Germany.

[8] Kavzoglu, T. and Mather, P. M. (2002). The role of feature selection in artificial network applications. *International Journal of Remote Sensing* 23(15), pp. 2919-2937.

[9] Liu, Q., Zhao, Z., Li, Y., Yu, X. and Wang, Y. (2013). A Novel Method of Feature Selection based on SVM. *Journal of Computers*, 8(8), pp. 2144-2149.

[10] Neagoe, V. E. and Stănăşilă, O. (1992). *The theory of Pattern Recognition*, Romanian Academy Press.

[11] Neagoe, V. E. and Stănăşilă, O. (1999). *Pattern Recognition and Neural Networks,* Matrix Press.

[12] Reyes-Aldasoro, C. C. and Bhalerao, A. (2006). The Bhattacharyya space for feature selection and its application to texture segmentation. *Pattern Recognition*, 39(5), pp. 812-826.

[13] Zhong, W., Lu, X. and Wu, J. (2017). Feature Selection for Cancer Classification Using Microarray Gene Expression Data. *Biostatistics and Biometrics Open Acc J.* 1(2).

BIOGRAPHICAL SKETCH

Iuliana- Florentina Iatan, PhD

The figure below illustrates the connections of my fundamental research field (in which I am specialized by my two PhDs), with some disciplines (from its adjacent fields) of Computer Science and Mathematics.

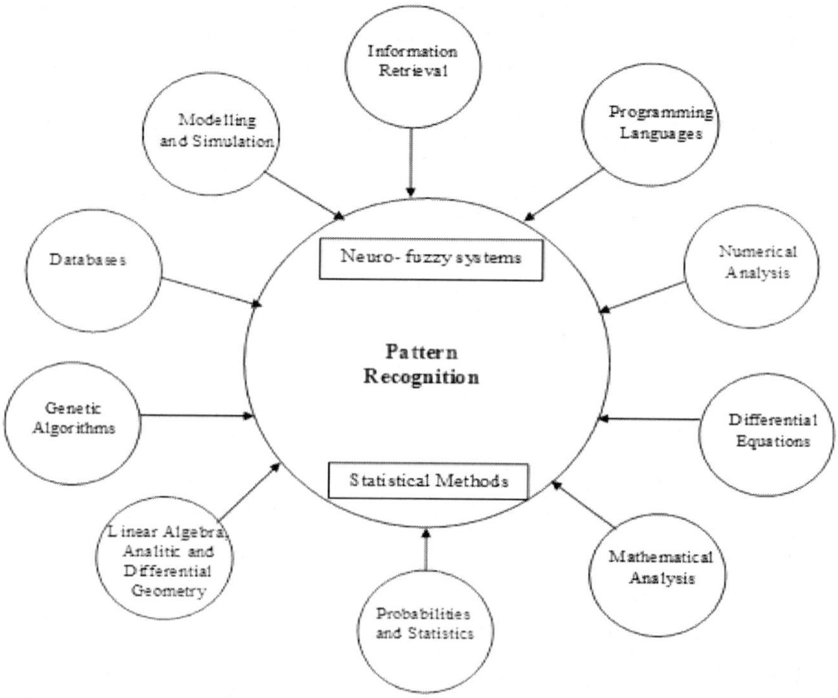

Figure 3. Connections of the Author's fundamental research field.

After I graduated my first PhD, in June 2003, from the University Politehnica of Bucharest, in a field of Artificial Intelligence, I wanted to deepen my scientific research on Pattern Recognition. I began a new PhD at the Faculty of Mathematics and Computer Science in the University of Bucharest. After graduating my second PhD, in October 2008, my competence domains have become Neuro-fuzzy systems, Probabilities, Statistics and Pattern Recognition, Information Retrieval, Genetic algorithms, Programming Languages, Databases, Numerical Analysis and Numerical Methods, Modelling and Simulation, Linear Algebra, Analytical and Differential Geometry, Mathematical Analysis, Differential Equations.

My scientific, professional and academic achievements deeply reflect on my activities performed as a postdoctoral researcher in the project Postdoctoral programme for training scientific researchers (POSDRU/89/1.5/S/58852, cofinanced by the European Social Fund within the Sectorial Operational Program Human Resources Development 2007-2013), in the field of Computer Science, of the Faculty of Mathematics and Computer Science in the University of Bucharest. I have also a wide experience as a member in the research teams within some national projects. Therefore, I am a researcher with a wide experience, proved by my both PhDs and by my postdoctoral project in the field of Computer Science, October 2010-March 2013. I have the experience of working as a researcher in a university from abroad. I have made two workouts (between April 1-July 31, 2012 and October 1- November 30, 2012) at the Institute of Informatics/ University of Amsterdam. I have been twice as a visiting professor to UvA, for one week and I made a stage of two weeks at the University of Dortmund, Department of Informatics I. I am always able to gain new knowledge in projects and research; my CV and my previous achievements prove that each of my steps is a step forward.

Relevant Scientific Papers (in the last 3 years)

1. Anastassiou, G. and Iatan, I. (2016). A Recurrent Neural Fuzzy Network. *Journal of Computational Analysis and Applications*, 20(2), pp. 213-225.

2. Anastassiou, G. and Iatan, I. (2016). A new approach of a possibility function based neural network, *Proceedings of the 3rd International Conference on Applied Mathematics & Approximation Theory- AMAT 2015,* May 28-31, 2015, Ankara-Turkey, Intelligent Mathematics II: Applied Mathematics and Approximation Theory, Vol. 441, Advances in Intelligent Systems and Computing, pp. 139-150, Springer.
3. Iatan, I. and Worring, M. (2016). A Fuzzy Neural Network for Determining Image Similarity, *Proceedings of Mathematics and Educational Symposium of Department of Mathematics and Computer Science,* Technical University of Civil Engineering Bucharest, The 2nd Edition, Bucharest, May 28, 2016, ed. MatrixRom, pp. 63-68.
4. Iatan I. (2017). *Issues in the Use of Neural Networks in Information Retrieval,* Springer.
5. Iatan, I. (2017). Studying the Equivalence between a Feed-Forward Neural Network and a Linear Regression Model, *The 14th Workshop of Scientific Communications, Department of Mathematics and Computer Science*, May 27, 2017, pp. 48-53, ed. MatrixRom.
6. Iatan, I. (2017). Dealing the Nonlinearity Associated with the Data Using Artificial Neural Networks. *Review of the Air Force Academy*, 2(34), pp. 15-22.
7. Iatan, I. and Giebel, S. (2018). A New Representation for the Fuzzy Systems. *Review of the Air Force Academy*, 2(37), pp. 45-52.

In: An Essential Guide to Fuzzy Systems
Editor: Michael Gr. Voskoglou ISBN: 978-1-53616-128-1

Chapter 8

Fuzzy Goursat Problems of Fractional Order with Generalized Hukuhara Derivatives

***Nguyen Thi Kim Son*[1,*], *Ha Thi Thanh Tam*[2†] and *Hoang Viet Long*[3‡]**
[1] Faculty of Natural Science, Hanoi Metropolitan University, Hanoi, Vietnam
[2] Department of Basic Sciences, University of Transport Technology, Hanoi, Vietnam
[3] Faculty of Information Technology, People's Police University of Technology and Logistics, Bac Ninh, Vietnam

Abstract

Fractional calculus emerged in 1695 in the sequel of the brilliant ideas of Gottfried Leibniz. Since then, this theory turned out to be very attractive to mathematicians. In recently times, fractional calculus has appeared as an important and useful tool in the modeling of phenomena, for example physics, mechanics, elasticity, dynamics, economics, biology, chemistry, etc [1, 22, 36]. Many different forms of fractional operators for real valued functions were proposed such as the Grunwald-Letnikov,

*Corresponding Author's Email: ntkson@daihocthudo.edu.vn
†Corresponding Author's Email: tamhtt@utt.edu.vn
‡Corresponding Author's Email: longhv08@gmail.com

Riemann-Liouville and Caputo fractional derivatives [3, 22, 37]. In which, Caputo's fractional derivative, originally introduced by Caputo and afterwards adopted in the theory of linear viscoelasticity, has many prominences. Problems studied under the Caputo's derivative allow the utilization of physically interpretable initial conditions [6, 17, 24, 25].

In the process of transforming real-world phenomena into initial or boundary problem for fractional differential equations, we are often interested in the uncertain transformation of initial conditions, boundary conditions or dependent functions of the model inputs into the model outputs. This may be due to the fact that initial data or boundary conditions have errors, dependent functions contain uncertain parameters, etc. Some mathematicians have used fuzzy fractional differential equations to overcome the above limitations. Since it was first studied in 2010, fuzzy fractional differential equations have had significant development steps. Studying the relative impacts of input variable uncertainties to output variables leads us to investigate the qualitative properties of solutions of these equations.

In this chapter, we study the well-known problem appeared in physics, e.g., Goursat problem for a fuzzy hyperbolic equation under the fractional Caputo g-derivative. We define the fractional integral and fractional Caputo g-derivatives for fuzzy-valued multivariable functions. Some properties between the fractional integral and the Caputo g-derivatives are established with some illustrated examples. The well-posedness of this problem is proven and some applications to real world problems are also investigated. In addition, qualitative properties of solutions of the problem including Ulam stability and Lyapunov stability are also presented in the last part of the chapter.

1. Introduction

In 2010, Agarwal et al. [5] introduced a new concept for fractional differential equations containing the uncertainty. This study laid the foundation for many later developments. In 2011, Arshad and Lupulescu [7] built the Riemann-Liouville integral and derivative concepts for fuzzy valued functions based on the Hukuhara difference and proved the existence and uniqueness of the probem

$$\begin{cases} {}^{RL}D^q u(t) = f(t, u(t)), t \in (0, a], q \in (0, 1) \\ \lim\limits_{t \to 0^+} t^{1-q} u(t) = u_0 \in E. \end{cases}$$

In 2015, in the article [1], the authors also employed fractional differential equations to determine the concentration of xylose in the acid hydrolysis of the Oil Palm Frond under uncertainty in parameters. In this case, fuzzy fractional differential equations are more suited to real-world situations than classical fractional differential equations. We also can find the model of the fuzzy fractional differential equations in some articles and their references [23, 27, 28, 36, 40].

There have been many studies on fuzzy fractional differential equations under different derivative concepts of fuzzy valued functions such as Hukuhara derivative [5, 7, 8], generalized Hukuhara derivative, Caputo derivative [6, 17, 29]. However, results on fuzzy fractional partial differential equations are limited. Therefore, in this chapter, we introduce fuzzy fractional partial differential equations with hyperbolic type. This problem is based on the concept of Caputo derivatives of fuzzy-valued two-variable functions presented in [25].

First, in section 2, we recall some notations on fuzzy numbers, fuzzy metric spaces that will be used throughout the chapter. In section 3, by using the Riemann - Liouville integral for real-valued two-variable functions and Stacking's theorem [31], we define the Riemann - Liouville integral concept for fuzzy-valued two-variable functions in Definition 3.10. The correctness of this concept is proved in Lemma 3.1. The concept of Caputo derivatives based on the generalized Hukuhara difference is given in Definition 3.11. Some examples are given to illustrate these concepts. In section 4, we investigate Goursat problem for the fractional fuzzy hyperbolic partial differential equation

$$\begin{cases} {}^{C}_{F}\mathcal{D}^q u(t, x) = f(t, x, u(t, x)) \\ u(t, 0) = \eta_1(t), \ u(0, x) = \eta_2(x), \end{cases}$$

where f, η_1, η_2 are given functions and will be presented in detail later.

Firstly, the solvability of the problem is considered in both finite and infinite domains. With the hypothesis of boundedness of the right hand side f, using Schauder fixed point theorem, we show that the existence of solutions of the problem in finite domain in Theorem 4.1. Results of solvability of the problem in infinite domains are given in Theorem 4.2 and Theorem 4.3 with the assumption of Lipschitz conditions of f with respect to third variable.

Next, in section 5, we study some qualitative properties of solutions of the problem. The first property is Ulam stability, including Hyers-Ulam stability and Hyers-Ulam-Rassias stability. The results are proven in Theorem 5.1 and Theorem 5.2. Different from the general stability, Ulam stability can guarantee the existence and even uniqueness of solutions, provided that there is an approximate solution via a given error. Therefore, Ulam stability not only establishes an important foundation for the existence and uniqueness of differential equations but also provides a theoretical basis for approximately solving differential equations. Obloza [33] is the first author studying Ulam stability of linear differential equations. Ulam stability of differential equations and partial differential equations is investigated in different approaches [2, 44]. However, there are few results on Ulam stability for fuzzy differential equations. So far, all the results are studied by Long [24, 27], Shen and Wang [38, 39].

When studying the asymptotic behavior of solutions of a dynamical system illustrated by a partial differential equation, an important issue is the stability of solutions near to the equilibrium point. In the physical sense, if solutions start near to the equilibrium point u_0 and stay near u_0 forever then u_0 is Lyapunov stable. It is considered in Theorem 5.3. Some studies of the stability of fuzzy differential equations can be mentioned as Cecconello et al. [10, 11, 12], Diamond [14], Ding et al. [16].

2. Generalized Metric Spaces of Fuzzy Numbers

In this section, we recall some basic knowledge about fuzzy numbers and operations on the set of fuzzy numbers with illustration examples.

Definition 2.1. *[15] A fuzzy subset $u : \mathbb{R} \to [0, 1]$ is called a fuzzy number if it satisfies following properties*

i) u is normal, that is there exist $x_0 \in \mathbb{R}$ such that $u(x_0) = 1$;

ii) u is fuzzy convex in meaning for $x, y \in \mathbb{R}$ and $0 < \lambda \leq 1$:

$$u(\lambda x + (1 - \lambda)y) \geq \min\{u(x), u(y)\};$$

iii) *u is upper semicontinuous on $\mathbb{R}$ (u is called upper semicontinuous at x_0 if for all $\varepsilon > 0$, there exists $\delta > 0$ such that for all x satisfying $|x - x_0| < \delta$ then $u(x) - u(x_0) < \varepsilon$);*

iv) $[u]^0 = \overline{\{x \in \mathbb{R} : u(x) > 0\}}$ *is a compact subset of* $\mathbb{R}$.

Denote E by the space of fuzzy numbers and $[u]^\alpha = \{x \in \mathbb{R} : u(x) \geq \alpha, \alpha \in (0,1]\}$ and $[u]^0 = cl(supp\ u)$ are level sets of fuzzy number u. It is proved that $[u]^\alpha = [u_\alpha^-, u_\alpha^+]$ and it is close compact interval in $\mathbb{R}$.

Example 2.1. *Let $a < b < c$. Consider the fuzzy number $u : \mathbb{R} \to [0,1]$ defined by the membership function*

$$u(x) = \begin{cases} \dfrac{x-a}{b-a} & \text{if } a \leq x \leq b \\ \dfrac{c-x}{c-b} & \text{if } b \leq x \leq c \\ 0 & \text{if } x \notin [a,c]. \end{cases}$$

Then u has level sets

$$[u]^\alpha = [(b-a)\alpha + a, (b-c)\alpha + c], \ \alpha \in [0,1].$$

Denote in short form $u = (a, b, c)$ and it is called a triangular fuzzy numbers.

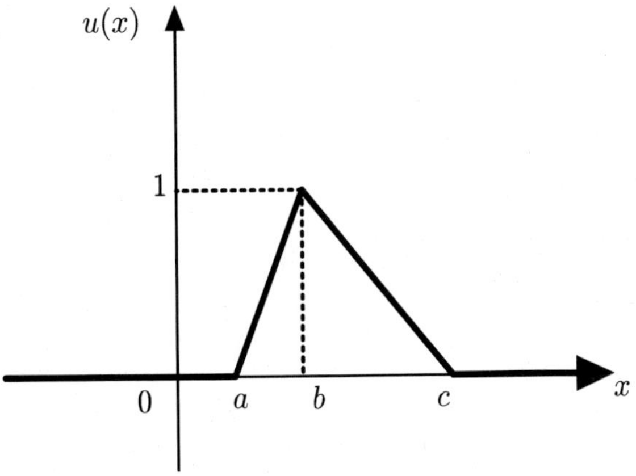

Figure 1. The triangular fuzzy number $u = (a, b, c)$.

Example 2.2. *Let $a < b < c < d$ and $u \in E$. Assume that u has the membership function*

$$u(x) = \begin{cases} \dfrac{x-a}{b-a} & \text{if } a \le x \le b \\ 1 & \text{if } b \le x \le c \\ \dfrac{d-x}{d-c} & \text{if } c \le x \le d \\ 0 & \text{if } x \notin [a,d]. \end{cases}$$

Level sets of u have the form

$$[u]^\alpha = [(b-a)\alpha + a, (c-d)\alpha + d], \ \alpha \in [0,1].$$

These numbers are called trapezoidal fuzzy numbers and for simplicity, we denote $u = (a, b, c, d)$.

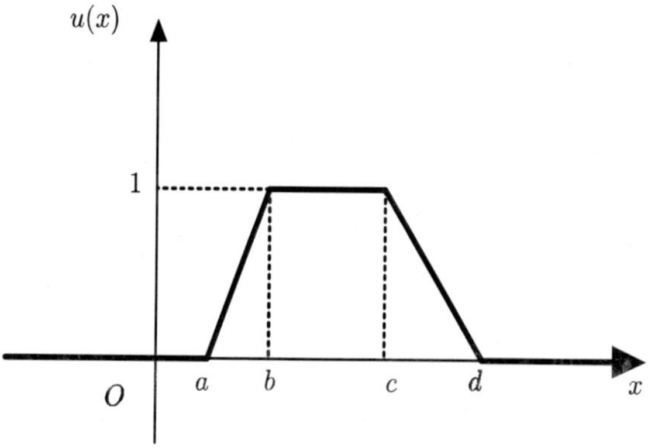

Figure 2. The trapezoidal fuzzy number $u = (a, b, c, d)$.

Theorem 2.1. *[31] (Stacking Theorem) If $u \in E$ and $[u]^\alpha$ are its level sets then:*

(i) *$[u]^\alpha$ are closed interval for all $\alpha \in [0,1]$;*

(ii) *if $0 \le \alpha_1 \le \alpha_2 \le 1$ then $[u]^{\alpha_2} \subseteq [u]^{\alpha_1}$;*

(iii) for any sequence $\{\alpha_n\}$ which is increasing monotone and converges from below to $\alpha \in (0,1]$, we have

$$\bigcap_{n=1}^{\infty} [u]^{\alpha_n} = [u]^{\alpha};$$

(iv) for any sequence $\{\alpha_n\}$ which is decreasing monotone and converges from above to 0, we have

$$cl\Big(\bigcup_{n=1}^{\infty} [u]^{\alpha_n}\Big) = [u]^0.$$

Theorem 2.2. *[31] (Negoita-Ralescu characterization Theorem) Assume that $\{M_\alpha : \alpha \in [0,1]\}$ is a family of subsets of $\mathbb{R}$ and satisfies four conditions*

(i) M_α is non-empty closed interval for any $\alpha \in [0,1]$;

(ii) if $0 \le \alpha_1 \le \alpha_2 \le 1$ then $M_{\alpha_2} \subseteq M_{\alpha_1}$;

(iii) for any sequence $\{\alpha_n\}$ which is increasing monotone and converges from below to $\alpha \in (0,1]$, we have

$$\bigcap_{n=1}^{\infty} M_{\alpha_n} = M_\alpha;$$

(iv) for any sequence $\{\alpha_n\}$ which is decreasing monotone and converges from above to 0, we have

$$cl\Big(\bigcup_{n=1}^{\infty} M_{\alpha_n}\Big) = M_0.$$

Then there exists a unique $u \in E$ such that $[u]^\alpha = M_\alpha$ for all $\alpha \in [0,1]$.

Denote $len[u]^\alpha$ the diameter of level sets of fuzzy number u. We have

$$len[u]^\alpha = u_\alpha^+ - u_\alpha^-.$$

Assume that u, v are two fuzzy numbers with level sets $[u]^\alpha = [u_\alpha^-, u_\alpha^+]$, $[v]^\alpha = [v_\alpha^-, v_\alpha^+]$. We define addition of two fuzzy numbers and scalar multiplication of the element $k \in \mathbb{R}$ with fuzzy numbers by level sets as follows.

The sum u and v denoted $u \oplus v$, is a fuzzy number defined by level sets

$$[u \oplus v]^\alpha = [u_\alpha^- + v_\alpha^-, u_\alpha^+ + v_\alpha^+].$$

For $k \in \mathbb{R}$, the product of k and u, is denoted by ku, is a fuzzy number defined by level sets

$$[ku]^\alpha = \begin{cases} [ku_\alpha^-, ku_\alpha^+], & k \geq 0, \\ [ku_\alpha^+, ku_\alpha^-], & k < 0. \end{cases}$$

The Hukuhara difference of u, v is defined as follows.

Definition 2.2. *[18] Let $u, v \in E$. If there exists $\omega \in E$ such that $u = v \oplus \omega$ then ω is called Hukuhara difference u and v, it is denoted by $u \ominus_H v$.*

Furthermore, if $u \ominus_H v$ exists then it is unique and $[u \ominus_H v]^\alpha = [u_\alpha^- - v_\alpha^-, u_\alpha^+ - v_\alpha^+]$ for all $0 \leq \alpha \leq 1$.

Proposition 2.1. *[42] The existence of Hukuhara difference $u \ominus_H v$ is guaranteed if and only if following conditions are satisfied*
(i) $len[u]^\alpha \geq len[v]^\alpha$ for all $0 \leq \alpha \leq 1$,
(ii) $u_\alpha^- - v_\alpha^-$ is increasing in α,
(iii) $u_\alpha^+ - v_\alpha^+$ is decreasing in α.

Proposition 2.2. *[26] Let $u, v, w \in E$. Following affirmations hold*
1) $(-1)(u \oplus v) = (-1)u \oplus (-1)v$.
2) *If $u \ominus_H v$ exists then $(-1)u \ominus_H (-1)v$ exists and $(-1)(u \ominus_H v) = (-1)u \ominus_H (-1)v$.*
3) *If $u \ominus_H (v \oplus w)$ exists then $u \ominus_H v \ominus_H w$ exists and $u \ominus_H (v \oplus w) = u \ominus_H v \ominus_H w$.*
4) *If $u \ominus_H v$ and $v \ominus_H w$ exists then $u \ominus_H (v \ominus_H w)$ exists and $u \ominus_H (v \ominus_H w) = (u \ominus_H v) \oplus w$.*

Remark 2.1. The Hukuhara difference has the property $u \ominus_H u = \hat{0}$. However this difference does not exist in the case that the diameter of the fuzzy number u is smaller than the diameter of the fuzzy number v. Therefore, a broader difference concept is given in [42] to overcome the limitations of Hukuhara difference.

Definition 2.3. *[42] Let $u, v \in E$, the generalized Hukuhara difference of u and v, denote by $u \ominus_g v$, is defined by the element $w \in E$ such that*

$$u \ominus_g v = w \Longleftrightarrow \begin{cases} (i) \;\; u = v \oplus w \\ (ii) \;\; v = u \oplus (-1)w. \end{cases}$$

Note that $u \ominus_g u = \hat{0}$, and if $u \ominus_H v$ exists then $u \ominus_g v = u \ominus_H v$. If (i) and (ii) in Definition 2.3 are simultaneously satisfied then w is a real number. It is easy to see that the generalized Hukuhara difference exists more cases than the Hukuhara difference, but it does not always exist in E. The following example shows that the difference $u \ominus_g v$ does not exist.

Example 2.3. Let the trapezoidal fuzzy number $u = (1, 2, 3, 5)$ and the triangular fuzzy number $v = (0, 3, 9)$. Then generalized Hukuhara difference $u \ominus_g v$ does not exist.

Indeed if we suppose the contrary the difference $u \ominus_g v$ exists then either $u_\alpha^- - v_\alpha^- \leq u_\alpha^+ - v_\alpha^+$ or $u_\alpha^- - v_\alpha^- \geq u_\alpha^+ - v_\alpha^+$ for all $\alpha \in [0, 1]$. Meanwhile, for $\alpha = 0$, we have

$$1 = u_0^- - v_0^- > u_0^+ - v_0^+ = -4$$

and for $\alpha = 1$ then $-1 = u_1^- - v_1^- < u_1^+ - v_1^+ = 0$. This is the contradiction. Therefore, $u \ominus_g v$ does not exist.

On E, a metric d_∞ is defined by

$$d_\infty(u, v) = \sup_{0 \leq \alpha \leq 1} \max\{|u_\alpha^- - v_\alpha^-|, |u_\alpha^+ - v_\alpha^+|\}.$$

Proposition 2.3. *[15, 20] Let $k \in \mathbb{R}$ and $u, v, w, e \in E$. It is clear that metric d_∞ satisfies some properties*

i) $d_\infty(ku, kv) = |k| d_\infty(u, v)$;

ii) $d_\infty(u \oplus v, w \oplus e) \leq d_\infty(u, w) + d_\infty(v, e)$;

iii) $d_\infty(u \oplus w, v \oplus w) = d_\infty(u, v)$;

iv) $d_\infty(u \ominus_H v, w \ominus_H e) \leq d_\infty(u, w) + d_\infty(v, e)$

and the space (E, d_∞) is a complete metric space.

3. Calculus of Fuzzy Metric Spaces

Definition 3.1. *Mapping*

$$\begin{aligned} f : U \subseteq \mathbb{R}^2 &\to E \\ (x, y) &\mapsto f(x, y) \end{aligned}$$

is called a fuzzy valued two-variable function.

Example 3.1. Let a triangular fuzzy number $T_0 = (0, 1, 2)$ with the membership function

$$T_0(t) = \begin{cases} t & \text{if } 0 \leq t \leq 1 \\ 2 - t & \text{if } 1 \leq t \leq 2 \\ 0 & \text{if } t \notin [0, 2]. \end{cases}$$

Consider the fuzzy value function

$$\begin{aligned} f : \Big[0, \frac{\pi}{2}\Big] \times [1, 5] &\longrightarrow E \\ (x, y) &\longmapsto f(x, y) = y \sin\Big(x + \frac{\pi}{6}\Big) T_0. \end{aligned}$$

Then $f(x, y) \in E$ is defined by the following membership function

$$f(x,y)(t) = \begin{cases} \dfrac{t}{y \sin\left(x + \frac{\pi}{6}\right)} & \text{if } 0 \leq t \leq y \sin\left(x + \frac{\pi}{6}\right) \\ \dfrac{2y \sin\left(x + \frac{\pi}{6}\right) - t}{y \sin\left(x + \frac{\pi}{6}\right)} & \text{if } y \sin\left(x + \frac{\pi}{6}\right) \leq t \leq 2y \sin\left(x + \frac{\pi}{6}\right) \\ 0 & \text{if } t \notin \left[0, 2y \sin\left(x + \frac{\pi}{6}\right)\right] \end{cases}$$

for all $(x, y) \in \big[0, \frac{\pi}{2}\big] \times [1, 5]$.

Definition 3.2. *Mapping $f : U \subseteq \mathbb{R}^2 \to E$ is called continuous at $(x_0, y_0) \in U$ if for any $\varepsilon > 0$, there exits $\delta > 0$ such that for each $(x, y) \in U$ satisfying $|x - x_0| + |y - y_0| < \delta$, we have*

$$d_\infty(f(x, y), f(x_0, y_0)) < \varepsilon.$$

The function f is continuous on U if it is continuous at all points in U.

Remark 3.1. *From definition of the metric d_∞, we see that $f : U \to E$ is continuous at (x_0, y_0) if and only if functions $f_\alpha^-, f_\alpha^+ : U \to \mathbb{R}$ are also continuous at (x_0, y_0) for all $\alpha \in [0, 1]$.*

Example 3.2. Consider the fuzzy function f is defined in Example 3.1. Then

$$[f(x,y)]^\alpha = \left[y \sin\left(x + \frac{\pi}{6}\right)\alpha, -y \sin\left(x + \frac{\pi}{6}\right)\alpha + 2y \sin\left(x + \frac{\pi}{6}\right)\right]$$

and f is continuous on $\left[0, \frac{\pi}{2}\right] \times [1, 5]$ because the functions $f_\alpha^-(x, y) = y \sin\left(x + \frac{\pi}{6}\right)\alpha$ and $f_\alpha^+(x, y) = -y \sin\left(x + \frac{\pi}{6}\right)\alpha + 2y \sin\left(x + \frac{\pi}{6}\right)$ are continuous on $\left[0, \frac{\pi}{2}\right] \times [1, 5]$ for all $\alpha \in [0, 1]$.

Denote $C(U, E)$ by the set of all fuzzy continuous functions from U to E. Let $U = [a, b] \times [c, d]$. On $C(U, E)$, we construct the metric H defined by

$$H(u, v) = \sup_{(x,y)\in U} d_\infty(u(x, y), v(x, y)).$$

Then $(C(U, E), H)$ is a complete metric space [32].

Definition 3.3. *[26] Mapping $f : U \times E \to E$ is called continuous at $(x_0, y_0, \varphi) \in U \times E$ if for $\varepsilon > 0$ arbitrary, there exists $\delta > 0$ such that for each $(x, y, u) \in U \times E$ satisfying $|x - x_0| + |y - y_0| + d_\infty(u, \varphi) < \delta$, we have*

$$d_\infty(f(x, y, u), f(x_0, y_0, \varphi)) < \varepsilon.$$

Mapping f is called continuous on $U \times E$ if f is continuous at all points $(x_0, y_0, \varphi) \in U \times E$.

Denote E_c by the space of all fuzzy numbers $u \in E$ such that the function $\alpha \mapsto [u]^\alpha$ is continuous with respect to Hausdorff metric on $[0, 1]$.

Definition 3.4. *[35]*

1) A subset $A \subset E_c$ is called to be compact-supported if there exists a compact set $K \subseteq \mathbb{R}$ such that $[u]^0 \subseteq K$ for all $u \in A$.

2) A subset $A \subset E_c$ is called to be level-equicontinuous at $\alpha_0 \in [0,1]$ if for all $\varepsilon > 0$ there exits $\delta > 0$ such that $|\alpha - \alpha_0| < \delta$ then $d_H([u]^\alpha, [u]^{\alpha_0}) < \varepsilon$, for all $u \in A$.

A subset A is called to be level-equicontinuous on $[0,1]$ if A to be level-equicontinuous at all points $\alpha \in [0,1]$.

3) A continuous function $f : U \subseteq \mathbb{R}^2 \times E_c \to E_c$ is called compact if for $I \subseteq U$ and a bounded subset $A \subseteq E_c$ then $f(I \times A)$ is relatively compact in E_c.

Theorem 3.1. *[35] Suppose that A is a compact-supported subset in E_c. Then the following assertions are equivalent:*

(a) A is a relatively compact subset of (E_c, d_∞).

(b) A is level-equicontinuous on $[0,1]$.

Remark 3.2. [21] We note that A is relatively compact in (E_c, d_∞) if and only if A is compact-supported in E_c and level-equicontinuous on $[0,1]$.

Denote by $\mathcal{K}_C$ the set of all nonempty compact convex subsets of $\mathbb{R}$ and $S(F)$ the set of all Lebesgue integrable selected functions of set-valued functions $F : [a,b] \to \mathcal{K}_C$.

Definition 3.5. *[13] The function $f : [a,b] \to E$ is called strongly measurable if the set-valued mapping $f_\alpha : [a,b] \to \mathcal{K}_C$ defined by $f_\alpha(x) = [f(x)]^\alpha$ is Lebesgue measurable for all $\alpha \in [0,1]$.*

The function $f : [a,b] \to E$ is called integrable bounded if there exists a integrable function $h : [a,b] \to [0,\infty)$ such that $d_\infty\big(f(x), \hat{0}\big) \leq h(x),\ \forall x \in [a,b]$.

A fuzzy function is strongly measurable and integrable bounded then it is integrable. Integral of f on $[a,b]$, denote by $\displaystyle\int_a^b f(x)\,dx$, is defined by level sets

$$\Big[\int_a^b f(x)dx\Big]^\alpha = \int_a^b [f(x)]^\alpha dx = \Big\{\int_a^b g(x)dx : g \in S\big([f(x)]^\alpha\big)\Big\}$$

for all $\alpha \in [0,1]$.

Denote by $L^1([a,b],E)$ the set of all integrable fuzzy functions on $[a,b]$.

Example 3.3. Consider the fuzzy function

$$f : [1,2] \to E$$
$$x \mapsto f(x)$$

where

$$f(x)(t) = \begin{cases} 1 & \text{if } t = 1 \\ x & \text{if } 0 \leq t < 1 \\ 0 & \text{otherwise.} \end{cases}$$

Then it is easy to see that f has level sets $[f(x)]^\alpha = [0,1]$ for all $x \in [1,2]$ and $\alpha \in [0,1]$. Therefore, we have

$$\Big[\int_1^2 f(x)dx\Big]^\alpha = \int_1^2 [f(x)]^\alpha dx = [0,1].$$

Proposition 3.1. *[19] If the function $f : [a,b] \to E$ is continuous then it is integrable. Moreover, for all $\alpha \in [0,1]$, we have*

$$\Big[\int_a^b f(x)dx\Big]^\alpha = \Big[\int_a^b f_\alpha^-(x)dx, \int_a^b f_\alpha^+(x)dx\Big]$$

where $[f(x)]^\alpha = [f_\alpha^-(x), f_\alpha^+(x)]$.

Theorem 3.2. *[13] Suppose $\lambda \in \mathbb{R}$ and $f, g \in L^1([a,b],E)$. Then*

i) $\int_a^b (f(x) \oplus g(x))dx = \int_a^b f(x)dx \oplus \int_a^b g(x)dx$

ii) $\int_a^b \lambda f(x)dx = \lambda \int_a^b f(x)dx$

iii) $d_\infty(f(x), g(x))$ is integrable and

$$d_\infty\Big(\int_a^b f(x)dx, \int_a^b g(x)dx\Big) \leq \int_a^b d_\infty(f(x), g(x))dx.$$

Definition 3.6. *[43] Let $x_0 \in (a,b)$. We say that $f : (a,b) \to E$ is generalized Hukuhara differentiable (or g-differentiable) at x_0 if there exists $w \in E$ such that for all h satisfying $x_0 + h \in (a,b)$, the difference $f(x_0+h) \ominus_g f(x_0)$ exists and*

$$\lim_{h \to 0} \frac{1}{h}\big(f(x_0+h) \ominus_g f(x_0)\big) = w.$$

Then w is called g-derivative of f at x_0, denote $f'_g(x_0)$.

Let $U \subset \mathbb{R}^2$, $f : U \to E$ and $(x_0, y_0) \in U$. When the function $x \mapsto f(x, y_0)$ is g-differentiable at x_0, we have the following definition.

Definition 3.7. *[4] Mapping $f : U \to E$ is called generalized Hukuhara differentiable (or g-differentiable) with respect to x at $(x_0, y_0) \in U$ if there exists $v \in E$ such that for all h satisfying $(x_0 + h, y_0) \in U$, the difference $f(x_0 + h, y_0) \ominus_g f(x_0, y_0)$ there exists and*

$$\lim_{h \to 0} \frac{1}{h}\big(f(x_0+h, y_0) \ominus_g f(x_0, y_0)\big) = v.$$

Then v is called a g-derivative of f with respect to x at (x_0, y_0), denote by $\frac{\partial f}{\partial x}(x_0, y_0)$, g-derivatives of f with respect to y and higher order of fuzzy partial derivative of f at $(x_0, y_0) \in U$ are defined similarly.

Definition 3.8. *[4] Assume that $f : U \to E$ is g-differentiable with respect to x at $(x_0, y_0) \in U$ and $[f(x,y)]^\alpha = [f^-_\alpha(x,y), f^+_\alpha(x,y)]$ for all $\alpha \in [0,1]$, $(x,y) \in U$. We say that*

- *f is g-differentiable in type 1 with respect to x at $(x_0, y_0) \in U$ if*

$$\Big[\frac{\partial f}{\partial x}(x_0, y_0)\Big]^\alpha = \Big[\frac{\partial f^-_\alpha}{\partial x}(x_0, y_0), \frac{\partial f^+_\alpha}{\partial x}(x_0, y_0)\Big], \ 0 \le \alpha \le 1,$$

- *f is g-differentiable in type 2 with respect to x at $(x_0, y_0) \in U$ if*

$$\Big[\frac{\partial f}{\partial x}(x_0, y_0)\Big]^\alpha = \Big[\frac{\partial f^+_\alpha}{\partial x}(x_0, y_0), \frac{\partial f^-_\alpha}{\partial x}(x_0, y_0)\Big], \ 0 \le \alpha \le 1.$$

Denote by

- $D_{xy}u(x,y) = \dfrac{\partial^2 u}{\partial x \partial y}(x,y)$ the second order generalized Hukuhara partial derivative of u with respect to x and y.
- $C^x_{(1)-g}(U,E)$ (or $C^y_{(1)-g}(U,E)$) the set of all continuous functions u which is differentiable in type 1 with respect to x (or y) on U.
- $C^x_{(2)-g}(U,E)$ (or $C^y_{(2)-g}(U,E)$) the set of all continuous functions u which is differentiable in type 2 with respect to x (or y) on U.
- $\mathcal{W}_1(U,E) = \Big\{ u | (u, u_x) \in C^x_{(k)-g}(U,E) \times C^y_{(k)-g}(U,E),\ k = 1,2 \Big\}.$
- $\mathcal{W}_2(U,E) = \Big\{ u | (u, u_x) \in C^x_{(1)-g}(U,E) \times C^y_{(2)-g}(U,E) \Big\}$

$$\bigcup \Big\{ u | (u, u_x) \in C^x_{(2)-g}(U,E) \times C^y_{(1)-g}(U,E) \Big\}.$$

Example 3.4. Consider the fuzzy function $f(x,y) = xe^{-y}.(1,2,3)$ for $(x,y) \in [0,1] \times [-2,0]$. Then we have

- f is g-differentiable in type 1 with respect to x and

$$\frac{\partial f}{\partial x}(x,y) = e^{-y}(1,2,3).$$

- f is g-differentiable in type 2 with respect to y and

$$\frac{\partial f}{\partial y}(x,y) = xe^{-y}(-3,-2,-1).$$

Proposition 3.2. *[9] Suppose that $f : [a,b] \to E$ is a continuous function and $\gamma \in E$.*

(i) The fuzzy function $F(t) = \gamma \oplus \int_a^t f(s)ds$ is g-differentiable in type 1 and $F'_g(t) = f(t)$, $t \in [a,b]$.

(ii) The fuzzy function $G(t) = \gamma \ominus_H \int_a^t f(s)ds$ is g-differentiable in type 2 and $G'_g(t) = (-1)f(t)$ such that the difference $\gamma \ominus_H \int_a^t f(s)ds$ exists, $t \in [a,b]$.

Proposition 3.3. *Let $f : [a,b] \to E$. Then we have following assertions.*

i) If f is g-differentiable in type 1 on $[a,b]$ and $f'_g(.)$ is integrable on $[a,b]$ then

$$\int_a^b f'_g(t)dt = f(b) \ominus_H f(a).$$

ii) If f is g-differentiable in type 2 on $[a,b]$ and $f'_g(.)$ is integrable on $[a,b]$ then

$$\int_a^b f'_g(t)dt = (-1)f(a) \ominus_H (-1)f(b).$$

Proof. i) If f is g-differentiable in type 1 then $[f'_g(t)]^\alpha = [(f_\alpha^-)'(t), (f_\alpha^+)'(t)]$ for all $t \in [a,b]$. This implies

$$\begin{aligned}\Big[\int_a^b f'_g(t)dt\Big]^\alpha &= \Big[\int_a^b (f_\alpha^-)'(t)dt, \int_a^b (f_\alpha^+)'(t)dt\Big] \\ &= \Big[f_\alpha^-(b) - f_\alpha^-(a), f_\alpha^+(b) - f_\alpha^+(a)\Big] \\ &= [f(b) \ominus_H f(a)]^\alpha, \ \forall \alpha \in [0,1].\end{aligned}$$

ii) If f is g-differentiable in type 2 then $[f'_g(t)]^\alpha = [(f_\alpha^+)'(t), (f_\alpha^-)'(t)]$ for all $t \in [a,b]$. Similarly, we have

$$\begin{aligned}\Big[\int_a^b f'_g(t)dt\Big]^\alpha &= \Big[\int_a^b (f_\alpha^+)'(t)dt, \int_a^b (f_\alpha^-)'(t)dt\Big] \\ &= \Big[f_\alpha^+(b) - f_\alpha^+(a), f_\alpha^-(b) - f_\alpha^-(a)\Big] \\ &= [(-1)f(a) \ominus_H (-1)f(b)]^\alpha,\end{aligned}$$

for all $\alpha \in [0,1]$. The proposition is proved. □

Definition 3.9. *[3] Let $q = (q_1, q_2) \in (0,1] \times (0,1]$, $J = [0,a] \times [0,b]$ and $u \in L^1(J, \mathbb{R})$. The Riemann - Liouville integral of order q of the real function u is defined by*

$$^{RL}I_{0^+}^q u(x,y) = \frac{1}{\Gamma(q_1)\Gamma(q_2)} \int_0^x \int_0^y (x-s)^{q_1-1}(y-t)^{q_2-1}u(s,t)dtds \quad (1)$$

provided that the integral on the right hand side is defined almost everywhere $(x,y) \in J$.

The fractional derivative of u in the Caputo sense with respect to x, y is defined by

$$^{C}D^{q}u(x,y) = {}^{RL}I_{0^+}^{1-q}\Big(\frac{\partial^2 u}{\partial x \partial y}(x,y)\Big),\ (x,y) \in J$$

provided that the integral on the right hand side is defined, where $1 - q = (1 - q_1, 1 - q_2)$.

Proposition 3.4. *[24] Let $q = (q_1, q_2) \in (0,1] \times (0,1]$. Consider the function $v(x,y) = e^{\lambda x},\ \lambda > 0,\ (x,y) \in [0,\infty) \times [0,b]$. For $\varepsilon > 0$ arbitrary, there exists $C > 0$ such that*

$$^{RL}I_{0^+}^{q} v(x,y) \leq \mathcal{L}(\lambda, b, q_1, q_2) e^{\lambda x}$$

where

$$\mathcal{L}(\lambda, b, q_1, q_2) = \frac{b^{q_2}}{\Gamma(q_1+1)\Gamma(q_2+1)} \max\Big\{\Big(\frac{C}{\lambda^{\frac{1}{1+\varepsilon}}}\Big)^{q_1}; 2\Big(\frac{C}{\lambda^{\frac{1}{1+\varepsilon}}}\Big)^{\frac{q_1}{2}} + \frac{1}{\lambda}\Big(\frac{C}{\lambda^{\frac{1}{1+\varepsilon}}}\Big)^{q_1}\Big\}$$

for all $(x,y) \in [0,\infty) \times [0,b]$.

Remark 3.3. *For each $k > 0$ given, we can choose $\lambda > 0$ big enough such that*

$$\mathcal{L}(\lambda, b, q_1, q_2) < k.$$

Lemma 3.1. *[25] Let $q = (q_1, q_2) \in (0,1] \times (0,1]$ and $u \in L^1(J, E)$, $[u(x,y)]^\alpha = [u_\alpha^-(x,y), u_\alpha^+(x,y)]$ for all $\alpha \in [0,1]$. If $u_\alpha^-, u_\alpha^+ \in L^1(J, \mathbb{R})$ and ${}^{RL}I_{0^+}^{q} u_\alpha^-(x,y)$, ${}^{RL}I_{0^+}^{q} u_\alpha^+(x,y)$ exist for all $\alpha \in [0,1]$, $(x,y) \in J$ then the family of closed interval*

$$G_\alpha(x,y) := \Big[{}^{RL}I_{0^+}^{q} u_\alpha^-(x,y), {}^{RL}I_{0^+}^{q} u_\alpha^+(x,y)\Big]$$

defined a fuzzy number $v(x,y) \in E$ such that $[v(x,y)]^\alpha = G_\alpha(x,y)$ for all $\alpha \in [0,1]$.

Proof. For each $(x,y) \in J$, because $u(x,y) \in E$, we set

$$G_\alpha = \frac{1}{\Gamma(q_1)\Gamma(q_2)}\Big[\int_0^x \int_0^y (x-s)^{q_1-1}(y-t)^{q_2-1} u_\alpha^-(s,t)dtds,$$
$$\int_0^x \int_0^y (x-s)^{q_1-1}(y-t)^{q_2-1} u_\alpha^+(s,t)dtds\Big].$$

Suppose $\alpha_1, \alpha_2 \in [0,1]$ and $\alpha_1 \leq \alpha_2$. Then $u^-_{\alpha_1}(s,t) \leq u^-_{\alpha_2}(s,t)$ and $u^+_{\alpha_1}(s,t) \geq u^+_{\alpha_2}(s,t)$. Hence $G_{\alpha_2} \subseteq G_{\alpha_1}$. For all $\alpha_n \in (0,1]$, because $u(0,0)^-(s,t) \leq u^-_{\alpha_n}(s,t) \leq u^-_1(s,t)$ and $u(0,0)^+(s,t) \geq u^+_{\alpha_n}(s,t) \geq u^+_1(s,t)$ we have:

$$\begin{cases} |u^-_{\alpha_n}(s,t)| \leq \max\{|u(0,0)^-(s,t)|, |u^-_1(s,t)|\} := g_1(s,t) \\ |u^+_{\alpha_n}(s,t)| \leq \max\{|u(0,0)^+(s,t)|, |u^+_1(s,t)|\} := g_2(s,t). \end{cases}$$

It is clear that for $i = 1, 2$, g_i are Lebesgue integrable on $[0,x] \times [0,y]$. Therefore, if $\{\alpha_n\}$ is a nondecreasing sequence and converges to $\alpha \in (0,1]$ then by the Lebesgues dominated convergence Theorem, we have

$$\lim_{n\to\infty} \frac{1}{\Gamma(q_1)\Gamma(q_2)} \int_0^x \int_0^y (x-s)^{q_1-1}(y-t)^{q_2-1} u^k_{\alpha_n}(s,t)dtds$$

$$= \frac{1}{\Gamma(q_1)\Gamma(q_2)} \int_0^x \int_0^y (x-s)^{q_1-1}(y-t)^{q_2-1} u^k_{\alpha}(s,t)dtds$$

where $u^k_{\alpha_n}$ denotes for $u^-_{\alpha_n}$ or $u^+_{\alpha_n}$. Hence $G_\alpha = \bigcap_{n\geq 1} G_{\alpha_n}$. By Theorem 2.2, we receive the result of this lemma. □

From above lemma, we can define fractional Riemann - Liouville integral for fuzzy functions as follows.

Definition 3.10. *[25] Assume that* $q = (q_1, q_2) \in (0,1] \times (0,1]$, $u \in L^1(J, E)$ *and* $[u(x,y)]^\alpha = [u^-_\alpha(x,y), u^+_\alpha(x,y)]$ *for all* $\alpha \in [0,1]$, $(x,y) \in J$. *Riemann - Liouville integral of order q of the fuzzy function u is denoted in the form*

$${}^{RL}_{F}\mathcal{I}^q_{0+}u(x,y) = \frac{1}{\Gamma(q_1)\Gamma(q_2)} \int_0^x \int_0^y (x-s)^{q_1-1}(y-t)^{q_2-1}u(s,t)dtds$$

and is defined by level sets

$$[{}^{RL}_{F}\mathcal{I}^q_{0+}u(x,y)]^\alpha = [{}^{RL}I^q_{0+}u^-_\alpha(x,y), {}^{RL}I^q_{0+}u^+_\alpha(x,y)], \alpha \in [0,1]$$

for all $(x,y) \in J$.

Example 3.5. Let $u : [0, 1] \times [0, 2] \to E$ be a fuzzy function defined by

$$u(x, y) = (x + \sqrt{y})C,$$

where C is a fuzzy number with level sets $[C]^\alpha = [c_\alpha^-, c_\alpha^+]$ for all $\alpha \in [0, 1]$. Then for $q \in (0, 1] \times (0, 1]$

$$[{}_F^{RL}\mathcal{I}_{0+}^q u(x, y)]^\alpha = [{}^{RL}I_{0+}^q c_\alpha^-(x + \sqrt{y}), {}^{RL}I_{0+}^q c_\alpha^+(x + \sqrt{y})].$$

We have

$$\begin{aligned}
{}^{RL}I_{0+}^q c_\alpha^-(x + \sqrt{y}) &= \frac{c_\alpha^-}{\Gamma(q_1)\Gamma(q_2)} \int_0^x \int_0^y (x - s)^{q_1-1}(y - t)^{q_2-1}(s + \sqrt{t})dsdt \\
&= \frac{c_\alpha^-}{\Gamma(q_1)\Gamma(q_2)} \Big[\int_0^x \int_0^y (x - s)^{q_1-1}(y - t)^{q_2-1} sdsdt \\
&\quad + \int_0^x \int_0^y (x - s)^{q_1-1}(y - t)^{q_2-1}\sqrt{t}dsdt \Big] \\
&= \frac{c_\alpha^-}{\Gamma(q_1)\Gamma(q_2)} \Big[\frac{x}{q_2} B(q_1, 2) + \frac{\sqrt{y}}{q_1} B(q_2, \frac{3}{2}) \Big] x^{q_1} y^{q_2},
\end{aligned}$$

$${}^{RL}I_{0+}^q c_\alpha^+(x + \sqrt{y}) = \frac{c_\alpha^+}{\Gamma(q_1)\Gamma(q_2)} \Big[\frac{x}{q_2} B(q_1, 2) + \frac{\sqrt{y}}{q_1} B(q_2, \frac{3}{2}) \Big] x^{q_1} y^{q_2}.$$

Therefore ${}_F^{RL}\mathcal{I}_{0+}^q u(x, y) = \frac{1}{\Gamma(q_1)\Gamma(q_2)} \Big[\frac{x}{q_2} B(q_1, 2) + \frac{\sqrt{y}}{q_1} B(q_2, \frac{3}{2}) \Big] x^{q_1} y^{q_2} C.$

Proposition 3.5. *[25] Let $p = (p_1, p_2), q = (q_1, q_2) \in (0, 1] \times (0, 1]$ such that $p + q = (p_1 + q_1, p_2 + q_2) \in (0, 1] \times (0, 1]$ and $u \in L^1(J, E)$. Then*

$$({}_F^{RL}\mathcal{I}_{0+}^p)({}_F^{RL}\mathcal{I}_{0+}^q)u = ({}_F^{RL}\mathcal{I}_{0+}^{p+q})u,$$

provided that the integrals in the right-hand side and the left-hand side are defined.

Definition 3.11. *[25] Let $q = (q_1, q_2) \in [0, 1) \times [0, 1)$ and $u \in \mathcal{W}_1(J, E) \cup \mathcal{W}_2(J, E)$. The Caputo g-derivative of order q of the fuzzy valued function u with respect to x, y is defined as follows*

$${}_F^C\mathcal{D}^q u(x, y) = {}_F^{RL}\mathcal{I}_{0+}^{1-q}(D_{xy}u(x, y)), \ (x, y) \in J$$

provided that the expression on the right-hand side is defined, where $1 - q = (1 - q_1, 1 - q_2)$.

- *If $u \in \mathcal{W}_1(J, E)$ then we say that u is Caputo g-differentiable in type 1 with respect to x, y on J.*

- *If $u \in \mathcal{W}_2(J, E)$ then we say that u is Caputo g-differentiable in type 2 with respect to x, y on J.*

Example 3.6. Consider the fuzzy function $u(x, y) = (x + \sqrt{y})C$ in Example 3.5. The g-derivative u with respect to x is calculated as follows,

$$\begin{aligned}\lim_{h\to 0} \frac{1}{h}\big(u(x+h, y) \ominus_g u(x, y)\big) &= \lim_{h\to 0} \frac{1}{h}\big[(x+h+\sqrt{y})C \ominus_g (x+\sqrt{y})C\big] \\ &= C.\end{aligned}$$

Therefore $\frac{\partial u}{\partial x}(x, y) = C$. Similarly we also have

$$\lim_{h\to 0} \frac{1}{h}\big(\frac{\partial u}{\partial x}(x+h, y) \ominus_g \frac{\partial u}{\partial x}(x, y)\big) = \lim_{h\to 0} \frac{1}{h}(C \ominus_g C) = \hat{0}.$$

This means $D_{xy}u(x, y) = \hat{0}$. Thus in this case $u \in \mathcal{W}_1([0, 1] \times [0, 2], E)$ and u is Caputo g-differentiable in type 1 and

$${}_F^C\mathcal{D}^q u(x, y) = \hat{0}.$$

Example 3.7. Consider the function

$$\begin{aligned} u : [0, 3] \times [0, \tfrac{1}{2}] &\to E \\ (x, y) &\mapsto (2x - 3xy)C, \end{aligned}$$

where $C = (0, 2, 3, 5)$. For $(x, y) \in [0, 3] \times [0, \frac{1}{2}]$, we have:

$$\begin{aligned}&\lim_{h\to 0} \frac{1}{h}\Big(u(x+h, y) \ominus_g u(x, y)\Big) \\ &\qquad = \lim_{h\to 0} \frac{1}{h}\Big((x+h)(2-3y)C \ominus_g x(2-3y)C\Big) = (2-3y)C.\end{aligned}$$

This means

$$\frac{\partial u}{\partial x}(x, y) = (2-3y)C.$$

Because $y \in [0, \frac{1}{2}]$, $2-3y > 0$. Then u is g-differentiable in type 1 with respect to x. Similarly, we obtain

$$D_{xy}u(x,y) = \lim_{h\to 0} \frac{1}{h}\Big(\frac{\partial u}{\partial x}(x,y) \ominus_g \frac{\partial u}{\partial x}(x,y+h)\Big) = (-3)C.$$

In this case, we see that $\frac{\partial u}{\partial x}$ is g-differentiable in type 2 with respect to y and therefore $u \in \mathcal{W}_2([0,3]\times[0,\frac{1}{2}], E)$. In addition, we have

$$\begin{aligned}
[{}^{C}_{F}\mathcal{D}^q u(x,y)]^\alpha &= [{}^{RL}_{F}\mathcal{I}^{1-q}_{0+} D_{xy}u(x,y)]^\alpha \\
&= \frac{1}{\Gamma(1-q_1)\Gamma(1-q_2)}\Big[\int_0^x\int_0^y (x-s)^{-q_1}(y-t)^{-q_2}(2\alpha-5)dtds, \\
&\quad \int_0^x\int_0^y (x-s)^{-q_1}(y-t)^{-q_2}(-2\alpha)dtds\Big] \\
&= \frac{1}{\Gamma(1-q_1)\Gamma(1-q_2)}\frac{x^{1-q_1}y^{1-q_2}}{(1-q_1)(1-q_2)}[2\alpha-5,-2\alpha].
\end{aligned}$$

This mean

$$[{}^{C}_{F}\mathcal{D}^q u(x,y)]^\alpha = \frac{1}{\Gamma(2-q_1)\Gamma(2-q_2)} x^{1-q_1}y^{1-q_2}[(-1)C]^\alpha.$$

Therefore, u is Caputo g-differentiable in type 2 and

$${}^{C}_{F}\mathcal{D}^q u(x,y) = \frac{-x^{1-q_1}y^{1-q_2}}{\Gamma(2-q_1)\Gamma(2-q_2)}C.$$

From Definition 3.11, it is easy to see that Caputo g-derivatives of order q of a fuzzy valued function has following properties.

Proposition 3.6. *Let $q = (q_1, q_2) \in [0,1)\times[0,1)$, $u, v \in \mathcal{W}_1(J,E) \cup \mathcal{W}_2(J,E)$, C is a constant in E and $k : J \to \mathbb{R}$. Then we have the following properties*

1. ${}^{C}_{F}\mathcal{D}^q(Ck(x,y)) = C\,{}^{C}D^q k(x,y)$.
2. ${}^{C}_{F}\mathcal{D}^q(u\oplus v)(x,y) = {}^{C}_{F}\mathcal{D}^q u(x,y) \oplus {}^{C}_{F}\mathcal{D}^q v(x,y)$.

4. Fuzzy Goursat Problems of Fractional Order under Caputo Generalized Hukuhara Derivatives

4.1. Goursat Problems in Finite Domains

In this part, by using a version of Schauder fixed point theorem in semilinear metric space, we prove the existence of solutions of the fractional fuzzy partial differential equation in hyperbolic form

$$ {}_F^C\mathcal{D}^q u(t,x) = f(t,x,u(t,x)),\ (t,x) \in \Omega_T^b = [0,T]\times[0,b] \tag{2} $$

with boundary conditions

$$ u(t,0) = \eta_1(t),\ t\in[0,T],\ u(0,x) = \eta_2(x),\ x\in[0,b], \tag{3} $$

where $f : \Omega_T^b \times C(\Omega_T^b, E_c) \to E_c$ and $\eta_1 \in C([0,T], E_c)$, $\eta_2 \in C([0,b], E_c)$ are given functions such that $\eta_1(0) = \eta_2(0) = u(0,0)$ and the differences $\eta_1(t) \ominus_H \eta_1(0)$, $\eta_2(x) \ominus_H \eta_1(0)$ exist for all $t \in [0,T], x \in [0,b]$.

Proposition 4.1. *[25] Suppose $u(.,.) \in \mathcal{W}_1(\Omega_T^b, E_c) \cup \mathcal{W}_2(\Omega_T^b, E_c)$ satisfies (2)-(3).*

1) If $u \in \mathcal{W}_1(\Omega_T^b, E_c)$ then u satisfies the following integral equation

$$ u(t,x) = \psi(t,x) \oplus {}_F^{RL}\mathcal{I}_{0+}^q f(t,x,u(t,x)) \text{ for all } (t,x)\in\Omega_T^b. \tag{4} $$

2) If $u \in \mathcal{W}_2(\Omega_T^b, E_c)$ then u satisfies the following integral equation

$$ u(t,x) = T_\psi^{f,q}[u](t,x) \text{ for all } (t,x)\in\Omega_T^b. \tag{5} $$

Where $\psi(t,x) = \eta_2(x) \oplus [\eta_1(t) \ominus_H \eta_1(0)]$ and

$$ T_\psi^{f,q}[u](t,x) = \psi(t,x) \ominus_H (-1){}_F^{RL}\mathcal{I}_{0+}^q f(t,x,u(t,x)),\ (t,x)\in\Omega_T^b. \tag{6} $$

Definition 4.1. *[25] The function $u \in C(\Omega_T^b, E_c)$ is called*

1) (1)-integral solution of problem (2)-(3) if it satisfies the equation (4),

2) (2)-integral solution of problem (2)-(3) if it satisfies the equation (5).

The existence of solutions of problem (2) - (3) is proved with some assumptions.

$(\mathbf{F_1})$ There exists $R > 0$ such that the function $f : \Omega_T^b \times B(\tilde{0}, R) \to E_c$ is compact, where $B(\tilde{0}, R) = \{u \in C(\Omega_T^b, E_c) : H(u, \tilde{0}) \leq R\}$ and $\tilde{0} \in C(\Omega_T^b, E_c)$ is defined by $\tilde{0}(t,x) = \hat{0}$.

$(\mathbf{F_2})$ ψ has compact-supported and $d_\infty(\psi(t,x), \hat{0}) \leq \frac{R}{2}$ for all $(t,x) \in \Omega_T^b$.

Theorem 4.1. *[25] Suppose that assumptions* $(\mathbf{F_1})$ - $(\mathbf{F_2})$ *are satisfied. Then problem (2)- (3) has at least (1)-integral solution on* Ω_T^b. *Moreover, if*

$$C_\psi^f(\Omega_T^b, E_c) = \left\{u \in C_\lambda^\infty(\Omega_T^b, E_c) : T_\psi^{f,q}[u](t,x) \in E_c, (t,x) \in \Omega_T^b\right\} \neq \emptyset$$

and for each $u \in C_\psi^f(\Omega_T^b, E_c)$, *we have* $T_\psi^{f,q}[u](t,x) \in C_\psi^f(\Omega_T^b, E_c)$, $(t,x) \in \Omega_T^b$ *then problem (2)- (3) has at least (2)-integral solution on* Ω_T^b.

Proof. It is clear that $B(\tilde{0}, R)$ is nonempty, bounded, closed, convex subset of $C(\Omega_T^b, E_c)$. For $u \in C(\Omega_T^b, E_c)$, set

$$P_1 u(t,x) = \psi(t,x) \oplus_F^{RL} \mathcal{I}_{0+}^q f(t,x,u(t,x)), \; (t,x) \in \Omega_T^b.$$

Because $f : \Omega_T^b \times B(\tilde{0}, R) \to E_c$ is compact, f is bounded. Then we assume

$$M_0 = \sup_{(t,x,\varphi) \in \Omega_T^b \times B(\tilde{0},R)} H(f(t,x,\varphi), \tilde{0}).$$

Moreover, since T^{q_1} is a positive exponential polynomial, there exists $T_1 > 0$ such that

$$\frac{M_0 T_1^{q_1} b^{q_2}}{\Gamma(q_1+1)\Gamma(q_2+1)} \leq \frac{R}{2}.$$

Set $T_* = \min\{T, T_1\}$, we have

$$\frac{M_0 T_*^{q_1} b^{q_2}}{\Gamma(q_1+1)\Gamma(q_2+1)} \leq \frac{R}{2}. \tag{7}$$

Consider the set $\Omega_* = [0, T_*] \times [0, b] \subset \Omega_T^b$ and $B_*(\tilde{0}, R) = \{u \in C(\Omega_*, E_c) : H(u, \tilde{0}) \leq R\}$.

Step 1. We shall prove $P_1(B_*(\tilde{0}, R)) \subset B_*(\tilde{0}, R)$.
Indeed, for each $u \in B_*(\tilde{0}, R)$, we have

$$\begin{aligned} d_\infty(P_1u(t,x),\hat{0}) &\leq d_\infty(\psi(t,x),\hat{0}) + d_\infty({}_F^{RL}\mathcal{I}_{0+}^q f(t,x,u(t,x)),\hat{0}) \\ &\leq d_\infty(\psi(t,x),\hat{0}) + \frac{M_0}{\Gamma(q_1)\Gamma(q_2)} \int_0^t \int_0^x (t-s)^{q_1-1}(x-z)^{q_2-1} dz ds. \end{aligned}$$

This implies

$$d_\infty(P_1u(t,x),\hat{0}) \leq d_\infty(\psi(t,x),\hat{0}) + \frac{M_0 T_*^{q_1} b_*^{q_2}}{\Gamma(q_1+1)\Gamma(q_2+1)}$$

for all $(t,x) \in \Omega_*$. From (7) and $d_\infty(\psi(t,x),\hat{0}) \leq \frac{R}{2}$, we have $H(P_1u,\tilde{0}) \leq R$. This means $P_1(u) \in B_*(\tilde{0}, R)$.
Step 2. We show that P_1 is continuous on $B_*(\tilde{0}, R)$.
Suppose $u_n \to u$ in $B_*(\tilde{0}, R)$. From the properties of the metric d_∞ in Proposition 2.3, we have

$$\begin{aligned} d_\infty(P_1u_n(t,x), P_1u(t,x)) &\leq d_\infty({}_F^{RL}\mathcal{I}_{0+}^q f(t,x,u_n(t,x)), {}_F^{RL}\mathcal{I}_{0+}^q f(t,x,u(t,x))) \\ &\leq \frac{1}{\Gamma(q_1)\Gamma(q_2)} \int_0^t \int_0^x (t-s)^{q_1-1}(x-z)^{q_2-1} \times \\ &\qquad \sup_{(s,z)\in[0,t]\times[0,x]} d_\infty\big(f(s,z,u_n(s,z)), f(s,z,u(s,z))\big) dz ds. \end{aligned}$$

Then

$$\begin{aligned} &d_\infty(P_1u_n(t,x), P_1u(t,x)) \\ &\quad\leq \frac{t^{q_1}x^{q_2}}{\Gamma(q_1+1)\Gamma(q_2+1)} \sup_{(t,x)\in\Omega_T^b} d_\infty\big(f(t,x,u_n(t,x)), f(t,x,u(t,x))\big) \\ &\quad\leq \frac{T^{q_1}b^{q_2}}{\Gamma(q_1+1)\Gamma(q_2+1)} \sup_{(t,x)\in\Omega_T^b} d_\infty\big(f(t,x,u_n(t,x)), f(t,x,u(t,x))\big). \end{aligned}$$

Because f is continuous, $\sup_{(t,x)\in\Omega_T^b} d_\infty\big(f(t,x,u_n(t,x)), f(t,x,u(t,x))\big) \overset{n\to\infty}{\longrightarrow} 0$.
Therefore, P_1 is continuous.

Step 3. We prove $P_1(B_*(\tilde{0}, R))$ is relatively compact in $C(\Omega_*, E_c)$.

To prove this, by Arzelà-Ascoli theorem , we need two parts
Part 1. $P_1(B_*(\tilde{0}, R))$ is a level-equicontinuous subset of $C(\Omega_*, E_c)$.
Fix $(t_1, x_1), (t_2, x_2) \in \Omega_*$ such that $t_1 < t_2, x_1 < x_2$ and $u \in B_*(\tilde{0}, R)$, we have

$$d_\infty(P_1u(t_2,x_2), P_1u(t_1,x_1)) \leq d_\infty[\psi(t_2,x_2), \psi(t_1,x_1)] + d_\infty\Big({}^{RL}_{F}\mathcal{I}^q_{0+} f(t_2,x_2,u(t_2,x_2)), {}^{RL}_{F}\mathcal{I}^q_{0+} f(t_1,x_1,u(t_1,x_1))\Big). \quad (8)$$

Furthermore, we also have

$$\begin{aligned} &d_\infty\Big({}^{RL}_{F}\mathcal{I}^q_{0+} f(t_2,x_2,u(t_2,x_2)), {}^{RL}_{F}\mathcal{I}^q_{0+} f(t_1,x_1,u(t_1,x_1))\Big) \\ &\leq \frac{1}{\Gamma(q_1)\Gamma(q_2)} d_\infty\Big(\int_0^{t_1}\int_0^{x_1}(t_2-s)^{q_1-1}(x_2-z)^{q_2-1} f(s,z,u(s,z))dzds, \\ &\qquad \int_0^{t_1}\int_0^{x_1}(t_1-s)^{q_1-1}(x_1-z)^{q_2-1} f(s,z,u(s,z))dzds\Big) \\ &\quad + d_\infty\Big(\int_{t_1}^{t_2}\int_0^{x_1}(t_2-s)^{q_1-1}(x_2-z)^{q_2-1} f(s,z,u(s,z))dzds, \hat{0}\Big) \\ &\quad + d_\infty\Big(\int_0^{t_1}\int_{x_1}^{x_2}(t_2-s)^{q_1-1}(x_2-z)^{q_2-1} f(s,z,u(s,z))dzds, \hat{0}\Big) \\ &\quad + d_\infty\Big(\int_{t_1}^{t_2}\int_{x_1}^{x_2}(t_2-s)^{q_1-1}(x_2-z)^{q_2-1} f(s,z,u(s,z))dzds, \hat{0}\Big). \end{aligned}$$

Because f is bounded we have the assertion

$$d_\infty\Big({}^{RL}_{F}\mathcal{I}^q_{0+} f(t_2,x_2,u(t_2,x_2)), {}^{RL}_{F}\mathcal{I}^q_{0+} f(t_1,x_1,u(t_1,x_1))\Big) \leq \frac{M_0}{\Gamma(q_1+1)\Gamma(q_2+1)}\Big[t_2^{q_1}x_2^{q_2} - t_1^{q_1}x_1^{q_2}\Big]. \quad (9)$$

Combine (8) and (9), it is easy to see that when $(t_1, x_1) \to (t_2, x_2)$ then

$$d_\infty(P_1u(t_2,x_2), P_1u(t_1,x_1)) \to 0$$

for all $u \in B(\tilde{0}, R)$. Thus, $P_1(B(\tilde{0}, R))$ is level-equicontinuous on $C(\Omega_*, E_c)$.

Part 2. $P_1(B(\tilde{0}, R))(t, x)$ is relatively compact in E_c. By Theorem 3.1, we have to prove two following conditions

a) $P_1(B(\tilde{0}, R))(t, x)$ is level-equicontinuous.

b) $P_1(B(\tilde{0}, R))(t, x)$ is a compact-supported subset of E_c.

Fix $(t, x) \in \Omega_*$. It is clear that $P_1(B(\tilde{0}, R))(t, x) \in E_c$ and if $\nu \in P_1(B(\tilde{0}, R))(t, x)$ then

$$\nu = \psi(t, x) \oplus_F^{RL} \mathcal{I}_{0+}^q f(t, x, u(t, x)) \text{ for } u \in B(\tilde{0}, R).$$

For each $(t, x) \in \Omega_*$, assume that m is a positive number such that

$$mt^{q_1} x^{q_2} < \frac{1}{2}.$$

Since f is compact, $f(\Omega_T^b \times B(\tilde{0}, R))$ is relatively compact in E_c. From Remark 3.2, we imply $f(\Omega_T^b \times B(\tilde{0}, R))$ is compact-supported and level-equicotinuous. Then for all $\varepsilon > 0$, there exits $\delta > 0$ such that for $\alpha, \beta \in [0, 1]$ satisfying $|\alpha - \beta| < \delta$ then

$$d_H([f(t, x, u(t, x))]^\alpha, [f(t, x, u(t, x))]^\beta) < \Gamma(q_1 + 1)\Gamma(q_2 + 1)m\varepsilon$$

and $d_H([\psi(t, x)]^\alpha, [\psi(t, x)]^\beta) \leq \frac{\varepsilon}{2}$ for all $(t, x) \in \Omega_*$.
Therefore, when $|\alpha - \beta| < \delta$, we have

$$\begin{aligned}
&d_H([\nu]^\alpha, [\nu]^\beta) = d_H([P_1 u(t, x)]^\alpha, [P_1 u(t, x)]^\beta) \\
&\leq d_H\big([\psi(t, x)]^\alpha, [\psi(t, x)]^\beta\big) \\
&+ \frac{1}{\Gamma(q_1)\Gamma(q_2)} d_H\Big(\Big[\int_0^t \int_0^x (t - s)^{q_1 - 1}(x - z)^{q_2 - 1} f(s, z, u(s, z))dzds\Big]^\alpha, \\
&\qquad \Big[\int_0^t \int_0^x (t - s)^{q_1 - 1}(x - z)^{q_2 - 1} f(s, z, u(s, z))dzds\Big]^\beta\Big) \\
&\leq \frac{\varepsilon}{2} + mt^{q_1} x^{q_2} \varepsilon = \varepsilon.
\end{aligned}$$

Therefore, $P_1(B(\tilde{0}, R))(t, x)$ is level-equicotinuous in E_c.

For each $(t, x) \in \Omega_*$, we have

$$\begin{aligned}
&\Big[\psi(t, x) \oplus \frac{1}{\Gamma(q_1)\Gamma(q_2)} \int_0^t \int_0^x (t - s)^{q_1 - 1}(x - z)^{q_2 - 1} f(s, z, u(s, z))dzds\Big]^0 \\
&= [\psi(t, x)]^0 + \frac{1}{\Gamma(q_1)\Gamma(q_2)} \int_0^t (t - s)^{q_1 - 1}[f(s, z, u(s, z))]^0 ds \int_0^x (x - z)^{q_2 - 1} dz.
\end{aligned}$$

Because ψ has compact-supported, there exits a compact subset $K_1 \subset \mathbb{R}$ such hat $[\psi(t,x)]^0 \subset K_1$.

In addition, because $f(\Omega_T^b \times B(\tilde{0}, R))$ has compact-supported, we suppose here exits $K_2 \subset \mathbb{R}$ such that $[f(t,x,u(t,x))]^0 \subset K_2$. Then, we have

$$\Big[\psi(t,x) \oplus \frac{1}{\Gamma(q_1)\Gamma(q_2)} \int_0^t \int_0^x (t-s)^{q_1-1}(x-z)^{q_2-1} f(s,z,u(s,z))dzds\Big]^0$$
$$\subseteq K_1 + \frac{K_2}{\Gamma(q_1)\Gamma(q_2)} \int_0^x (x-z)^{q_2-1}dz = K_1 + \frac{K_2 t^{q_1} x^{q_2}}{\Gamma(q_1+1)\Gamma(q_2+1)}.$$

This means there exits a compact set $K_0 \subseteq \mathbb{R}$ such that

$$\Big[\psi(t,x) \oplus \frac{1}{\Gamma(q_1)\Gamma(q_2)} \int_0^t \int_0^x (t-s)^{q_1-1}(x-z)^{q_2-1} f(s,z,u(s,z))dads\Big]^0 \subseteq K_0.$$

or $P_1(B(\tilde{0}, R))(t,x)$ has compact-supported for each $(t,x) \in \Omega_*$. Therefore, P_1 is compact.

By Schauder fixed point theorem, P_1 has at least a fixed point u in $B(\tilde{0}, R)$ and u is a (1)-integral solution of the problem. By the hypothesis $C_\psi^f(\Omega_T^b, E_c) \neq \emptyset$ and for each $u \in C_\psi^f(\Omega_T^b, E_c)$ then $T_\psi^{f,q}[u](t,x) \in C_\psi^f(\Omega_T^b, E_c)$ for all $(t,x) \in \Omega_T^b$, we can define the mapping

$$P_2(u(t,x)) = T_\psi^{f,q}[u](t,x)$$

for all $(t,x) \in \Omega_T^b$. Next use Lemma 2.3 and repeat steps as proof P_1, we obtain P_2 is compact and by Schauder fixed point theorem, P_2 has at least a fixed point u in $B(\tilde{0}, R)$ and u is a (2)-integral solution of the problem. □

4.2. Goursat Problems in Infinite Domains

As we know, the space (E, d_∞) is not Banach space, this is difficult for us to study differential equation classes on spaces of fuzzy numbers. In this part, by constructing suitable metric spaces, we prove the existence of integral solutions of problem (2) - (3) when the time variable tends to infinity.

Consider the fractional order fuzzy partial differential equation in hyperbolic form

$$^C_F\mathcal{D}^q u(t,x) = f(t,x,u(t,x)),\ \Omega_\infty^b = [0,\infty) \times [0,b] \tag{10}$$

with boundary conditions

$$u(t,0) = \eta_1(t),\ t \in [0,\infty),\ u(0,x) = \eta_2(x),\ x \in [0,b] \tag{11}$$

where $f : \Omega^b_\infty \times C(\Omega^b_\infty, E) \to E, \eta_1 \in C([0,\infty), E), \eta_2 \in C([0,b], E)$ are given functions such that $\eta_1(0) = \eta_2(0) = u(0,0)$ and the differences $\eta_1(t) \ominus_H \eta_1(0), \eta_2(x) \ominus_H \eta_1(0)$ exist for $t \in [0,\infty), x \in [0,b]$.

Proposition 4.2. *[41] Let $u(.,.) \in \mathcal{W}_1(\Omega^b_\infty, E) \cup \mathcal{W}_2(\Omega^b_\infty, E)$ satisfy (10) - (11).*

1) If $u \in \mathcal{W}_1(\Omega^b_\infty, E)$ then u satisfies the integral equation

$$u(t,x) = \psi(t,x) \oplus_F^{RL} \mathcal{I}^q_{0+} f(t,x,u(t,x)) \text{ for all } (t,x) \in \Omega^b_\infty. \tag{12}$$

2) If $u \in \mathcal{W}_2(\Omega^b_\infty, E)$ then u satisfies the integral equation

$$u(t,x) = T^{f,q}_\psi[u](t,x) \text{ for all } (t,x) \in \Omega^b_\infty, \tag{13}$$

where $T^{f,q}_\psi[u]$ is defined in (6).

Definition 4.2. *[41] The function $u \in C(\Omega^b_\infty, E)$ is called*

1) (1)-integral solution of problem (10)- (11) if it satisfies the integral equation (12),

2) (2)-integral solution of problem (10)- (11) if it satisfies the integral equation (13).

Following hypotheses will be held in considering the solvability of problems (10)- (11).

($\mathbf{A_1}$) The function $f : \Omega^b_\infty \times C(\Omega^b_\infty, E) \to E$ satisfies the Lipschitz condition with respect to the third variable, i.e., there exists a positive real number L such that

$$d_\infty(f(t,x,\phi_1), f(t,x,\phi_2)) \le L DL_\infty(\phi_1,\phi_2)$$

for all $(t,x) \in \Omega^b_\infty, \phi_1,\phi_2 \in C(\Omega^b_\infty, E)$.

($\mathbf{A_2}$) There exist positive real numbers M_1, c_1 such that

$$d_\infty(f(t,x,\hat{0}),\hat{0}) \le M_1 e^{c_1 t},\ \ (t,x) \in \Omega^b_\infty.$$

$(\mathbf{A_3})$ For all $(t,x) \in \Omega^b_\infty$, there exist positive real numbers M_2, M_3 and c_2 such that $d_\infty(\eta_1(t), \hat{0}) \le M_2 e^{c_2 t}$, $d_\infty(\eta_2(x), \hat{0}) \le M_3$.

When using Banach fixed point theorem, it is important to build a appropriate complete metric space such that the mapping from this space into itself is contraction mapping. In this part, we will point out problem (10) - (11) has a unique (1)-integral solution in the space $C^\infty_\lambda(\Omega^b_\infty, E)$ consisting of all functions $u \in C(\Omega^b_\infty, E)$ with the metric $H_\lambda(u,v)$ is defined by

$$H_\lambda(u,v) = \sup_{(t,x)\in\Omega^b_\infty} \left\{ d_\infty(u(t,x), v(t,x)) e^{-\lambda t} \right\}$$

such that $\sup_{(t,x)\in\Omega^b_\infty} \left\{ d_\infty(u(t,x), \hat{0}) e^{-\lambda t} \right\} < \infty$, for each $\lambda > 0$.

Theorem 4.2. *[41] Under assumptions $f \in C(\Omega^b_\infty \times C(\Omega^b_\infty, E), E)$ and hypotheses $(\mathbf{A_1}) - (\mathbf{A_3})$ hold then there exists a (1)-integral solution of problem (10) - (11) on Ω^b_∞ and this solution is unique in $C^\infty_\lambda(\Omega^b_\infty, E)$ for $\lambda > 0$ big enough.*

Proof. For $u \in C^\infty_\lambda(\Omega^b_\infty, E)$, we set

$$P_3(u(t,x)) = \psi(t,x) \oplus^{AL}_F \mathcal{I}^c_{0+} f(t,x,u(t,x)),\ (t,x) \in \Omega^b_\infty.$$

Because $u \in C^\infty_\lambda(\Omega^b_\infty, E)$ there exits $\rho_0 > 0$ such that

$$d_\infty(u(t,x), \hat{0}) \le \rho_0 e^{\lambda t}, \forall\, (t,x) \in \Omega^b_\infty.$$

Combine with the condition $(\mathbf{A_1})$ and $(\mathbf{A_2})$, we have

$$\begin{aligned} d_\infty(f(t,x,u(t,x)), \hat{0}) &\le d_\infty(f(t,x,u(t,x)), f(t,x,\hat{0})) + d_\infty(f(t,x,\hat{0}), \hat{0}) \\ &\le L\rho_0 e^{\lambda t} + M_1 e^{c_1 t}. \end{aligned}$$

For all $\lambda > c_1 > 0$, we have

$$\begin{aligned} & d_\infty({}^{RL}_F\mathcal{I}^q_{0+} f(t,x,u(t,x)), \hat{0}) \\ & \le \frac{1}{\Gamma(q_1)\Gamma(q_2)} \int_0^t \int_0^x (t-s)^{q_1-1} (x-z)^{q_2-1} d_\infty(f(s,z,u(s,z)), \hat{0}) dz ds \\ & \le \frac{1}{\Gamma(q_1)\Gamma(q_2)} \int_0^t \int_0^x (t-s)^{q_1-1} (x-z)^{q_2-1} \left[L\rho_0 e^{\lambda s} + M_1 e^{c_1 s} \right] ds dz \\ & \le \frac{L\rho_0 + M_1}{\Gamma(q_1)\Gamma(q_2)} \int_0^t \int_0^x (t-s)^{q_1-1} (x-z)^{q_2-1} e^{\lambda s} ds dz. \end{aligned}$$

Apply Proposition 3.4, one obtains

$$d_\infty({}_F^{RL}\mathcal{I}_{0+}^q f(t,x,u(t,x)),\hat{0}) \le (L\rho_0 + M_1)\mathcal{L}(\lambda,b,q_1,q_2)e^{\lambda t}. \quad (14)$$

If $\lambda \ge \max\{c_1,c_2\}$ then from the hypothesis ($\mathbf{A_3}$) and (14), we imply

$$\begin{aligned} d_\infty(P_3u(t,x),\hat{0})e^{-\lambda t} &\le 2M_2e^{(c_2-\lambda)t} + M_3e^{-\lambda t} + (L\rho_0+M_1)\mathcal{L}(\lambda,b,q_1,q_2) \\ &\le 2M_2 + M_3 + (L\rho_0+M_1)\mathcal{L}(\lambda,b,q_1,q_2) < \infty \quad (15) \end{aligned}$$

for all $(t,x)\in\Omega_\infty^b$. This show that $P_3(u)\in C_\lambda^\infty(\Omega_\infty^b,E)$.

For all $u,v\in C_\lambda^\infty(\Omega_\infty^b,E)$, from Lemma 3.4, we have

$$\begin{aligned} d_\infty(P_3u(t,x),P_3v(t,x)) &\le d_\infty\left({}_F^{RL}\mathcal{I}_{0+}^q f(t,x,u(t,x)),{}_F^{RL}\mathcal{I}_{0+}^q f(t,x,v(t,x))\right) \\ &\le \frac{L}{\Gamma(q_1)\Gamma(q_2)}H_\lambda(u,v)\int_0^t\int_0^x (t-s)^{q_1-1}(x-z)^{q_2-1}e^{\lambda s}dsdz \\ &\le L\mathcal{L}(\lambda,b,q_1,q_2)H_\lambda(u,v)e^{\lambda t}. \quad (16) \end{aligned}$$

Thus $H_\lambda(P_3u,P_3v)\le L\mathcal{L}(\lambda,b,q_1,q_2)H_\lambda(u,v)$. Set

$$\lambda_0 := \inf\{\lambda > 0 \text{ such that } L\mathcal{L}(\lambda,b,q_1,q_2) < 1\}$$

and $\lambda_1 = \max\{\lambda_0,c_1,c_2\}$. Then if we choose $\lambda\ge\lambda_1$ big enough then $L\mathcal{L}(\lambda,b,q_1,q_2)<1$. Therefore, P_3 is a contraction mapping.

On the other hand, because $(C_\lambda^\infty(\Omega_\infty^b,E),H_\lambda)$ is a complete metric space for $\lambda>0$ (see [41]), by Banach fixed point Theorem, there exists a unique fuzzy function u is defined on Ω_∞^b satisfying the equation (12), it is a (1)-integral solution of problem (10) - (11) on Ω_∞^b. This completes the proof. □

Set

$$C_{\lambda,\psi}^{\infty,f}(\Omega_\infty^b,E) = \left\{u\in C_\lambda^\infty(\Omega_\infty^b,E) : T_\psi^{f,q}[u](t,x)\in E,\, (t,x)\in\Omega_\infty^b\right\}$$

for each $\lambda>0$.

To prove the existence of the (2)-integral solution of the problem, we assume that

($\mathbf{A_4}$) $C_{\lambda,\psi}^{\infty,f}(\Omega_\infty^b,E)\ne\emptyset$ and if $u\in C_{\lambda,\psi}^{\infty,f}(\Omega_\infty^b,E)$ then

$$T_\psi^{f,q}[u](t,x)\in C_{\lambda,\psi}^{\infty,f}(\Omega_\infty^b,E),\ \forall(t,x)\in\Omega_\infty^b.$$

Theorem 4.3. *[41] Suppose that $f \in C(\Omega^b_\infty \times C(\Omega^b_\infty, E), E)$ and the assumptions $(\mathbf{A_1})$ - $(\mathbf{A_4})$ hold. Then there exists a (2)-integral solution of problem (10) (11) on Ω^b_∞ and this solution is unique in $C^{\infty,f}_{\lambda,\psi}(\Omega^b_\infty, E)$ for $\lambda > 0$ big enough.*

Proof. Consider the mapping $P_4 : C^{\infty,f}_{\lambda,\psi}(\Omega^b_\infty, E) \to C^{\infty,f}_{\lambda,\psi}(\Omega^b_\infty, E)$ defined by

$$P_4(u(t,x)) = T^{f,q}_{\psi}[u](t,x).$$

By utilizing similar method in (16), we have

$$d_\infty(P_4u(t,x),\ P_4v(t,x)) \leq L\mathcal{L}(\lambda, b, q_1, q_2)H_\lambda(u,v)e^{\lambda t}.$$

This implies that

$$H_\lambda(P_4u, P_4v) \leq L\mathcal{L}(\lambda, b, q_1, q_2)H_\lambda(u,v).$$

For $\lambda \geq \lambda_1$, we have $L\mathcal{L}(\lambda, b, q_1, q_2) < 1$. Therefore, P_4 is a contraction mapping. Furthermore, since $C^{\infty,f}_{\lambda,\psi}(\Omega^b_\infty, E)$ is a complete metric space for each positive real number λ (see [41]), by Banach fixed point Theorem, P_4 has a unique fixed point $u \in C^{\infty,f}_{\lambda,\psi}(\Omega^b_\infty, E)$. □

5. Some Qualitative Properties of Solutions of Fuzzy Fractional Goursat Problems

5.1. Ulam Stability

In this part, we present the concepts and prove Ulam stability of the problem

$$^C_F\mathcal{D}^q u(t,x) = f(t,x,u(t,x)),\ (t,x) \in \Omega^b_\infty = [0,\infty) \times [0,b] \quad (9)$$
$$u(t,0) = \eta_1(t),\ \ t \in [0,\infty),\ u(0,x) = \eta_2(x),\ x \in [0,b] \quad (10)$$

where $f : \Omega^b_\infty \times C(\Omega^b_\infty, E) \to E, \eta_1 \in C([0,\infty), E), \eta_2 \in C([0,b], E)$ are given functions such that $\eta_1(0) = \eta_2(0) = u(0,0)$ and the differences $\eta_1(t) \ominus_H \eta_1(0), \eta_2(x) \ominus_H \eta_1(0)$ exist for $t \in [0,\infty), x \in [0,b]$.

a. Hyers-Ulam Stability

For all $\varepsilon > 0$, we consider the following inequation

$$d_\infty({}^C_F\mathcal{D}^q v(t,x), f(t,x,v(t,x))) \leq \varepsilon,\ (t,x) \in \Omega^b_\infty. \tag{17}$$

Definition 5.1.

1. *The function* $v \in \mathcal{W}_1(\Omega^b_\infty, E)$ *is called a solution in type 1 of (17) if there exists a function* $h_1 \in C(\Omega^b_\infty, E)$ *such that*

   i. $d_\infty(h_1(t,x), \hat{0}) \leq \varepsilon,\ (t,x) \in \Omega^b_\infty;$

   ii. ${}^C_F\mathcal{D}^i v(t,x) = f(t,x,v(t,x)) \oplus h_1(t,x), (t,x) \in \Omega^b_\infty.$

2. *The function* $v \in \mathcal{W}_2(\Omega^b_\infty, E)$ *is called a solution in type 2 of (17) if there exists a function* $h_2 \in C(\Omega^b_\infty, E)$ *such that*

   i. $d_\infty(h_2(t,x), \hat{0}) \leq \varepsilon,\ (t,x) \in \Omega^b_\infty;$

   ii. ${}^C_F\mathcal{D}^i v(t,x) = f(t,x,v(t,x)) \oplus h_2(t,x), (t,x) \in \Omega^b_\infty.$

Proposition 5.1.

1. *If v is a solution in type 1 of (17) then it satisfies the following integral inequation*

$$\Gamma(q_1+1)\Gamma(q_2+1) d_\infty\big(v(t,x), v(t,0) \oplus v(0,x) \ominus_H v(0,0) \oplus {}^{RL}_F\mathcal{I}^q_{0+} f(t,x,v(t,x))\big) \leq \varepsilon t^{q_1} x^{q_2} \tag{18}$$

   for all $(t,x) \in \Omega^b_\infty$.

2. *If v is a solution in type 2 of (17) then it satisfies the following integral inequation*

$$\Gamma(q_1+1)\Gamma(q_2+1) d_\infty\big(v(t,x), v(t,0) \oplus v(0,x) \ominus_H v(0,0) \ominus_H (-1) {}^{RL}_F\mathcal{I}^q_{0+} f(t,x,v(t,x))\big) \leq \varepsilon t^{q_1} x^{q_2} \tag{19}$$

   for all $(t,x) \in \Omega^b_\infty$.

Proof.

1. Suppose $v \in \mathcal{W}_1(\Omega^b_\infty, E)$ is a solution in type 1 of the inequation (17). Then there exists the function $h_1 \in C(\Omega^b_\infty, E)$ such that

$$ {}^C_F\mathcal{D}^i v(t,x) = f(t,x,v(t,x)) \oplus h_1(t,x)\ (t,x) \in \Omega^b_\infty. $$

This means

$$ {}^{AL}_F\mathcal{I}^{1-q}_{0+} D_{TX} v(t,x) = f(t,x,v(t,x)) \oplus h_1(t,x). $$

Hence

$$ {}^{AL}_F\mathcal{I}^{q}_{0+}\,{}^{AL}_F\mathcal{I}^{1-q}_{0+} D_{TX} v(t,x) = {}^{AL}_F\mathcal{I}^{q}_{0+}[f(t,x,v(t,x)) \oplus h_1(t,x)]. $$

From Proposition 3.5, we imply that

$$ \int_0^t \int_0^x D_{sz} v(s,z)dzds = {}^{RL}_F\mathcal{I}^{q}_{0+}[f(t,x,v(t,x)) \oplus h_1(t,x)]. $$

Because $v \in \mathcal{W}_1(\Omega^b_\infty, E)$, utilizing similar argument to Proposition 4.1, we get

$$ v(t,x) = v(t,0) \oplus v(0,x) \ominus_H v(0,0) \oplus {}^{RL}_F\mathcal{I}^{q}_{0+}[f(t,x,v(t,x)) \oplus h_1(t,x)] \tag{20} $$

for all $(t,x) \in \Omega^b_\infty$. Then we have

$$ \begin{aligned} d_\infty(v(t,x), v(t,0) \oplus v(0,x) \ominus_H v(0,0) \oplus {}^{RL}_F\mathcal{I}^{q}_{0+} f(t,x,v(t,x))) \\ \leq d_\infty({}^{RL}_F\mathcal{I}^{q}_{0+} h_1(t,x), \hat{0}) \\ \leq \frac{\varepsilon}{\Gamma(q_1)\Gamma(q_2)} \int_0^t \int_0^x (t-s)^{q_1-1}(x-z)^{q_2-1} dzds \\ \leq \frac{\varepsilon t^{q_1} x^{q_2}}{\Gamma(q_1+1)\Gamma(q_2+1)}. \end{aligned} $$

2. Suppose that $v \in \mathcal{W}_2(\Omega^b_\infty, E)$ is a solution in type 2 of the inequation (17). By proving the same as above, we obtain

$$ v(t,x) = v(t,0) \oplus v(0,x) \ominus_H v(0,0) \ominus_H (-1){}^{RL}_F\mathcal{I}^{q}_{0+}[f(t,x,v(t,x)) \oplus h_2(t,x)]. $$

Then apply Proposition 2.3, we also have the following estimation

$$
\begin{aligned}
d_\infty(v(t,x), v(t,0)\oplus v(0,x) \ominus_H v(0,0) \ominus_H (-1)\,{}^{RL}_{F}\mathcal{I}^q_{0+} f(t,x,v(t,x))) \\
&\leq d_\infty({}^{RL}_{F}\mathcal{I}^q_{0+} h_2(t,x), \hat{0}) \\
&\leq \frac{\varepsilon}{\Gamma(q_1)\Gamma(q_2)} \int_0^t \int_0^x (t-s)^{q_1-1}(x-z)^{q_2-1} dz ds \\
&\leq \frac{\varepsilon t^{q_1} x^{q_2}}{\Gamma(q_1+1)\Gamma(q_2+1)}
\end{aligned}
$$

for all $(t,x) \in \Omega^b_\infty$. The proof is complete. □

Definition 5.2.

1. *The function $v \in C(\Omega^b_\infty, E)$ is called an integral solution in type 1 of (17) if it satisfies (18).*

2. *The function $v \in C(\Omega^b_\infty, E)$ is called an integral solution in type 2 of (17) if it satisfies (19).*

Definition 5.3. *problem (10) - (11) is called Hyers-Ulam stability in type k ($k = 1, 2$) if there exists $c_o > 0$ such that for arbitrary $\varepsilon > 0$ and for each integral solution v in type k of the inequation (17), there exists a (k)-integral solution of problem (10) - (11) such that $H_\lambda(u, v) \leq c_o \varepsilon$.*

Theorem 5.1.

1. *Assume that $f \in C(\Omega^b_\infty \times C(\Omega^b_\infty, E), E)$ and hypotheses $(\mathbf{A_1}) - (\mathbf{A_3})$ hold. Then problem (10) - (11) is Hyers-Ulam stable in type 1.*

2. *Assume that $f \in C(\Omega^b_\infty \times C(\Omega^b_\infty, E), E)$ and hypotheses $(\mathbf{A_1}) - (\mathbf{A_4})$ are satisfied. Then problem (10) - (11) is Hyers-Ulam stable in type 2.*

Proof. Suppose that $v \in C(\Omega^b_\infty, E)$ is an integral solution in type 1 of (17). Because hypotheses $(\mathbf{A_1}) - (\mathbf{A_3})$ hold, by Theorem 4.2, we see that problem (10) - (11) with boundary conditions

$$\eta_1(t) = v(t,0),\ t \in [0,\infty),\ \eta_2(x) = v(0,x),\ x \in [0,b]$$

has unique a (1)-integral solution $u(t,x)$ satisfying

$$u(t,x) = v(t,0) \oplus v(0,x) \ominus_H v(0,0) \oplus_F^{AL} \mathcal{I}_{0+}^q f(t,x,u(t,x))$$

for all $(t,x) \in \Omega_\infty^b$. Then we have

$$\begin{aligned} &d_\infty(u(t,x), v(t,x)) \\ &\quad \le \frac{\varepsilon t^{q_1} x^{q_2}}{\Gamma(q_1+1)\Gamma(q_2+1)} + d_\infty\big({}_F^{RL}\mathcal{I}_{0+}^q f(t,x,u(t,x)), {}_F^{RL}\mathcal{I}_{0+}^q f(t,x,v(t,x))\big). \end{aligned}$$

Using similar assertions as (16), we get

$$d_\infty(u(t,x), v(t,x)) \le \frac{\varepsilon t^{q_1} x^{q_2}}{\Gamma(q_1+1)\Gamma(q_2+1)} + L\mathcal{L}(\lambda, b, q_1, q_2) H_\lambda(u,v) e^{\beta t}. \tag{21}$$

Multiplying both sides of the inequation (21) by $e^{-\lambda t}$, it follows that

$$e^{-\lambda t} d_\infty(u(t,x), v(t,x)) \le \frac{\varepsilon t^{q_1} x^{q_2}}{e^{\beta t}\Gamma(q_1+1)\Gamma(q_2+1)} + L\mathcal{L}(\lambda, b, q_1, q_2) H_\lambda(u,v). \tag{22}$$

It is easy to see that for $\lambda \ge q_1$, we have

$$\max_{t\in[0,\infty)} \frac{t^{q_1}}{e^{\lambda t}} = \frac{q_1^{q_1}}{e^{\lambda q_1}}.$$

Taking supremum both sides of the inequality (22), one obtains

$$H_\lambda(u,v) \le \frac{\varepsilon q_1^{q_1} b^{q_2}}{e^{\lambda q_1}\Gamma(q_1+1)\Gamma(q_2+1)} + L\mathcal{L}(\lambda, b, q_1, q_2) H_\lambda(u,v).$$

From Remark 3.3, we can choose $\lambda \ge 0$ big enough such that

$$c_o := \frac{q_1^{q_1} b^{q_2}}{e^{\lambda q_1}\Gamma(q_1+1)\Gamma(q_2+1)\Big[1 - L\mathcal{L}(\lambda, b, q_1, q_2)\Big]} > 0.$$

Therefore $H_\lambda(u,v) \le c_o\varepsilon$. Then problem (10)- (11) is Hyers-Ulam stable in type 1.

In addition, when the condition $(\mathbf{A_4})$ holds, because the conditions of Theorem 4.3 are satisfied, problem (10) - (11) with following conditions

$$\eta_1(t) = v(t,0),\ t \in [0,\infty),\ \eta_2(x) = v(0,x),\ x \in [0,b]$$

has unique a (2)-integral solution $u(t,x)$ satisfying

$$u(t,x) = v(t,0) \oplus v(0,x) \ominus_H v(0,0) \ominus_H (-1)\,{}^{AL}_{F}\mathcal{I}^{q}_{0+} f(t,x,u(t,x)),\ \forall\,(t,x) \in \Omega^b_\infty.$$

From the properties $iv)$ of the metric d_∞ in Proposition 2.3, we have

$$\begin{aligned} & d_\infty(u(t,x), v(t,x)) \\ & \le \frac{\varepsilon t^{q_1} x^{q_2}}{\Gamma(q_1+1)\Gamma(q_2+1)} + d_\infty\big({}^{RL}_{F}\mathcal{I}^{q}_{0+} f(t,x,u(t,x)), {}^{RL}_{F}\mathcal{I}^{q}_{0+} f(t,x,v(t,x))\big). \end{aligned}$$

Continuing the same arguments as in proof of Hyers-Ulam stability in type 1, we can also point out that for $\lambda > 0$ big enough, there exists

$$c_o := \frac{q_1^{q_1} b^{q_2}}{e^{\lambda q_1}\Gamma(q_1+1)\Gamma(q_2+1)\Big[1 - L\mathcal{L}(\lambda, b, q_1, q_2)\Big]} > 0$$

such that $H_\lambda(u,v) \le c_o\varepsilon$. This means problem (10) -(11) is Hyers-Ulam stable in type 2. □

b. Hyers-Ulam-Rassias Stability

For $\Phi \in L^1(\Omega^b_\infty, [0,\infty)) \cap L^\infty(\Omega^b_\infty, [0,\infty))$, we consider the following inequation

$$d_\infty({}^{C}_{F}\mathcal{D}^q v(t,x), f(t,x,v(t,x))) \le \Phi(t,x),\ (t,x) \in \Omega^b_\infty. \tag{23}$$

Definition 5.4.

1. *The function $v \in \mathcal{W}_1(\Omega^b_\infty, E)$ is a solution in type 1 of the inequation (23) if there exists the function $h_3 \in C(\Omega^b_\infty, E)$ such that*

   i. $d_\infty(h_3(t,x), \hat{0}) \le \Phi(t,x),\ (t,x) \in \Omega^b_\infty,$

   ii. ${}^{C}_{F}\mathcal{D}^q v(t,x) = f(t,x,v(t,x)) \oplus h_3(t,x),\ (t,x) \in \Omega^b_\infty.$

2. *The function $v \in \mathcal{W}_2(\Omega^b_\infty, E)$ is a solution in type 1 of the inequation (23) if there exists the function $h_4 \in C(\Omega^b_\infty, E)$ such that*

   i. $d_\infty(h_4(t,x), \hat{0}) \leq \Phi(t,x)$, $(t,x) \in \Omega^b_\infty$,

   ii. ${}^C_F\mathcal{D}^q v(t,x) = f(t,x,v(t,x)) \oplus h_4(t,x)$, $(t,x) \in \Omega^b_\infty$.

By proving similar to Proposition 5.1, we get the following result.

Proposition 5.2.

1. *If v is an integral solution in type 1 of (23) then it satisfies following integral inequation for all $(t,x) \in \Omega^b_\infty$*

$$d_\infty(v(t,x), v(t,0) \oplus v(0,x) \ominus_H v(0,0) \oplus^{RL}_F \mathcal{I}^q_{0+} f(t,x,v(t,x)) \\ \leq^{RL} I^q_{0+} \Phi(t,x). \tag{24}$$

2. *If v is an integral solution in type 2 of (23) then it satisfies following integral inequation for all $(t,x) \in \Omega^b_\infty$*

$$d_\infty(v(t,x), v(t,0) \oplus v(0,x) \ominus_H v(0,0) \ominus_H (-1)^{RL}_F\mathcal{I}^q_{0+} f(t,x,v(t,x))) \\ \leq^{RL} I^q_{0+} \Phi(t,x). \tag{25}$$

Definition 5.5.

1. *The function $v \in C(\Omega^b_\infty, E)$ is called an integral solution in type 1 of (23) if it satisfies (24).*

2. *The function $v \in C(\Omega^b_\infty, E)$ is called an integral solution in type 2 of (23) if it satisfies (25).*

Definition 5.6. *problem (10) - (11) is called Hyers-Ulam-Rassias stable in type k ($k = 1,2$) with respect to Φ if there exists a real number $c_{f,\Phi} > 0$ such that for each integral solution v in type k of the inequation (23), there exists a (k)-integral solution of problem (10) - (11) such that*

$$H_\lambda(u,v) \leq c_{f,\Phi} \sup_{(t,x)\in\Omega^b_\infty} \Phi(t,x).$$

Theorem 5.2. *Assume that $f \in C(\Omega^b_\infty \times C(\Omega^b_\infty, E), E)$, the hypotheses $(\mathbf{A_1}) - (\mathbf{A_3})$ are fulfilled and for each the function $\Phi \in L^1(\Omega^b_\infty, [0, +\infty)) \cap L^\infty(\Omega^b_\infty, [0, \infty))$, there exists $m_\Phi > 0$, $\nu \geq 0$ such that*

$$^{RL}I^q_{0+}\Phi(t, x) \leq m_\Phi e^{\nu t}\Phi(t, x), \ \forall (t, x) \in \Omega^b_\infty. \tag{26}$$

Then problem (10) - (11) is Hyers-Ulam-Rassias stable in type 1. Moreover, if condition $(\mathbf{A_4})$ is satisfied then problem (10) - (11) is Hyers-Ulam-Rassias stable in type 2.

Proof. We shall show that problem (10) - (11) is Hyers-Ulam-Rassias stable in type 2. Hyers-Ulam-Rassias stability in type 1 of the problem is similarly proved based on the results of Proposition 2.3 and Theorem 4.2.

Assume that $v \in C(\Omega^b_\infty, E)$ is an integral solution in type 2 of the inequation (23). Because the hypotheses $(\mathbf{A_1}) - (\mathbf{A_4})$ hold, by Theorem 4.3, we see that problem (10)-(11) with the conditions

$$\eta_1(t) = v(t, 0), \, t \in [0, \infty), \, \eta_2(x) = v(0, x), \, x \in [0, b]$$

has unique a (2)-integral solution $u(t, x)$ satisfying

$$u(t, x) = v(t, 0) \oplus v(0, x) \ominus_H v(0, 0) \ominus_H (-1)^{AL}_F\mathcal{I}^q_{0+}f(t, x, u(t, x))$$

for all $(t, x) \in \Omega^b_\infty$. Then we have

$$\begin{aligned} &d_\infty(u(t, x), v(t, x)) \\ &\quad\leq d_\infty(v(t, 0) \oplus v(0, x) \ominus_H v(0, 0) \ominus_H (-1)^{RL}_F\mathcal{I}^q_{0+}f(t, x, u(t, x)), v(t, x)) \\ &\quad\leq^{RL} I^q_{0+}\Phi(t, x) + d_\infty\big({}^{RL}_F\mathcal{I}^q_{0+}f(t, x, u(t, x)), {}^{RL}_F\mathcal{I}^q_{0+}f(t, x, v(t, x))\big). \end{aligned}$$

Proving similar to (16), we get

$$d_\infty(u(t, x), v(t, x)) \leq^{RL} I^q_{0+}\Phi(t, x) + L\mathcal{L}(\lambda, b, q_1, q_2)H_\lambda(u, v)e^{\lambda t}. \tag{27}$$

From the hypothesis (26), for all $\lambda > \nu$, we see that

$$e^{-\lambda t}d_\infty(u(t, x), v(t, x)) \leq m_\Phi\Phi(t, x) + L\mathcal{L}(\lambda, b, q_1, q_2)H_\lambda(u, v).$$

By Remark 3.3, we can choose $\lambda \geq 0$ such that

$$c_{f,\Phi} = \frac{m_\Phi}{1 - L\mathcal{L}(\lambda, b, q_1, q_2)} > 0.$$

This shows that $H_\lambda(u, v) \leq c_{f,\Phi} \sup_{(t,x)\in\Omega^b_\infty} \Phi(t, x)$. Hence problem (10) - (11) is Hyers-Ulam-Rassias stable in type 2. □

.2. Lyapunov Stability

n this part, we investigate problem (10)-(11) via the hypotheses $f(t,x,\hat{0}) \equiv \hat{0}$, $f : \Omega^b_\infty \times C(\Omega^b_\infty, E) \to E$, $\eta_1 \in C([0,\infty), E)$, $\eta_2 \in C([0,b], E)$ are given unctions such that $\eta_1(0) = \eta_2(0) = u(0,0)$ and the differences $\eta_1(t) \ominus_H \eta_1(0)$, $\eta_2(x) \ominus_H \eta_1(0)$ exist for all $t \geq 0, x \in [0,b]$.

In this case, fuzzy number $\hat{0}$ is called a equilibrium point of the dynamic ystem determined by equation (10).

Definition 5.7. *[41] The equilibrium point $\hat{0}$ of problem (10) - (11) is called stability with respect to t if for all $\varepsilon > 0$, there exists $\delta > 0$ such that if*

$$d_\infty(\eta_1(t), \hat{0}) < \delta, \;\; t \geq 0, \;\; d_\infty(\eta_2(x), \hat{0}) < \delta, \;\; x \in [0,b]$$

hen all solutions $u(t,x)$ of problem (10) - (11) satisfying

$$d_\infty(u(t,x), \hat{0}) < \varepsilon \text{ for all } (t,x) \in \Omega^b_\infty.$$

To prove the stability of problem (10) - (11), we suppose the function f satisfying the condition

($\mathbf{G}_1$) There exists $\beta > 1$ such that

$$d_\infty(f(t,x,\phi_1(t,x)), f(t,x,\phi_2(t,x))) \leq L(t+1)^{-\beta} d_\infty(\phi_1(t,x), \phi_2(t,x))$$

for all $\phi_1, \phi_2 \in C(\Omega^b_\infty, E)$, $(t,x) \in \Omega^b_\infty$. Following lemma is proved in [34].

Lemma 5.1. *[34] Suppose $r(t,x), a(t,x)$ v $b(t,x)$ are nonnegative continuous functions for all $t, x \geq 0$. If*

$$r(t,x) \leq a(t,x) + \int_0^t \int_0^x b(s,z) r(s,z) dads, \; t, x \geq 0$$

then

$$r(t,x) \leq \max_{(s,z)\in[0,t]\times[0,x]} a(s,z) \exp\left(\int_0^t \int_0^x b(s,z) dads \right) \text{ for all } t, x \geq 0.$$

Theorem 5.3. *[41] Under the assumptions $f \in C(\Omega^b_\infty \times C(\Omega^b_\infty, E), E)$ and the conditions ($\mathbf{G}_1$), ($\mathbf{A}_3$) are satisfied, the equilibrium point $\hat{0}$ of problem (10) - (11 is stable.*

Proof. From the hypothesis ($\mathbf{G_1}$), there exists $\beta > 1$ such that

$$d_\infty(f(t,x,u(t,x)), f(t,x,v(t,x))) \le L(t+1)^{-\beta} d_\infty(u(t,x), v(t,x))$$

for all $(t,x) \in \Omega^b_\infty$. This implies

$$d_\infty(f(t,x,u(t,x)), f(t,x,v(t,x))) < L d_\infty(u(t,x), v(t,x)), t \ge 0, x \in [0,b].$$

Therefore the condition ($\mathbf{A_1}$) is satisfied. By Theorem 4.2, for each condition (11), the equation (10) has unique a (1)-integral solution on Ω^b_∞.

Assume that $u(t,x)$ is a (1)-integral solution of the problem. From the property ii) of the metric d_∞ in Proposition 2.3 and the hypothesis ($\mathbf{G_1}$), we have

$$\begin{aligned} d_\infty(u(t,x), \hat{0}) &\le d_\infty(\psi(t,x), \hat{0}) + d_\infty({}^{RL}_{F}\mathcal{I}^q_{0+} f(t,x,u(t,x)), \hat{0}) \\ &\le d_\infty(\psi(t,x), \hat{0}) \\ &\quad + \frac{L}{\Gamma(q_1)\Gamma(q_2)} \int_0^t \int_0^x (t-s)^{q_1-1}(x-z)^{q_2-1}(s+1)^{-\beta} d_\infty(u(s,z), \hat{0}) dz ds. \end{aligned}$$

Apply Lemma 5.1, we receive

$$\begin{aligned} d_\infty(u(t,x), \hat{0}) \le &\max_{(s,z)\in[0,t]\times[0,x]} d_\infty(\psi(s,z), \hat{0}) \\ &\times \exp\Big(\frac{L}{\Gamma(q_1)\Gamma(q_2)} \int_0^t \int_0^x (t-s)^{q_1-1}(x-z)^{q_2-1}(s+1)^{-\beta} dz ds\Big) \end{aligned}$$

Besides, for $\beta > 1$, we have

$$\begin{aligned} &\int_0^t (t-s)^{q_1-1}(s+1)^{-\beta} ds \\ &\qquad = \int_0^{\frac{t}{2}} (t-s)^{q_1-1}(s+1)^{-\beta} ds + \int_{\frac{t}{2}}^t (t-s)^{q_1-1}(s+1)^{-\beta} ds \\ &\qquad \le \left(\frac{t}{2}\right)^{q_1-1} \int_0^{\frac{t}{2}} (s+1)^{-\beta} ds + \left(\frac{t}{2}+1\right)^{-\beta} \int_{\frac{t}{2}}^t (t-s)^{q_1-1} ds \\ &\qquad \le \left(\frac{t}{2}\right)^{q_1-1} \frac{1-\left(\frac{t}{2}+1\right)^{1-\beta}}{\beta-1} + \left(\frac{t}{2}+1\right)^{-\beta} \frac{\left(\frac{t}{2}\right)^{q_1-1}}{q_1}. \end{aligned} \tag{28}$$

Moreover, for all $t \geq 0, \beta > 1, q_1 \in (0,1)$, it is easy to see that

$$\frac{1-\left(\frac{t}{2}+1\right)^{1-\beta}}{\beta-1} \leq \frac{1}{1-q_1}\left(\frac{t}{2}\right)^{1-q_1}. \tag{29}$$

Combine (28) and (29), one obtains

$$\int_0^t (t-s)^{q_1-1}(s+1)^{-\beta}ds \leq \frac{1}{1-q_1}+\frac{1}{q_1}, \text{ for } \beta > 1. \tag{30}$$

This implies

$$d_\infty(u(t,x),\hat{0}) \leq \max_{(s,z)\in[0,t]\times[0,x]} d_\infty(\psi(s,z),\hat{0}) \exp\left(\frac{Lb^{q_2}}{\Gamma(q_1+1)\Gamma(q_2+1)(1-q_1)}\right).$$

Thus for all $\varepsilon > 0$, there exists $\delta = \varepsilon \exp\left(-\frac{Lb^{q_2}}{\Gamma(q_1+1)\Gamma(q_2+1)(1-q_1)}\right) > 0$, such that the functions η_1, η_2 satisfying

$$\max_{s\in[0,t]} d_\infty(\eta_1(s),\hat{0}) < \frac{\delta}{3} \text{ v } \max_{z\in[0,x]} d_\infty(\eta_2(z),\hat{0}) < \frac{\delta}{3}.$$

This means

$$\max_{(s,z)\in[0,t]\times[0,x]} d_\infty(\psi(t,x),\hat{0}) \leq \max_{s\in[0,t]} d_\infty(\eta_1(s),\hat{0}) + \max_{z\in[0,x]} d_\infty(\eta_2(z),\hat{0}) + d_\infty(\eta_1(0),\hat{0}) < \delta.$$

Then $d_\infty(u(t,x),\hat{0}) < \varepsilon, \ \forall\ (t,x) \in \Omega^b_\infty$. Thus the equilibrium point $\hat{0}$ of problem (10)- (11) is stable. □

6. Applications

Example 6.1. Consider the fractional fuzzy partial differential equation

$$^{C}_{F}\mathcal{D}^{\frac{1}{2}}u(t,x) = \sqrt{tx}C, \quad (t,x) \in \Omega^1_4 = [0,4]\times[0,1] \tag{31}$$

with boundary conditions

$$u(t,0) = tC,\ x \in [0,1], u(0,x) = xC,\ t \in [0,1],\ u(0,0) = \hat{0} \tag{32}$$

where C is a fuzzy number having the level sets $[C]^\alpha = [c^-_\alpha, c^+_\alpha]$.

For $(t, x) \in \Omega_4^1$, we set

$$f(t, x, u(t, x)) = \sqrt{TX}C.$$

Choose $R = \frac{d_\infty(C, \hat{0})}{8}$. It is easy to see that $f : \Omega_4^1 \times B(\tilde{0}, R) \to E_c$ is a compact mapping. Furthermore, we have $\psi(t, x) = (t + x)C$ and ${}^{AL}_{F}\mathcal{I}^{\frac{1}{2}}_{0+} f(t, x, u(t, x)) = \pi TXC$. This implies $[\psi(t, x)]^\alpha = (t + x)[c^-_\alpha, c^+_\alpha]$ and

$$[(-1)^{AL}_{F}\mathcal{I}^{\frac{1}{2}}_{0+} f(t, x, u(t, x))]^\alpha = \Big[-\pi TXc^+_\alpha, -\pi TXc^-_\alpha\Big]$$

Then we have $Len[\psi(t, x)]^\alpha = (t + x)(c^+_\alpha - c^-_\alpha)$ and

$$Len[(-1)^{AL}_{F}\mathcal{I}^{\frac{1}{2}}_{0+} f(t, x, u(t, x))]^\alpha = \pi TX(c^+_\alpha - c^-_\alpha).$$

Therefore

$$Len[\psi(t, x)]^\alpha \geq Len[(-1)^{AL}_{F}\mathcal{I}^{\frac{1}{2}} f(t, x, u(t, x))]^\alpha, \ (t, x) \in \Omega_4^1.$$

Thus the Hukuhara difference $\psi(t, x) \ominus_H (-1)^{AL}_{F}\mathcal{I}^{\frac{1}{2}}_{0+} f(t, x, u(t, x))$ exists for all $(t, x) \in \Omega_4^1$. Similarly, we also point out there exists the difference

$$\psi(t, x) \ominus_H (-1)^{AL}_{F}\mathcal{I}^{\frac{1}{2}}_{0+} f(t, x, \nu(t, x)),$$

where $\nu(t, x) = \psi(t, x) \ominus_H (-1)^{AL}_{F}\mathcal{I}^{\frac{1}{2}}_{0+} f(t, x, u(t, x)), (t, x) \in \Omega_4^1$. From above remark, we see that the hypotheses of Theorem 4.1 are satisfied and therefore problem (31)-(32) has at least a (2)-integral solution on $\Omega_* \subset \Omega_4^1$.

Example 6.2. Consider the fractional fuzzy partial differential equation

$$^{C}_{F}\mathcal{D}^q u(t, x) = \frac{e^x + \sin x}{\Gamma(2 - q_1)\Gamma(2 - q_2)} u(t, x) \oplus \frac{\sqrt{tx} + (e^x + \sin x)(tx + 1)}{\Gamma(2 - q_1)\Gamma(2 - q_2)} C \tag{33}$$

for all $(t, x) \in \Omega_\infty^{\frac{\pi}{2}}$ with boundary conditions

$$u(t, 0) = C, t \in [0, +\infty), \ u(0, x) = C, x \in [0, \frac{\pi}{2}], \tag{34}$$

where $q = (\frac{1}{2}, \frac{1}{2})$, $\Omega_\infty^{\frac{\pi}{2}} = [0, +\infty) \times [0, \frac{\pi}{2}]$ and C is a fuzzy number.

For all $(t,x) \in \Omega_\infty^{\frac{\pi}{2}}$, we set

$$f(t,x,u(t,x)) = \frac{e^x + \sin x}{\Gamma(2-q_1)\Gamma(2-q_2)}u(t,x) \oplus \frac{\sqrt{tx} + (e^x + \sin x)(tx+1)}{\Gamma(2-q_1)\Gamma(2-q_2)}C.$$

It is clear that f satisfies Lipschitz condition for Lipschitz constant $L = \frac{e^{\frac{\pi}{2}}+1}{\Gamma(2-q_1)\Gamma(2-q_2)}$ and $f(t,x,\hat{0}) = \frac{\sqrt{tx} + (e^x + \sin x)(tx+1)}{\Gamma(2-q_1)\Gamma(2-q_2)}C$. Therefore we have

$$\begin{aligned} d_\infty(f(t,x,\hat{0}),\hat{0}) &\leq \frac{\sqrt{tx} + (e^x + \sin x)(tx+1)}{\Gamma(2-q_1)\Gamma(2-q_2)} d_\infty(C,\hat{0}) \\ &\leq \frac{\sqrt{\frac{\pi}{2}t} + (e^{\frac{\pi}{2}}+1)(\frac{\pi}{2}t+1)}{\Gamma(2-q_1)\Gamma(2-q_2)} d_\infty(C,\hat{0}) \\ &\leq \frac{\sqrt{\frac{\pi}{2}} + (e^{\frac{\pi}{2}}+1)}{\Gamma(2-q_1)\Gamma(2-q_2)} d_\infty(C,\hat{0}) e^{\frac{\pi}{2}t}. \end{aligned}$$

This shows that the condition $(\mathbf{A_2})$ holds with $c_1 = \frac{\pi}{2}$ and

$$M_1 = \frac{\sqrt{\frac{\pi}{2}} + (e^{\frac{\pi}{2}}+1)}{\Gamma(2-q_1)\Gamma(2-q_2)} d_\infty(C,\hat{0}).$$

Moreover, because $u(t,0) = \eta_1(t) = C,\ u(0,x) = \eta_2(x) = C$, the condition $(\mathbf{A_3})$ is satisfied. By Theorem 4.2, problem (33)-(34) has a unique (1)-integral solution on $\Omega_\infty^{\frac{\pi}{2}}$.

Example 6.3. Consider the following fractional fuzzy hyperbolic partial differential equation

$$ {}^C_F\mathcal{D}^q u(t,x) = \frac{1}{e^{t+x+5}+t+2}u(t,x),\ (t,x) \in \Omega_\infty^3 = [0,\infty) \times [0,3] \quad (35)$$

with conditions

$$u(t,0) = (t+1)C,\ t \in [0,\infty),\ u(0,x) = e^x C,\ x \in [0,3] \quad (36)$$

where $q = (q_1,q_2) \in [0,1) \times [0,1)$ and C is a fuzzy number.

For all $(t, x) \in \Omega^3_\infty$, set

$$f(t, x, u(t, x)) = \frac{1}{e^{t+x+5} + t + 2} u(t, x).$$

It is easy to see that

$$d_\infty(f(t, x, u(t, x)), f(t, x, v(t, x))) \leq \frac{1}{e^5 + 2} d_\infty(u(t, x), v(t, x))$$

and $f(t, x, \hat{0}) = \hat{0}$. Hence, f satisfies the hypotheses $(\mathbf{A_1})$ and $(\mathbf{A_2})$. Moreover, for all $(t, x) \in \Omega^3_\infty$, we have

$$d_\infty(u(t, 0), \hat{0}) \leq e^t d_\infty(C, \hat{0}) \text{ v } d_\infty(u(0, x), \hat{0}) \leq e^3 d_\infty(C, \hat{0}).$$

Thus the condition $(\mathbf{A_3})$ holds with $M_2 = d_\infty(C, \hat{0}), M_3 = e^3 d_\infty(C, \hat{0})$ and $c_5 = 1$. By Theorem 4.2, problem (35) -(36) has unique a (1)-integral solution on Ω^3_∞. In addition, by Theorem 5.1, problem (35)-(36) is Hyers-Ulam stable in type 1.

On other hand, if we choose $\Phi(t, x) = (t + 1)^{-a} e^{bx}$, $a > 1$, $b > 0$, for all $(t, x) \in \Omega^3_\infty$ then we have

$${}^{RL}I^q \Phi(t, x) = \frac{1}{\Gamma(q_1)\Gamma(q_2)} \int_0^t (t - s)^{q_1 - 1}(s + 1)^{-a} ds \int_0^x (x - z)^{q_2 - 1} e^{bz} dz.$$

From Proof of Lemma 3.4, we imply for $\varepsilon > 0$ arbitrary, there exists $C > 0$ such that

$$\int_0^x (x - z)^{q_2 - 1} e^{bz} dz \leq \frac{e^{bx}}{q_2} \Big[2 \Big(\frac{C}{b^{\frac{1}{1+\varepsilon}}} \Big)^{\frac{q_2}{2}} + \frac{1}{b} \Big(\frac{C}{b^{\frac{1}{1+\varepsilon}}} \Big)^{q_2} \Big].$$

Furthermore, by (30), one has the estimation

$$\int_0^t (t - s)^{q_1 - 1}(s + 1)^{-a} ds < \frac{1}{1 - q_1} + \frac{1}{q_1}.$$

Therefore

$$\begin{aligned} &{}^{RL}I^q \Phi(t, x) \\ &\leq \frac{1}{\Gamma(q_1 + 1)\Gamma(q_2 + 1)(1 - q_1)} \Big[2 \Big(\frac{C}{b^{\frac{1}{1+\varepsilon}}} \Big)^{\frac{q_2}{2}} + \frac{1}{b} \Big(\frac{C}{b^{\frac{1}{1+\varepsilon}}} \Big)^{q_2} \Big] (t + 1)^a \Phi(t, x). \end{aligned}$$

Thus there exist

$$m_\Phi = \frac{1}{\Gamma(q_1+1)\Gamma(q_2+1)(1-q_1)}\left[2\left(\frac{C}{b^{\frac{1}{1+\varepsilon}}}\right)^{\frac{q_2}{2}} + \frac{1}{b}\left(\frac{C}{b^{\frac{1}{1+\varepsilon}}}\right)^{q_2}\right] > 0$$

and $\nu = a$ such that ${}^{RL}I^q\Phi(t,x) \le m_\Phi e^{\nu t}\Phi(t,x)$. This shows that Φ satisfying the condition (26). By Theorem 5.2, problem (35) -(36) is Hyers-Ulam-Rassias stable in type 1 with respect to $\Phi(t,x) = (t+1)^{-a}e^{bx}$, $a > 1, b > 0, (t,x) \in \Omega^3_\infty$.

Example 6.4. *As an application of our results, we consider the fuzzy fractional partial differential equations*

$$ {}^{C}_{gH}D^q u(t,x) = \frac{e^{2019x} + x^5 + x^3 + \log_2 x + 2018}{(t+1)^3} u(t,x), \ (t,x) \in \Omega^1_\infty \tag{37}$$

with the conditions

$$u(t,0) = (t+1)e^t C, \ t \ge 0, \ u(0,x) = (x^2 + e^x + 23)C, \ x \in [0,1]. \tag{38}$$

where $q = (q_1, q_2) \in (0,1) \times (0,1)$ *and* $C = (3,4,2)$.

For all $(t,x) \in \Omega^1_\infty$, we set

$$f(t,x,u(t,x)) = \frac{e^{2019x} + x^5 + x^3 + \log_2 x + 1}{(t+1)^3} u(t,x).$$

It is clear that for $u \in C(\Omega^1_\infty, E)$, $f(t,x,u(t,x))$ is a continuous function and the condition $(\mathbf{G_1})$ is satisfied for $\beta = 3, L = e^{2019} + 2020$.

Moreover, since $\eta_1(t) = (t+1)e^t C$, $\eta_2(x) = (x^2+e^x+23)C$ for all $(t,x) \in \Omega^1_\infty$, we imply that the condition $(\mathbf{A_3})$ satisfies with $M_2 = d_\infty(C,\hat{0}), c_2 = 2$ and $M_3 = (e+24)d_\infty(C,\hat{0})$.

Therefore, by Theorem 5.3, the equilibrium point $\hat{0}$ of problem (37)-(38) is stable.

Conclusion

In this chapter, we study Goursat problem for fuzzy fractional partial differential equations. The achieved results include:

- Define the fractional derivative for fuzzy-valued two-variable functions. Based on the concept of generalized Hukuhara derivative and the definition of Riemann-Liouville integral for real functions, we define the Caputo g-derivatives for fuzzy functions with some illustrative examples.

- Prove the existence and uniqueness of integral solutions of the problem in finite and infinite domains with or without hypotheses of Lipschitz of the right hand side f. The results are obtained in both cases of Caputo g-derivatives.

- Prove some results on the stability of the problems such as

  + Ulam stability.
  + Lyapunov stability.

Besides the achieved results in this chapter, some related open issues need to be further studied.

1. Study fractional order fuzzy partial differential equations with distribution delay.

2. Study the nonlocal boundary value problem of fractional order fuzzy partial differential equations.

3. Study fuzzy evolution equations by semigroup methods [40].

References

[1] Ahmadian, A., Salahshour, S., Baleanu, D. (2015), Amirkhani, H., Yunus, R., Tau method for the numerical solution of a fuzzy fractional kinetic model and its application to the Oil Palm Frond as a promising source of xylose, *J. Comput. Phys.* 294, 562-584.

[2] Abbas, S., Benchohra, M., Petrusel, A. (2014), Ulam stability for partial fractional differential inclusions via Picard operators theory, *Electron. J. Qual. Theory Differ. Equ.*, 51, 1 - 13.

[3] Abbas, S., Benchohra, M. , N'Guérékata, G.M. (2012), *Topics in Fractional Differential Equations*, Springer, New York.

[4] Allahviranloo, T., Gouyandeh, Z., Armand, A., Hasanoglu, A. (2015), On fuzzy solutions for heat equation based on generalized Hukuhara differentiability, *Fuzzy Sets Syst.* 265, 1-23.

[5] Agarwal, R.P., Lakshmikantham, V., Nieto, J.J. (2010), On the concept of solution for fractional differential equations with uncertainty, *Nonlinear Anal.* 72, 2859-2862.

[6] Allahviranloo, T., Gouyandeh, Z., Armand, A. (2014), Fuzzy fractional differential equations under generalized fuzzy Caputo derivative, *J. Intell. Fuzzy Syst.*, 26, No. 3, 1481-1490.

[7] Arshad, S., Lupulescu, V. (2011), On the fractional differential equations with uncertainty, *Nonlinear Anal.*, 74, No.11, 3685 - 3693.

[8] Arshad, S. (2013), On existence and uniqueness of solution of fuzzy fractional differential equations, *Iran. J. Fuzzy Syst.* 6, 137-151.

[9] Bede, B., Stefanini, L. (2013), Generalized differentiability of fuzzy-valued functions, *Fuzzy Sets Syst.*, 230, 119-141.

[10] Cecconello, M.S., Leite, J., Bassanezi, R.C. (2017), Asymptotic analysis of continuous fuzzy flows, *Comp. Appl. Math.*, 36, No.4, 1681 -1697.

[11] Cecconello, M.S., Leite, J., Bassanezi, R.C., Brandão, A.J. (2015), Invariant and attractor sets for fuzzy dynamical systems, *Fuzzy Sets Syst.*, 265, 99-109.

[12] Cecconello, M.S., Bassanezi, R.C., Brandão, A.J., Leite, J. (2014), On the stability of fuzzy dynamical systems, *Fuzzy Sets Syst.*, 248, 106-121.

[13] Diamond, P., Kloeden, P. (1994), *Metric Spaces of Fuzzy Sets: Theory and Applications*, World Scientific, Singapore.

[14] Diamond, P. (2000), Stability and periodicity in fuzzy differential equations, *IEEE Trans. Fuzzy Syst.*, 8, No.5, 583-590.

[15] Diamond, P., Kloeden, P. (2000), *Metric Topology of Fuzzy Numbers and Fuzzy Analysis*. In: Dubois D., Prade H. (eds) Fundamentals of Fuzzy Sets. The Handbooks of Fuzzy Sets Series, vol 7. Springer, Boston, MA.

[16] Ding, Z., Kandel, A. (1997), Existence and stability of fuzzy differential equations, *J. Fuzzy Math.*, 5, 681-697.

[17] Hoa, N.V. (2015), Fuzzy fractional functional differential equations under Caputo gH-differentiability, *Commun. Nonlinear Sci. Numer. Simul.*, 22, No.1-3, 1134-1157.

[18] Hukuhara, M. (1967), Integration des applications measurables dont la valeur est un compact convexe, *Funkcialaj Ekvacioj*, 10, 205-223.

[19] Kaleva, O. (1987), Fuzzy differential equations, *Fuzzy Sets Syst.*, 24, 301-317.

[20] Khastan, A., Neito, J.J., Rodríguez-López, R. (2014), Fuzzy delay differential equations under generalized differentiability, *Inf. Sci.*, 275, 145 - 167.

[21] Khastan, A., Nieto, J.J., Rodríguez-López, R. (2014), Schauder fixed-point theorem in semilinear spaces and its application to fractional differential equations with uncertainty, *Fixed Point Theory Appl.*, 2014: 21, 1687-1812.

[22] Kilbas, A.A., Srivastava, H.M., Trujillo, J.J. (2006), *Theory and Applications of Fractional Differential Equations*, Elsevier Science, Amsterdam.

[23] Long, H.V. (2018), On random fuzzy fractional partial integro-differential equations under Caputo generalized Hukuhara differentiability, *Computational and Applied Mathematics* 37 (3), 2738-2765.

[24] Long, H.V., Son, N.T.K., Tam, H.T.T., Yao, J.C. (2017), Ulam stability for fractional partial integro-differential equation with uncertainty, *Acta Mathematica Vietnamica*, 42(4), 675-700.

[25] Long, H.V., Son, N.T.K., Tam, H.T.T. (2017), The solvability of fuzzy fractional partial differential equations under Caputo gH-differentiability, *Fuzzy Sets Syst.* 309, 35-63.

[26] Long, H.V., Son, N.T.K., Tam, H.T.T. (2015), Global existence of solutions to fuzzy partial hyperbolic functional differential equations with generalized Hukuhara derivatives, *J. Intell. Fuzzy Syst.* 29, 939-954.

[27] Long, H.V., Thao, H.T.P. (2018), Hyers-Ulam stability for nonlocal fractional partial integro-differential equation with uncertainty, *Journal of Intelligent and Fuzzy Systems* 34(1), 233-244.

[28] Malinowski, M.T. (2015), Random fuzzy fractional integral equations - Theoretical foundations, *Fuzzy Sets Syst.*, 265, 39-62.

[29] Mazandarani, M., Kamyad, A.V. (2013), Modified fractional Euler method for solving fuzzy fractional initial value problem, *Commun. Nonlinear Sci. Numer. Simul.* 18, 12-21.

[30] Miller, K. S. and Ross, B. (1993), *An Introduction to the Fractional Calculus and Differential Equations*, John Wiley & Sons, New York.

[31] Negoita, C., Ralescu, D. (1975), *Application of Fuzzy Sets to System Analysis*, Wiley, New York.

[32] Nieto , J.J. (1999), The Cauchy problem for continuous fuzzy differential equations, *Fuzzy Sets Syst.* 102, 259-262.

[33] Obloza, M. (1993), Hyers stability of the linear differential equation, *Rocznik Nauk.-Dydakt. Prace Mat.*, 13, 259-270.

[34] Pachpatte, B. G. (1998), *Inequalities for Differential and Integral Equations*, Academic Press, Inc., San Diego, CA.

[35] Romn-Flores, H., Roas-Medar, M. (2002), Embedding of level-continuous fuzzy sets on Banach spaces, *Inf. Sci.*, 144, No.1-4, 227-247.

[36] Salahshour, S., Ahmadian, A., Senu, N., Baleanu, D., Agarwal, P. (2015), On analytical solutions of the fractional differential equation with uncertainty, Application to the Basset Problem, *Entropy*, 17, No.2, 885-902.

[37] Samko, S.G., Kilbas, A.A., Marichev, O.I. (1993), *Fractional integrals and derivatives, Theory and Applications*, Gordon and Breach Science Publishers, Singapore.

[38] Shen, Y. (2015), On the Ulam stability of first order linear fuzzy differential equations under generalized differentiability, *Fuzzy Sets Syst.*, 280, 27-57.

[39] Shen, Y., Wang, F. (2016), A fixed point approach to the Ulam stability of fuzzy differential equations under generalized differentiability, *J. Intell. Fuzzy Syst.*, 30, No.6, 3253-3260.

[40] Son, N.T.K. (2018), A foundation on semigroups of operators defined on the set of triangular fuzzy numbers and its application to fuzzy fractional evolution equations, *Fuzzy Sets and Systems* 347, 1-28.

[41] Son, N.T.K., Tam, H.T.T. (2018), On the stability and global attractivity of solutions of fuzzy fractional partial differential equations, *J. Intell. Fuzzy Syst.* 35, No.3, 3797-3806.

[42] Stefanini, L. (2010), A generalization of Hukuhara difference and division for interval and fuzzy arithmetric, *Fuzzy Sets Syst.*, 161, 1564-1584.

[43] Stefanini, L., Bede, B. (2009), Generalized Hukuhara differentiability of interval-valued functions and interval differential equations, *Nonlinear Anal.*, 71, No.3-4, 1311-1328.

[44] Zada, A., Shah, O., Shah, R. (2015), Hyers-Ulam stability of nonautonomous systems in terms of boundedness of Cauchy problems, *Appl. Math. Comput.*, 271, 512-518.

ISBN: 978-1-53616-128-1

Chapter 9

A Foundation on Neutrosophic Calculus

Nguyen Thi Kim Son[1,*]***, Nguyen Phuong Dong***[2†] ***and Hoang Viet Long***[3‡]
[1] Faculty of Natural Science, Hanoi Metropolitan University, Hanoi, Vietnam
[2] Department of Mathematics, Hanoi Pedagogical University 2, Hanoi, Vietnam
[3] Faculty of Information Technology, People's Police University of Technology and Logistics, Bac Ninh, Vietnam

Abstract

This chapter is devoted to introducing some preliminaries on the space of single valued neutrosophic numbers (SVNNs) and granular calculus related to single valued neutrosophic functions. Firstly, we introduce some different types of SVNNs and their parametric form representations. Next, by using horizontal membership functions approach, we establish basic arithmetic operations on the space of SVNNs, that allows us to directly introduce the terms of neutrosophic function in usual mathematical formulas. Especially, thanks to the foundation of the metric defined on space of SVNNs induced from Hamming distance, we can define some

*Corresponding Author's Email: ntkson@daihocthudo.edu.vn
†Corresponding Author's Email: nguyenphuongdong@hpu2.edu.vn
‡Corresponding Author's Email: longhv08@gmail.com

background on the limit, derivative and integral of single valued neutrosophic functions. To illustrate for the obtained results, we represent some numerical examples.

1. Introduction

Uncertainty theory plays an important role in modeling real-world problems. However, there is a basic question regarding how we can use the uncertainty concepts in our mathematical modeling. Motivated by this reason, many researchers have defined and introduced many approaches to investigate the uncertainty theory, but it should be noted that there is no unique reorientation of uncertain parameters. Thus, the problems with uncertainty are always attractive topics for scientists and mathematicians. Here, we give some information about uncertainty and show how they differ from each other by using following flowchart (see Fig.1).

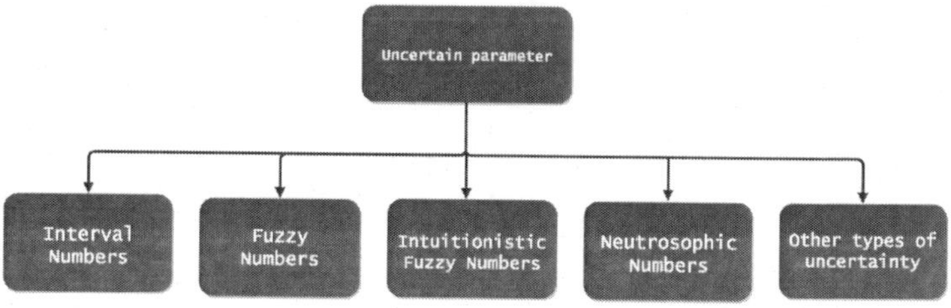

Figure 1. Flowchart for different uncertainty parameters.

Neutrosophic set (NS) and neutrosophic logic were first introduced by Smarandache [1], which are really extension of appeared earlier logic in the philosophical and mathematical aspects. Neutrosophic theory studies the objects whose values are neither true nor false, but in between, that can be called by neutral, indeterminate, unclear or contradictory quantities. Smarandache [2] laid the first attempt to study neutrosophic precalculus and neutrosophic calculus based on the existing calculus of interval analysis. Neutrosophic algebraic structures and neutrosophic cognitive maps were investigated in [3]. Neutrosophic measure, neutrosophic probability and statistics were studied in [4, 5, 6, 7]. Application of neutrosophic theory in decision making seems to be very successful. Ye [8, 9] proposed a multi-attribute decision making method using the correlation coefficient under single-valued neutrosophic environment. [10] further developed clustering method and decision making methods by similarity measures of SVNS. Meanwhile, [11] presented cross entropy measures of SVNNs and applied them to decision making (for more details, see [12, 13]).

Based on the horizontal membership functions approach, Mazandarani et. al. [14] introduced the concept of granular differentiability and applied to fuzzy differential systems, which can be considered as a particular scenario of neutrosophic dynamic systems. However, neutrosophic set theory and neutrosophic dynamic systems are still in the beginning stage of development. On the other hand, there are only some literature that have attained the first step of defining the distance between neutrosophic sets and neutrosophic numbers [10, 13] or some fundamental concepts in neutrosophic calculus [5, 2]. Until now, the studies on analysis structures and topological structures on the space of neutrosophic set and neutrosophic numbers have almost never appeared due to the disadvantages coming from the semilinearity structures of the space of neutrosophic numbers. To the best of our knowledge, there is rarely a suitable derivative concept defined for the class of neutrosophic functions. Hence, it leads to big challenges for researchers in studying the analysis properties as well as constructing dynamical systems on this space.

2. The Space of Neutrosophic Numbers

Let us recall from [1, 5] that a neutrosophic set (NS) $\mathcal{A}$ is defined in the universal of discourse X and can be represented in following form

$$\mathcal{A} = \{\langle x, \mathtt{T}_{\mathcal{A}}(x), \mathtt{I}_{\mathcal{A}}(x), \mathtt{F}_{\mathcal{A}}(x)\rangle : x \in X\}$$

where $\mathtt{T}_{\mathcal{A}} : X \to ^{-}]0,1[^{+}$ is denoted for the truth membership function representing the degree of confidence, $\mathtt{I}_{\mathcal{A}} : X \to ^{-}]0,1[^{+}$ is called the indeterminacy membership function which represents the degree of uncertainty and $\mathtt{F}_{\mathcal{A}} : X \to ^{-}]0,1[^{+}$ is called the falsity membership function which represents the degree of scepticism such that the following relation holds

$$0^{-} \leq \mathtt{T}_{\mathcal{A}}(x) + \mathtt{I}_{\mathcal{A}}(x) + \mathtt{F}_{\mathcal{A}}(x) \leq 3^{+}.$$

The most popular kind of neutrosophic sets is the single valued neutrosophic set, which is a neutrosophic set $\mathcal{A}$ with x is a single valued independent variable (see [16]), whose the truth, indeterminacy and falsity membership functions exhibit the following relation

$$0 \leq \mathtt{T}_{\mathcal{A}}(x) + \mathtt{I}_{\mathcal{A}}(x) + \mathtt{F}_{\mathcal{A}}(x) \leq 3.$$

Next, we give the concept of neutrosophic number, which was firstly introduced in [1]

Definition 2.1. *A single valued NS $\mathcal{A}$ defined on the universal set of real numbers $\mathbb{R}$ is said to be a SVNN or neutrosophic number (NN) for short if it satisfies following properties*

(i) $\mathcal{A}$ is neut-convex, i.e., the following inequality are fulfilled

$$\begin{aligned}
\mathtt{T}_{\mathcal{A}}\left(\lambda x_1 + (1-\lambda)x_2\right) &\geq \min\left\{\mathtt{T}_{\mathcal{A}}(x_1), \mathtt{T}_{\mathcal{A}}(x_2)\right\},\\
\mathtt{I}_{\mathcal{A}}\left(\lambda x_1 + (1-\lambda)x_2\right) &\leq \max\left\{\mathtt{I}_{\mathcal{A}}(x_1), \mathtt{I}_{\mathcal{A}}(x_2)\right\},\\
\mathtt{F}_{\mathcal{A}}\left(\lambda x_1 + (1-\lambda)x_2\right) &\leq \max\left\{\mathtt{F}_{\mathcal{A}}(x_1), \mathtt{F}_{\mathcal{A}}(x_2)\right\},
\end{aligned}$$

for each $\lambda \in [0, 1]$ and x_1, $x_2 \in \mathbb{R}$.

(ii) $\mathcal{A}$ is neut-normal, i.e., there exist three elements a_0, b_0 and $c_0 \in \mathbb{R}$ such that $\mathtt{T}_{\mathcal{A}}(a_0) = 1$, $\mathtt{I}_{\mathcal{A}}(b_0) = 1$ and $\mathtt{F}_{\mathcal{A}}(c_0) = 1$.

Definition 2.2. *[5] The (α, β, γ) - cut (or level set) of a SVNN $\mathcal{A}$, denoted by $\mathcal{A}_{(\alpha,\beta,\gamma)}$, is defined by*

$$\mathcal{A}_{(\alpha,\beta,\gamma)} = \{x \in X : \mathtt{T}_{\mathcal{A}}(x) \geq \alpha, \mathtt{I}_{\mathcal{A}}(x) \leq \beta, \mathtt{F}_{\mathcal{A}}(x) \leq \gamma\},$$

where α, β, $\gamma \in [0, 1]$ such that $\alpha + \beta + \gamma \leq 3$.

3. Single Valued Triangular Neutrosophic Number

In this section, we consider a special type of SVNN with a wide application in many aspects of engineering, namely single valued triangular neutrosophic number.

Definition 3.1. *A single valued triangular NN is given by*

$$\mathcal{A} = \langle [(p_1, q_1, r_1); \alpha], [(p_2, q_2, r_2); \beta], [(p_3, q_3, r_3); \gamma] \rangle,$$

where α, β, $\gamma \in [0, 1]$ *and the truth membership function* $\mathtt{T}_{\mathcal{A}} : \mathbb{R} \to [0, \alpha]$, *the indeterminacy membership function* $\mathtt{I}_{\mathcal{A}} : \mathbb{R} \to [\beta, 1]$ *and the falsity membership function* $\mathtt{F}_{\mathcal{A}} : \mathbb{R} \to [\gamma, 1]$ *satisfy following condition*

$$0 \leq \mathtt{T}_{\mathcal{A}}(x) + \mathtt{I}_{\mathcal{A}}(x) + \mathtt{F}_{\mathcal{A}}(x) \leq 3 \textit{ for all } x \in \mathcal{A}.$$

We denote by $\mathcal{E}$ the set of all single valued triangular NNs. Then, based on the dependence between quantities: the truth, the indeterminacy and the falsity, we can classify the set $\mathcal{E}$ of single valued triangular NNs into three of the following types:

3.1. Linear Single Valued Triangular Neutrosophic Number

Here, we will introduce some types of linear single valued triangular neutrosophic numbers.

a. Single Valued Triangular NN of Type 1: Here, the quantities of the truth, indeterminacy and falsity are not dependent. Then, a linear single valued triangular neutrosophic number of type 1 is defined as

$$\mathcal{A} = \langle p_1, q_1, r_1; p_2, q_2, r_2; p_3, q_3, r_3 \rangle,$$

with respect to the membership functions

$$\mathtt{T}_{\mathcal{A}}(x) = \begin{cases} \frac{x-p_1}{q_1-p_1} & p_1 \leq x \leq q_1, \\ 1 & x = q_1, \\ \frac{r_1-x}{r_1-q_1} & q_1 \leq x \leq r_1, \\ 0 & \text{otherwise}, \end{cases}$$

$$\mathtt{I}_{\mathcal{A}}(x) = \begin{cases} \frac{q_2-x}{q_2-p_2} & p_2 \leq x \leq q_2, \\ 0 & x = q_2, \\ \frac{x-q_2}{r_2-q_2} & q_2 \leq x \leq r_2, \\ 1 & \text{otherwise}, \end{cases}$$

$$\mathtt{F}_{\mathcal{A}}(x) = \begin{cases} \frac{q_3-x}{q_3-p_3} & p_3 \leq x \leq q_3, \\ 0 & x = q_3, \\ \frac{x-q_3}{r_3-q_3} & q_3 \leq x \leq r_3, \\ 1 & \text{otherwise}, \end{cases}$$

Then, the parametric form of $\mathcal{A}$ is given as

$$\mathcal{A}_{(\alpha,\beta,\gamma)} = \left[\mathtt{T}^-_{\mathcal{A}}(\alpha), \mathtt{T}^+_{\mathcal{A}}(\alpha); \mathtt{I}^-_{\mathcal{A}}(\beta), \mathtt{I}^+_{\mathcal{A}}(\beta); \mathtt{F}^-_{\mathcal{A}}(\gamma), \mathtt{F}^+_{\mathcal{A}}(\gamma)\right], \qquad (1)$$

where the parameters α, β, $\gamma \in [0, 1]$ satisfy $0 \leq \alpha + \beta + \gamma \leq 3$ and

$$\begin{aligned}
\mathtt{T}^-_{\mathcal{A}}(\alpha) &= p_1 + \alpha(q_1 - p_1), \\
\mathtt{T}^+_{\mathcal{A}}(\alpha) &= r_1 - \alpha(r_1 - q_1), \\
\mathtt{I}^-_{\mathcal{A}}(\beta) &= q_2 - \beta(q_2 - p_2), \\
\mathtt{I}^+_{\mathcal{A}}(\beta) &= q_2 + \beta(r_2 - q_2), \\
\mathtt{F}^-_{\mathcal{A}}(\gamma) &= q_3 - \gamma(q_3 - p_3), \\
\mathtt{F}^+_{\mathcal{A}}(\gamma) &= q_3 + \gamma(r_3 - q_3).
\end{aligned}$$

To illustrate for above result, let us consider following example

Example 3.1. *Consider a single valued triangular neutrosophic number* $\mathcal{A} = (8, 14, 20; 12, 16, 22; 10, 15, 24)$. *By using the formula (1), we obtain the parametric form of* $\mathcal{A}$ *is as follows*

$$\mathcal{A}_{(\alpha,\beta,\gamma)} = [8 + 6\alpha, 20 - 6\alpha; 16 - 4\beta, 16 + 6\beta; 15 - 5\gamma, 15 + 9\gamma],$$

for all α, $\beta, \gamma \in [0, 1]$.

In Table 1, we present some values of $\mathtt{T}^-_{\mathcal{A}}(\alpha)$, $\mathtt{T}^+_{\mathcal{A}}(\alpha)$, $\mathtt{I}^-_{\mathcal{A}}(\beta)$, $\mathtt{I}^+_{\mathcal{A}}(\beta)$, $\mathtt{F}^-_{\mathcal{A}}(\gamma)$, $\mathtt{F}^+_{\mathcal{A}}(\gamma)$ *of the neutrosophic number* $\mathcal{A}$ *at some concrete levels and then, we also show the graphical representation of this number in Figure 2.*

Table 1. Values of $\mathtt{T}^-_{\mathcal{A}}(\alpha), \mathtt{T}^+_{\mathcal{A}}(\alpha), \mathtt{I}^-_{\mathcal{A}}(\beta), \mathtt{I}^+_{\mathcal{A}}(\beta), \mathtt{F}^-_{\mathcal{A}}(\gamma), \mathtt{F}^+_{\mathcal{A}}(\gamma)$

α, β, γ	$\mathtt{T}^-_{\mathcal{A}}(\alpha)$	$\mathtt{T}^+_{\mathcal{A}}(\alpha)$	$\mathtt{I}^-_{\mathcal{A}}(\beta)$	$\mathtt{I}^+_{\mathcal{A}}(\beta)$	$\mathtt{F}^-_{\mathcal{A}}(\gamma)$	$\mathtt{F}^+_{\mathcal{A}}(\gamma)$
0	8	20	16	16	15	15
0.1	8.6	19.4	15.6	16.6	14.5	15.9
0.2	9.2	18.8	15.2	17.2	14	16.8
0.3	9.8	18.2	14.8	17.8	13.5	17.7
0.4	10.4	17.6	14.4	18.4	13	18.6
0.5	11	17	14	19	12.5	19.5
0.6	11.6	16.4	13.6	19.6	12	20.4
0.7	12.2	15.8	13.2	20.2	11.5	21.3
0.8	12.8	15.2	12.8	20.8	11	22.2
0.9	13.4	14.6	12.4	21.4	10.5	23.1
1	14	14	12	22	10	24

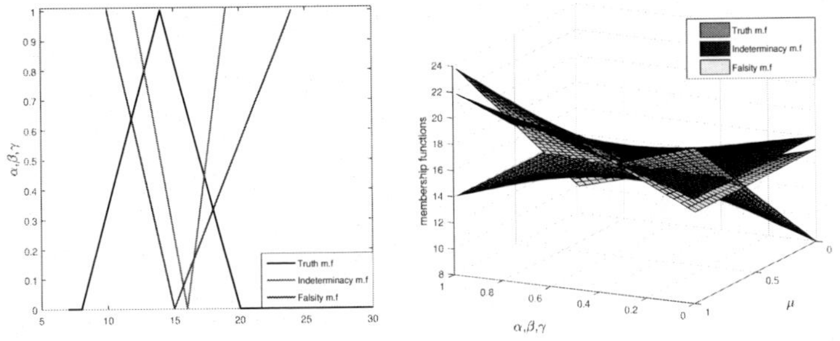

Figure 2. Membership functions of NN of type 1 $\mathcal{A}$.

b. Single Valued Triangular NN of Type 2: In this type of number, two quantities: indeterminacy membership function and falsity membership function are dependent. Then, a linear single valued triangular NN of type 2 is defined by $\mathcal{A} = \langle p_1, q_1, r_1; p_2, q_2, r_2; \beta_{neu}; \gamma_{neu}\rangle$, whose membership functions are given by

$$\mathtt{T}_{\mathcal{A}}(x) = \begin{cases} \frac{x-p_1}{q_1-p_1} & p_1 \le x \le q_1, \\ 1 & x = q_1, \\ \frac{r_1-x}{r_1-q_1} & q_1 \le x \le r_1, \\ 0 & \text{otherwise,} \end{cases}$$

$$\mathtt{I}_{\mathcal{A}}(x) = \begin{cases} \frac{q_2-x+\beta_{neu}(x-p_2)}{q_2-p_2} & p_2 \le x \le q_2, \\ \beta_{neu} & x = q_2, \\ \frac{x-q_2+\beta_{neu}(r_2-x)}{r_2-q_2} & q_2 \le x \le r_2, \\ 1 & \text{otherwise,} \end{cases}$$

$$\mathtt{F}_{\mathcal{A}}(x) = \begin{cases} \frac{q_2-x+\gamma_{neu}(x-p_2)}{q_2-p_2} & p_2 \le x \le q_2, \\ \gamma_{neu} & x = q_2, \\ \frac{x-q_2+\gamma_{neu}(r_2-x)}{r_2-q_2} & q_2 \le x \le r_2, \\ 1 & \text{otherwise,} \end{cases}$$

Thus, we can give the parametric form of $\mathcal{A}$ as follows

$$\mathcal{A}_{(\alpha,\beta,\gamma)} = \left[\mathtt{T}^-_{\mathcal{A}}(\alpha), \mathtt{T}^+_{\mathcal{A}}(\alpha); \mathtt{I}^-_{\mathcal{A}}(\beta), \mathtt{I}^+_{\mathcal{A}}(\beta); \mathtt{F}^-_{\mathcal{A}}(\gamma), \mathtt{F}^+_{\mathcal{A}}(\gamma)\right],$$

where

$$\begin{aligned}
\mathrm{T}^{-}_{\mathcal{A}}(\alpha) &= p_1 + \alpha(q_1 - p_1), \\
\mathrm{T}^{+}_{\mathcal{A}}(\alpha) &= r_1 - \alpha(r_1 - q_1), \\
\mathrm{I}^{-}_{\mathcal{A}}(\beta) &= \frac{q_2 - \beta_{neu}p_2 - \beta(q_2 - p_2)}{1 - \beta_{neu}}, \\
\mathrm{I}^{+}_{\mathcal{A}}(\beta) &= \frac{q_2 - \beta_{neu}r_2 + \beta(r_2 - q_2)}{1 - \beta_{neu}}, \\
\mathrm{F}^{-}_{\mathcal{A}}(\gamma) &= \frac{q_2 - \gamma_{neu}p_2 - \gamma(q_2 - p_2)}{1 - \gamma_{neu}}, \\
\mathrm{F}^{+}_{\mathcal{A}}(\gamma) &= \frac{q_2 - \gamma_{neu}r_2 + \gamma(r_2 - q_2)}{1 - \gamma_{neu}},
\end{aligned}$$

where $\alpha \in [0, 1]$, $\beta \in [\beta_{neu}, 1]$ and $\gamma \in [\gamma_{neu}, 1]$ satisfy

$$0 \leq \alpha + \beta + \gamma \leq 3.$$

Now, we will illustrate the obtained result by following example

Example 3.2. *Let* $\mathcal{A} = (8, 14, 20; 12, 16, 22; 0.5; 0.6)$ *be a SVNN. As a result, we have its parametric form is given by*

$$\mathcal{A}_{(\alpha,\beta,\gamma)} = [8 + 6\alpha, 20 - 6\alpha; 20 - 8\beta, 10 + 12\beta; 22 - 10\gamma, 7 + 15\gamma],$$

where $\alpha \in [0, 1]$, $\beta \in [0.5, 1]$, $\gamma \in [0.6, 1]$.

In Table 2, some values of $\mathrm{T}^{-}_{\mathcal{A}}(\alpha)$, $\mathrm{T}^{+}_{\mathcal{A}}(\alpha)$, $\mathrm{I}^{-}_{\mathcal{A}}(\beta)$, $\mathrm{I}^{+}_{\mathcal{A}}(\beta)$, $\mathrm{F}^{-}_{\mathcal{A}}(\gamma)$, $\mathrm{F}^{+}_{\mathcal{A}}(\gamma)$ *are given and then, the graphical representation of the neutrosophic number* $\mathcal{A}$ *is shown in Figure 3.*

Table 2. Values of $\mathrm{T}^{-}_{\mathcal{A}}(\alpha)$, $\mathrm{T}^{+}_{\mathcal{A}}(\alpha)$, $\mathrm{I}^{-}_{\mathcal{A}}(\beta)$, $\mathrm{I}^{+}_{\mathcal{A}}(\beta)$, $\mathrm{F}^{-}_{\mathcal{A}}(\gamma)$, $\mathrm{F}^{+}_{\mathcal{A}}(\gamma)$

α, β, γ	$\mathrm{T}^{-}_{\mathcal{A}}(\alpha)$	$\mathrm{T}^{+}_{\mathcal{A}}(\alpha)$	$\mathrm{I}^{-}_{\mathcal{A}}(\beta)$	$\mathrm{I}^{+}_{\mathcal{A}}(\beta)$	$\mathrm{F}^{-}_{\mathcal{A}}(\gamma)$	$\mathrm{F}^{+}_{\mathcal{A}}(\gamma)$
0	8	20				
0.1	8.6	19.4				
0.2	9.2	18.8				
0.3	9.8	18.2				
0.4	10.4	17.6				
0.5	11	17	16	16		
0.6	11.6	16.4	17.2	19.6	16	16
0.7	12.2	15.8	18.4	20.2	15	17.5
0.8	12.8	15.2	19.6	20.8	14	19
0.9	13.4	14.6	20.8	21.4	13	20.5
1	14	14	12	22	12	22

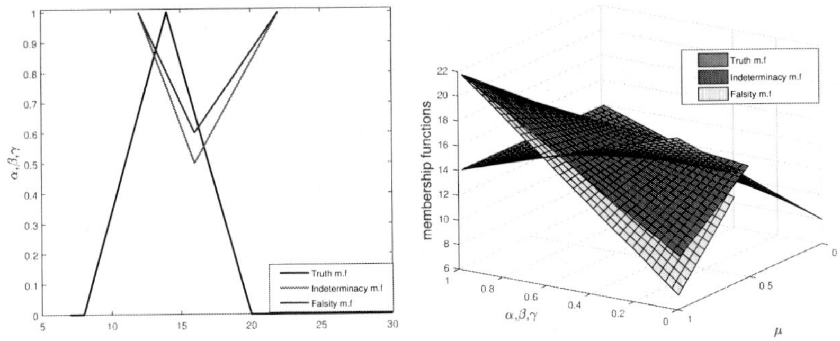

Figure 3. Membership functions of NN of type 2 $\mathcal{A}$.

c. Single Valued Triangular NN of Type 3: In this type of neutrosophic numbers, the quantities: the truth, indeterminacy and falsity membership function are dependent. Then, a single valued triangular NN of type 3 is defined as $\mathcal{A} = \langle p_1, q_1, r_1; \alpha_{neu}; \beta_{neu}; \gamma_{neu} \rangle$, whose membership functions are defined as follows

$$\mathtt{T}_{\mathcal{A}}(x) = \begin{cases} \alpha_{neu}\frac{x-p_1}{q_1-p_1} & p_1 \le x \le q_1, \\ \alpha_{neu} & x = q_1, \\ \alpha_{neu}\frac{r_1-x}{r_1-q_1} & q_1 \le x \le r_1, \\ 0 & \text{otherwise}, \end{cases}$$

$$\mathtt{I}_{\mathcal{A}}(x) = \begin{cases} \frac{q_1-x+\beta_{neu}(x-p_1)}{q_1-p_1} & p_1 \le x \le q_1, \\ \beta_{neu} & x = q_1, \\ \frac{x-q_1+\beta_{neu}(r_1-x)}{r_1-q_1} & q_1 \le x \le r_1, \\ 1 & \text{otherwise}, \end{cases}$$

$$\mathtt{F}_{\mathcal{A}}(x) = \begin{cases} \frac{q_1-x+\gamma_{neu}(x-p_1)}{q_1-p_1} & p_1 \le x \le q_1, \\ \gamma_{neu} & x = q_1, \\ \frac{x-q_1+\gamma_{neu}(r_1-x)}{r_1-q_1} & q_1 \le x \le r_1, \\ 1 & \text{otherwise}, \end{cases}$$

Here, the parametric form of the number $\mathcal{A}$ is

$$\mathcal{A}_{(\alpha,\beta,\gamma)} = \left[\mathtt{T}^-_{\mathcal{A}}(\alpha), \mathtt{T}^+_{\mathcal{A}}(\alpha); \mathtt{I}^-_{\mathcal{A}}(\beta), \mathtt{I}^+_{\mathcal{A}}(\beta); \mathtt{F}^-_{\mathcal{A}}(\gamma), \mathtt{F}^+_{\mathcal{A}}(\gamma)\right],$$

where

$$\begin{aligned}
\mathtt{T}^-_{\mathcal{A}}(\alpha) &= p_1 + \frac{\alpha}{\alpha_{neu}}(q_1 - p_1),\\
\mathtt{T}^+_{\mathcal{A}}(\alpha) &= r_1 - \frac{\alpha}{\alpha_{neu}}(r_1 - q_1),\\
\mathtt{I}^-_{\mathcal{A}}(\beta) &= \frac{q_1 - \beta_{neu}p_1 - \beta(q_1 - p_1)}{1 - \beta_{neu}},\\
\mathtt{I}^+_{\mathcal{A}}(\beta) &= \frac{q_1 - \beta_{neu}r_1 + \beta(r_1 - q_1)}{1 - \beta_{neu}},\\
\mathtt{F}^-_{\mathcal{A}}(\gamma) &= \frac{q_1 - \gamma_{neu}p_1 - \gamma(q_1 - p_1)}{1 - \gamma_{neu}},\\
\mathtt{F}^+_{\mathcal{A}}(\gamma) &= \frac{q_1 - \gamma_{neu}r_1 + \gamma(r_1 - q_1)}{1 - \gamma_{neu}},
\end{aligned}$$

where $\alpha \in [0, \alpha_{neu}]$, $\beta \in [\beta_{neu}, 1]$ and $\gamma \in [\gamma_{neu}, 1]$ satisfy

$$0 \le \alpha + \beta + \gamma \le 3.$$

Finally, we will illustrate the obtained result by following example

Example 3.3. *Let $\mathcal{A} = (14, 16, 22; 0.5; 0.8; 0.7)$ be a SVNN. Then, we have that the parametric form of $\mathcal{A}$ is given as*

$$\mathcal{A}_{(\alpha,\beta,\gamma)} = \left[14 + 4\alpha, 22 - 12\alpha; 24 - 10\beta, -8 + 30\beta; \frac{62}{3} - \frac{20}{3}\gamma, 2 + 20\gamma\right],$$

for $\alpha \in [0, 0.5]$, $\beta \in [0.8, 1]$, $\gamma \in [0.7, 1]$.

In Table 3, some values of $\mathtt{T}^-_{\mathcal{A}}(\alpha)$, $\mathtt{T}^+_{\mathcal{A}}(\alpha)$, $\mathtt{I}^-_{\mathcal{A}}(\beta)$, $\mathtt{I}^+_{\mathcal{A}}(\beta)$, $\mathtt{F}^-_{\mathcal{A}}(\gamma)$, $\mathtt{F}^+_{\mathcal{A}}(\gamma)$ are given and then, we present graphical representation of the number $\mathcal{A}$ in Figure 4.

Table 3. Values of $\mathtt{T}^-_{\mathcal{A}}(\alpha)$, $\mathtt{T}^+_{\mathcal{A}}(\alpha)$, $\mathtt{I}^-_{\mathcal{A}}(\beta)$, $\mathtt{I}^+_{\mathcal{A}}(\beta)$, $\mathtt{F}^-_{\mathcal{A}}(\gamma)$, $\mathtt{F}^+_{\mathcal{A}}(\gamma)$

α, β, γ	$\mathtt{T}^-_{\mathcal{A}}(\alpha)$	$\mathtt{T}^+_{\mathcal{A}}(\alpha)$	$\mathtt{I}^-_{\mathcal{A}}(\beta)$	$\mathtt{I}^+_{\mathcal{A}}(\beta)$	$\mathtt{F}^-_{\mathcal{A}}(\gamma)$	$\mathtt{F}^+_{\mathcal{A}}(\gamma)$
0	14	22				
0.1	14.4	20.8				
0.2	14.8	19.6				
0.3	15.2	18.4				
0.4	15.6	17.2				
0.5	16	16				
0.6						
0.7					16	16
0.8			16	16	15.333	18
0.9			15	19	14.667	20
1			14	22	14	22

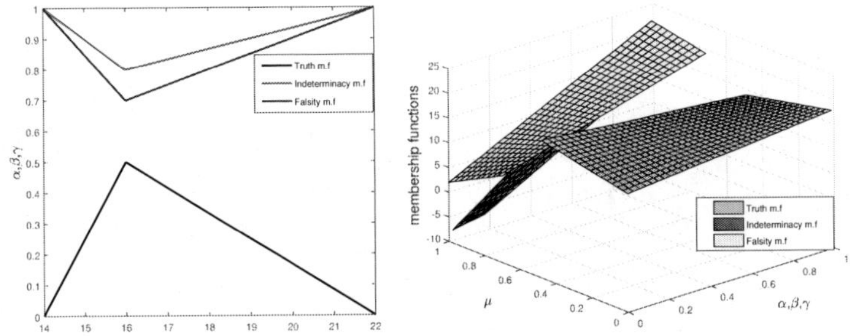

Figure 4. Membership functions of NN of type 3 $\mathcal{A}$.

Remark 3.1. *From the definition of the truth, the indeterminacy and the falsity membership functions, we can see that*

- *The truth membership function* $\mathtt{T}_{\mathcal{A}}(x)$ *increases with a constant rate for each* $x \in [p_1, q_1]$ *and decreases for each* $x \in [q_1, r1]$.
- *The indeterminacy membership function* $\mathtt{I}_{\mathcal{A}}(x)$ *decreases for each* $x \in [p_2, q_2]$ *and increases for each* $x \in [q_2, r2]$, *while the falsity membership function* $\mathtt{F}_{\mathcal{A}}(x)$ *decreases in the interval* $[p_3, q_3]$ *and increases in the interval* $[q_3, r3]$.

3.2. Non-Linear Single Valued Triangular Neutrosophic Number

The generalized form of a non-linear single valued triangular neutrosophic number can be given as an element

$$\mathcal{A} = \langle p_1, q_1, r_1; p_2, q_2, r_2; p_3, q_3, r_3; \alpha_{neu}; \beta_{neu}; \gamma_{neu} \rangle$$

whose truth, indeterminacy, and falsity membership functions are defined as follows

$$\mathrm{T}_{\mathcal{A}}(x) = \begin{cases} \alpha_{neu}\left(\frac{x-p_1}{q_1-p_1}\right)^{a_1} & p_1 \le x \le q_1, \\ \alpha_{neu} & x = q_1, \\ \alpha_{neu}\left(\frac{r_1-x}{r_1-q_1}\right)^{a_2} & q_1 \le x \le r_1, \\ 0 & \text{otherwise,} \end{cases}$$

$$\mathrm{I}_{\mathcal{A}}(x) = \begin{cases} \beta_{neu}\left(\frac{q_2-x}{q_2-p_2}\right)^{b_1} & p_2 \le x \le q_2, \\ \beta_{neu} & x = q_2, \\ \beta_{neu}\left(\frac{x-q_2}{r_2-q_2}\right)^{b_2} & q_2 \le x \le r_2, \\ 1 & \text{otherwise,} \end{cases}$$

$$\mathrm{F}_{\mathcal{A}}(x) = \begin{cases} \gamma_{neu}\left(\frac{q_3-x}{q_3-p_3}\right)^{c_1} & p_3 \le x \le q_3, \\ \gamma_{neu} & x = q_3, \\ \gamma_{neu}\left(\frac{x-q_3}{r_3-q_3}\right)^{c_2} & q_3 \le x \le r_3, \\ 1 & \text{otherwise,} \end{cases}$$

To illustrate for the concept of non-linear single valued triangular neutrosophic number, let us consider following example

Example 3.4. *In this example, we consider a non-linear single valued triangular neutrosophic number* $\mathcal{A}$

$$\mathcal{A} = (10, 15, 20; 14, 16, 22; 12, 15, 19),$$

where the parameter $a_1 = a_2 = 2$, $b_1 = b_2 = 1$ *and* $c_1 = c_2 = \frac{1}{2}$. *Here, for simplicity, we only consider the case* $\alpha_{neu} = \beta_{neu} = \gamma_{neu} = 1$.

Then, we can present the truth, the indeterminacy and the falsity member-

hip functions of the number $\mathcal{A}$ as follows

$$\mathtt{T}_{\mathcal{A}}(x) = \begin{cases} \left(\dfrac{x-10}{5}\right)^2 & 10 \leq x \leq 15, \\ 1 & x = 15, \\ \left(\dfrac{20-x}{5}\right)^2 & 15 \leq x \leq 20, \\ 0 & \textit{otherwise}, \end{cases}$$

$$\mathtt{I}_{\mathcal{A}}(x) = \begin{cases} \dfrac{16-x}{2} & 14 \leq x \leq 16, \\ 0 & x = 16, \\ \dfrac{x-16}{6} & 16 \leq x \leq 22, \\ 1 & \textit{otherwise}, \end{cases}$$

$$\mathtt{F}_{\mathcal{A}}(x) = \begin{cases} \left(\dfrac{15-x}{3}\right)^{\frac{1}{2}} & 12 \leq x \leq 15, \\ 0 & x = 15, \\ \left(\dfrac{x-15}{4}\right)^{\frac{1}{2}} & 15 \leq x \leq 19, \\ 1 & \textit{otherwise}, \end{cases}$$

The parametric form of $\mathcal{A}$ is given as

$$\mathcal{A}_{(\alpha,\beta,\gamma)} = \left[\mathtt{T}^-_{\mathcal{A}}(\alpha), \mathtt{T}^+_{\mathcal{A}}(\alpha); \mathtt{I}^-_{\mathcal{A}}(\beta), \mathtt{I}^+_{\mathcal{A}}(\beta); \mathtt{F}^-_{\mathcal{A}}(\gamma), \mathtt{F}^+_{\mathcal{A}}(\gamma)\right],$$

where the parameters α, β, $\gamma \in [0,1]$ satisfy $0 \leq \alpha+\beta+\gamma \leq 3$ and

$$\begin{aligned} \mathtt{T}^-_{\mathcal{A}}(\alpha) &= 10 + 5\sqrt{\alpha}, \\ \mathtt{T}^+_{\mathcal{A}}(\alpha) &= 20 - 5\sqrt{\alpha}, \\ \mathtt{I}^-_{\mathcal{A}}(\beta) &= 16 - 2\beta, \\ \mathtt{I}^+_{\mathcal{A}}(\beta) &= 16 + 6\beta, \\ \mathtt{F}^-_{\mathcal{A}}(\gamma) &= 15 - 3\gamma^2, \\ \mathtt{F}^+_{\mathcal{A}}(\gamma) &= 15 + 4\gamma^2. \end{aligned}$$

Finally, the graphical representation of $\mathcal{A}$ can be shown in Figure 5.

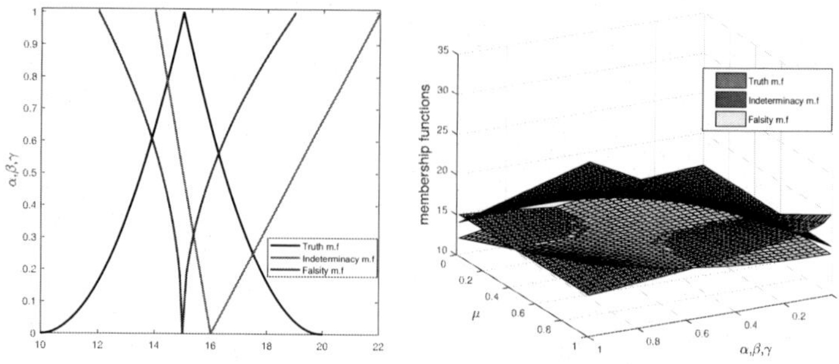

Figure 5. Membership functions of non-linear single valued triangular neutrosophic number $\mathcal{A}$.

Remark 3.2. *In special case, if* $a_1 = a_2 = b_1 = b_2 = c_1 = c_2 = 1$, *the non-linear single valued triangular neutrosophic number will be converted into the linear single valued triangular neutrosophic number and additionally, the type of this linear single valued triangular neutrosophic number is determined by the dependent of three quantities: the truth, the indeterminacy and the falsity membership functions.*

4. Granular Presentation of Single Valued Triangular Neutrosophic Number

Here, the concept of arithmetic operations on the space of SVNNs is defined via their horizontal membership functions. This original idea was introduced by Piegat et al. (see [17, 18]) and developed for granular differentiability of fuzzy-valued functions by Mazandarani et al. [14]. Especially, thanks to this approach, we can define the granular difference between neutrosophic numbers - one important step to define further differentiability of neutrosophic functions as well as neutrosophic differential equations and other applications.

4.1. Horizontal Membership Function

Definition 4.1. *[19] Let $\mathcal{A}$ be a SVNN whose parametric form is given by*

$$\mathcal{A}_{(\alpha,\beta,\gamma)} = \left[\mathtt{T}^-_{\mathcal{A}}(\alpha), \mathtt{T}^+_{\mathcal{A}}(\alpha); \mathtt{I}^-_{\mathcal{A}}(\beta), \mathtt{I}^+_{\mathcal{A}}(\beta); \mathtt{F}^-_{\mathcal{A}}(\gamma), \mathtt{F}^+_{\mathcal{A}}(\gamma)\right].$$

Then, we can represent the horizontal membership function (HMF) of the number $\mathcal{A}$ as an element $A^{gr}(\alpha,\beta,\gamma,\mu)$, where the function

$$A^{gr} : [0, \alpha_{neu}] \times [\beta_{neu}, 1] \times [\gamma_{neu}, 1] \times [0,1]^3 \to [a_1, c_1] \times [a_2, c_2] \times [a_3, c_3],$$

maps an element $(\alpha,\beta,\gamma,\mu)$ into $(x_\alpha(\mu_1), x_\beta(\mu_2), x_\gamma(\mu_3)) \in \mathbb{R}^3$.

Here, the notion "gr" represents for the granular information that are contained in $(x_\alpha, x_\beta, x_\gamma) \in [a_1, c_1] \times [a_2, c_2] \times [a_3, c_3]$ and $\mu \in [0,1]^3$ standing for μ_1, μ_2, μ_3 is called relative-distance-measure (RDM for short) variables.

In particular, we have $A^{gr}(\alpha,\beta,\gamma,\mu) = (x_\alpha(\mu_1), x_\beta(\mu_2), x_\gamma(\mu_3))$, where

$$\begin{aligned}
x_\alpha(\mu_1) &:= \mathtt{T}^{gr}_{\mathcal{A}}(\alpha,\mu_1) = \mathtt{T}^-_{\mathcal{A}}(\alpha) + \left(\mathtt{T}^+_{\mathcal{A}}(\alpha) - \mathtt{T}^-_{\mathcal{A}}(\alpha)\right)\mu_1,\\
x_\beta(\mu_2) &:= \mathtt{I}^{gr}_{\mathcal{A}}(\beta,\mu_2) = \mathtt{I}^-_{\mathcal{A}}(\beta) + \left(\mathtt{I}^+_{\mathcal{A}}(\beta) - \mathtt{I}^-_{\mathcal{A}}(\beta)\right)\mu_2,\\
x_\gamma(\mu_3) &:= \mathtt{F}^{gr}_{\mathcal{A}}(\gamma,\mu_3) = \mathtt{F}^-_{\mathcal{A}}(\gamma) + \left(\mathtt{F}^+_{\mathcal{A}}(\gamma) - \mathtt{F}^-_{\mathcal{A}}(\gamma)\right)\mu_3.
\end{aligned}$$

Proposition 4.1. *[19] The horizontal membership function of a neutrosophic number $\mathcal{A}$ is denoted by $\mathcal{H}(\mathcal{A}) \triangleq A^{gr}(\alpha,\beta,\gamma,\mu)$. Moreover, the $(\alpha,\beta,\gamma)-$cuts of the number $\mathcal{A}$ can be obtained by using following inverse transformation*

$$\begin{aligned}
\mathcal{A}_{(\alpha,\beta,\gamma)} &:= \mathcal{H}^{-1}(A^{gr}(\alpha,\beta,\gamma,\mu))\\
&= \left\{\left[\inf_{\xi\geq\alpha}\min_{\mu_1}\mathtt{T}^{gr}_{\mathcal{A}}(\xi,\mu_1), \sup_{\xi\geq\alpha}\max_{\mu_1}\mathtt{T}^{gr}_{\mathcal{A}}(\xi,\mu_1)\right],\right.\\
&\qquad \left[\inf_{\xi\geq\beta}\min_{\mu_2}\mathtt{I}^{gr}_{\mathcal{A}}(\xi,\mu_2), \sup_{\xi\geq\beta}\max_{\mu_2}\mathtt{I}^{gr}_{\mathcal{A}}(\xi,\mu_2)\right],\\
&\qquad \left.\left[\inf_{\xi\geq\gamma}\min_{\mu_3}\mathtt{F}^{gr}_{\mathcal{A}}(\xi,\mu_3), \sup_{\xi\geq\gamma}\max_{\mu_3}\mathtt{F}^{gr}_{\mathcal{A}}(\xi,\mu_3)\right]\right\}.
\end{aligned}$$

For two arbitrary neutrosophic numbers, one of our aims is to determine whenever they are said to be equal, that is the detail of following definition.

Definition 4.2. *Two neutrosophic numbers $\mathcal{A}$, $\tilde{\mathcal{A}}$ are said to be equal and written by $\mathcal{A} = \tilde{\mathcal{A}}$ if and only if $\mathcal{H}(\mathcal{A}) = \mathcal{H}(\tilde{\mathcal{A}})$, i.e.,*

$$\begin{cases} \mathrm{T}^{gr}_{\mathcal{A}}(\alpha, \mu_1) = \mathrm{T}^{gr}_{\tilde{\mathcal{A}}}(\alpha, \mu_1) \\ \mathrm{I}^{gr}_{\mathcal{A}}(\beta, \mu_2) = \mathrm{I}^{gr}_{\tilde{\mathcal{A}}}(\beta, \mu_2) \\ \mathrm{F}^{gr}_{\mathcal{A}}(\gamma, \mu_3) = \mathrm{F}^{gr}_{\tilde{\mathcal{A}}}(\gamma, \mu_3) \end{cases}$$

for all $(\alpha, \beta, \gamma) \in [0, \alpha_{neu}] \times [\beta_{neu}, 1] \times [\gamma_{neu}, 1]$ and $(\mu_1, \mu_2, \mu_3) \in [0, 1]^3$.

4.2. Arithmetic Operations

Based on horizontal membership function approach, we can define the arithmetic operations on the set of neutrosophic numbers as follows:

Definition 4.3. *We denote by "$\otimes$" one of three arithmetic operations on the space $\mathcal{E}$: addition, subtraction or multiplication operation. Then, the following equality holds*

$$\mathcal{H}(\mathcal{A}_1 \otimes \mathcal{A}_2) \triangleq \mathcal{H}(\mathcal{A}_1) \odot \mathcal{H}(\mathcal{A}_2),$$

where the notion "$\odot$" denotes for the vector addition, vector subtraction or scalar multiplication operations in the space of $\mathbb{R}^3$.

It should be noted that the difference in this sense is called granular difference (gr-difference) and denoted by $\ominus^{gr}$.

Example 4.1. *For two SVNNs*

$$\mathcal{A} = (5, 10, 15; 3, 6, 9; 10, 16, 22) \text{ and } \tilde{\mathcal{A}} = (4, 6, 8; 1, 3, 5; 9, 11, 13),$$

we immediately get the parametric representations of the numbers $\mathcal{A}$ and $\tilde{\mathcal{A}}$ are as follows

$$\mathcal{A}_{(\alpha,\beta,\gamma)} = \{[5 + 5\alpha, 15 - 5\alpha], [6 - 3\beta, 6 + 3\beta], [16 - 6\gamma, 16 + 6\gamma]\},$$
$$\tilde{\mathcal{A}}_{(\alpha,\beta,\gamma)} = \{[4 + 2\alpha, 8 - 2\alpha], [3 - 2\beta, 3 + 2\beta], [11 - 2\gamma, 11 + 2\gamma]\}.$$

From Definition 4.1, we can present the neutrosophic numbers $\mathcal{A}$ and $\tilde{\mathcal{A}}$ in following parametric forms

$$A^{gr}(\alpha,\beta,\gamma,\mu)=\begin{pmatrix}5+5\alpha+(10-10\alpha)\mu_1;\\ 6-3\beta+6\beta\mu_2;\\ 16-6\gamma+12\gamma\mu_3\end{pmatrix}, \qquad (2)$$

$$\tilde{A}^{gr}(\alpha,\beta,\gamma,\tilde{\mu})=\begin{pmatrix}4+2\alpha+(4-4\alpha)\tilde{\mu}_1;\\ 3-2\beta+4\beta\tilde{\mu}_2;\\ 11-2\gamma+4\gamma\tilde{\mu}_3\end{pmatrix}.$$

Then, by employing Definition 4.3, all the arithmetic operations on the space of SVNNs can be easily done via their horizontal membership functions.

(i) The addition: Since the representation (2), we have that

$$A^{gr}(\alpha,\beta,\gamma,\mu)+\tilde{A}^{gr}(\alpha,\beta,\gamma,\mu)=\begin{pmatrix}9+7\alpha+(14-14\alpha)\mu_1;\\ 9-5\beta+10\beta\mu_2;\\ 27-8\gamma+16\gamma\mu_3\end{pmatrix}.$$

Next, by using Proposition 4.1 and $(\alpha,\beta,\gamma)-$ cuts representation theorem, we obtain that

$$\begin{aligned}\mathcal{A}+\tilde{\mathcal{A}}&=\mathcal{H}^{-1}\left(A^{gr}(\alpha,\beta,\gamma,\mu)+\tilde{A}^{gr}(\alpha,\beta,\gamma,\mu)\right)\\&=(9,17,24;4,9,14;19,27,35).\end{aligned}$$

(ii) The subtraction: Since the granular representation (2), we have that

$$A^{gr}(\alpha,\beta,\gamma,\mu)-\tilde{A}^{gr}(\alpha,\beta,\gamma,\mu)=\begin{pmatrix}1+2\alpha+(6-6\alpha)\mu_1;\\ 3-\beta+2\beta\mu_2;\\ 5-4\gamma+8\gamma\mu_3\end{pmatrix}.$$

Then, by using Proposition 4.1 and $(\alpha,\beta,\gamma)-$ cuts representation theorem, we obtain that

$$\begin{aligned}\mathcal{A}\ominus^{gr}\tilde{\mathcal{A}}&=\mathcal{H}^{-1}\left(A^{gr}(\alpha,\beta,\gamma,\mu)-\tilde{A}^{gr}(\alpha,\beta,\gamma,\mu)\right)\\&=(-3,4,11;-2,3,8;-3,5,13).\end{aligned}$$

(iii) *The multiplication with a scalar $\lambda = 0.5$: From the granular representation (2), we get*

$$0.5 \cdot A^{gr}(\alpha,\beta,\gamma,\mu) = \begin{pmatrix} 2.5 + 2.5\alpha + (5 - 5\alpha)\mu_1; \\ 3 - 1.5\beta + 3\beta\mu_2; \\ 8 - 3\gamma + 6\gamma\mu_3 \end{pmatrix}.$$

Next, by using Proposition 4.1 and $(\alpha,\beta,\gamma)-$ cuts representation theorem, we obtain that

$$\lambda\mathcal{A} = \mathcal{H}^{-1}\left(0.5 \cdot A^{gr}(\alpha,\beta,\gamma,\mu)\right) = (2.5, 5, 7.5; 1.5, 3, 4.5; 5, 8, 11).$$

(iv) *The multiplication: By the granular representation (2), we have*

$$\begin{aligned} &A^{gr}(\alpha,\beta,\gamma,\mu) \times \tilde{A}^{gr}(\alpha,\beta,\gamma,\mu) \\ &= \begin{pmatrix} 20 + 30\alpha + 10\alpha^2 + 20(1-\alpha)(2+3\alpha)\mu_1 + 40(1-\alpha)^2\mu_1^2; \\ 18 - 21\beta + 6\beta^2 + 6\beta(7-4\beta)\mu_2 + 24\beta^2\mu_2^2; \\ 176 - 98\gamma + 12\gamma^2 + 4\gamma(49 - 12\gamma)\mu_3 + 48\gamma^2\mu_3^2 \end{pmatrix}. \end{aligned}$$

Then, by using Proposition 4.1 and $(\alpha,\beta,\gamma)-$ cuts representation theorem, we obtain that

$$\begin{aligned} \mathcal{A} \cdot \tilde{\mathcal{A}} &= \mathcal{H}^{-1}\left(A^{gr}(\alpha,\beta,\gamma,\mu) \times \tilde{A}^{gr}(\alpha,\beta,\gamma,\mu)\right) \\ &= (20, 60, 120; 3, 18, 45; 90, 176, 286). \end{aligned}$$

Especially, we have the spare of $\mathcal{A}$ is given by

$$\mathcal{A}^2 := \mathcal{A} \cdot \mathcal{A} = (25, 100, 225; 9, 36, 81; 100, 256, 484).$$

Definition 4.4. *[19] Let $f : [a,b] \to \mathcal{E}$ be a function including n distinct SVNNs $\mathcal{A}_1, \mathcal{A}_2, \ldots, \mathcal{A}_n$. Then, the horizontal membership function of the function f at a point $t \in [a,b]$ is denoted by $\mathcal{H}(f(t)) \triangleq f^{gr}(t,\alpha,\beta,\gamma,\mu_f)$ and given as follows*

$$f^{gr} : [a,b] \times [0,\alpha_{neu}] \times [\beta_{neu},1] \times [\gamma_{neu},1] \times [0,1]^{3n} \to \mathbb{R}^3,$$

where $\mu_f \triangleq (\mu_{1,\mathcal{A}_1}, \ldots, \mu_{1,\mathcal{A}_n}, \mu_{2,\mathcal{A}_1}, \ldots, \mu_{2,\mathcal{A}_n}, \mu_{3,\mathcal{A}_1}, \ldots, \mu_{3,\mathcal{A}_n})$.

Remark 4.1. *For a neutrosophic function f defined on $[a,b] \subset \mathbb{R}$, due to the value of $f(t)$ belongs to $\mathcal{E}$, we can rewrite $f(t)$ as below*

$$f(t) = \left\{ \langle x, \mathtt{T}_{f(t)}(x), \mathtt{I}_{f(t)}(x), \mathtt{F}_{f(t)}(x) \rangle : x \in \mathbb{R} \right\},$$

for each $t \in [a,b]$ and its horizontal membership function is given by

$$f^{gr}(t,\alpha,\beta,\gamma,\mu_f) = \left(\mathtt{T}^{gr}_{f(t)}(\alpha,\mu_{1,f}), \mathtt{I}^{gr}_{f(t)}(\beta,\mu_{2,f}), \mathtt{F}^{gr}_{f(t)}(\gamma,\mu_{3,f}) \right).$$

Example 4.2. *Let two single valued triangular neutrosophic numbers*

$$\mathcal{A} = (5,10,15;3,6,9;10,16,22), \text{ and } \tilde{\mathcal{A}} = (4,6,8;1,3,5;9,11,13).$$

First of all, by Definition 4.1, we have their respective horizontal membership function are as follows

$$A^{gr}(\alpha,\beta,\gamma,\mu) = (5 + 5\alpha + 10(1-\alpha)\mu_1; 6 - 3\beta + 6\beta\mu_2; 16 - 6\gamma + 12\gamma\mu_3)$$
$$\tilde{A}^{gr}(\alpha,\beta,\gamma,\tilde{\mu}) = (4 + 2\alpha + 4(1-\alpha)\tilde{\mu}_1; 3 - 2\beta + 4\beta\tilde{\mu}_2; 11 - 2\gamma + 4\gamma\tilde{\mu}_3)$$

where $\alpha,\beta,\gamma \in [0,1]$ and $\mu = (\mu_1,\mu_2,\mu_3)$, $\tilde{\mu} = (\tilde{\mu}_1,\tilde{\mu}_2,\tilde{\mu}_3) \in [0,1]^3$.

Next, we consider a neutrosophic function

$$f : [0,2\pi] \to \mathcal{E},\ t \mapsto f(t) = \mathcal{A}\sin t + \tilde{\mathcal{A}}\sin 2t,$$

whose horizontal membership function of $f(t)$ can be given as follows

$$\begin{aligned}
\mathcal{H}(f(t)) &= A^{gr}(\alpha,\beta,\gamma,\mu)\sin t + \tilde{A}^{gr}(\alpha,\beta,\gamma,\tilde{\mu})\sin 2t \\
&= ([5 + 5\alpha + 10(1-\alpha)\mu_1]\sin t + [4 + 2\alpha + 4(1-\alpha)\tilde{\mu}_1]\sin 2t; \\
&\quad [6 - 3\beta + 6\beta\mu_2]\sin t + [3 - 2\beta + 4\beta\tilde{\mu}_2]\sin 2t; \\
&\quad [16 - 6\gamma + 12\gamma\mu_3]\sin t + [11 - 2\gamma + 4\gamma\tilde{\mu}_3]\sin 2t),
\end{aligned}$$

where α, β, $\gamma \in [0,1]$ and μ_i, $\tilde{\mu}_i \in [0,1]$ ($i = 1,2,3$). Then, its graphical representation can be shown as below (see Figure 6).

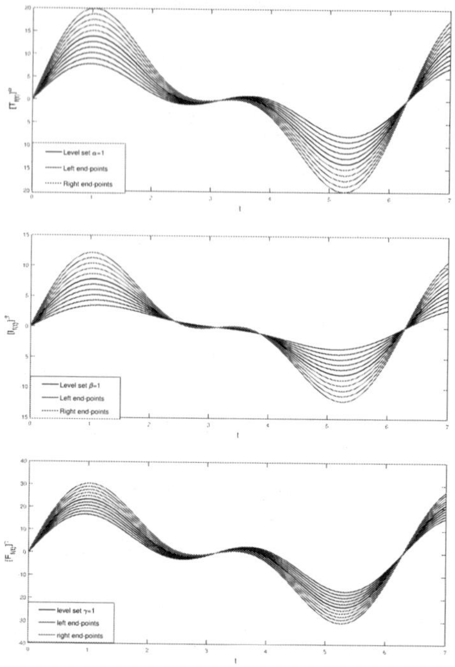

Figure 6. The (α, β, γ)- cuts of neutrosophic function $f(t)$ in Example 4.2

4.3. Neutrosophic Partially Ordered Space

Definition 4.5. *In space $\mathcal{E}$, a partial ordering $\preceq$ can be defined as follows*

$$\mathcal{A}_1 \preceq \mathcal{A}_2 \textit{ if and only if } \begin{cases} \mathtt{T}^-_{\mathcal{A}_1}(\alpha) \leq \mathtt{T}^-_{\mathcal{A}_2}(\alpha) \textit{ and } \mathtt{T}^+_{\mathcal{A}_1}(\alpha) \leq \mathtt{T}^+_{\mathcal{A}_2}(\alpha) \\ \mathtt{I}^-_{\mathcal{A}_1}(\beta) \geq \mathtt{I}^-_{\mathcal{A}_2}(\beta) \textit{ and } \mathtt{I}^+_{\mathcal{A}_1}(\beta) \geq \mathtt{I}^+_{\mathcal{A}_2}(\beta) \\ \mathtt{F}^-_{\mathcal{A}_1}(\gamma) \geq \mathtt{F}^-_{\mathcal{A}_2}(\gamma) \textit{ and } \mathtt{F}^+_{\mathcal{A}_1}(\gamma) \geq \mathtt{F}^+_{\mathcal{A}_2}(\gamma) \end{cases}$$

for all α, β, $\gamma \in [0, 1]$, where $\mathcal{A}_1$, $\mathcal{A}_2 \in \mathcal{E}$ have following parametric forms

$$\mathcal{A}_{1,(\alpha,\beta,\gamma)} = \left[\mathtt{T}^-_{\mathcal{A}_1}(\alpha), \mathtt{T}^+_{\mathcal{A}_1}(\alpha); \mathtt{I}^-_{\mathcal{A}_1}(\beta), \mathtt{I}^+_{\mathcal{A}_1}(\beta); \mathtt{F}^-_{\mathcal{A}_1}(\gamma), \mathtt{F}^+_{\mathcal{A}_2}(\gamma)\right],$$

$$\mathcal{A}_{2,(\alpha,\beta,\gamma)} = \left[\mathtt{T}^-_{\mathcal{A}_2}(\alpha), \mathtt{T}^+_{\mathcal{A}_2}(\alpha); \mathtt{I}^-_{\mathcal{A}_2}(\beta), \mathtt{I}^+_{\mathcal{A}_2}(\beta); \mathtt{F}^-_{\mathcal{A}_2}(\gamma), \mathtt{F}^+_{\mathcal{A}_2}(\gamma)\right].$$

Remark 4.2. *We can check that the partial ordering $\preceq$ is well-defined. Indeed, this relation satisfies all axioms of a partially ordered relation*

(i) *Reflexivity: By the definition of the relation, it is obvious that* $\mathcal{A} \preceq \mathcal{A}$ *for all* $\mathcal{A} \in \mathcal{E}$.

(ii) *Antisymmetry: For two arbitrary numbers* $\mathcal{A}_1, \mathcal{A}_2 \in \mathcal{E}$, *let us assume that* $\mathcal{A}_1 \preceq \mathcal{A}_2$ *and* $\mathcal{A}_2 \preceq \mathcal{A}_1$, *i.e.,*

$$\begin{cases} \mathtt{T}^-_{\mathcal{A}_1}(\alpha) \leq \mathtt{T}^-_{\mathcal{A}_2}(\alpha) \text{ and } \mathtt{T}^+_{\mathcal{A}_1}(\alpha) \leq \mathtt{T}^+_{\mathcal{A}_2}(\alpha) \\ \mathtt{I}^-_{\mathcal{A}_1}(\beta) \geq \mathtt{I}^-_{\mathcal{A}_2}(\beta) \text{ and } \mathtt{I}^+_{\mathcal{A}_1}(\beta) \geq \mathtt{I}^+_{\mathcal{A}_2}(\beta) \\ \mathtt{F}^-_{\mathcal{A}_1}(\gamma) \geq \mathtt{F}^-_{\mathcal{A}_2}(\gamma) \text{ and } \mathtt{F}^+_{\mathcal{A}_1}(\gamma) \geq \mathtt{F}^+_{\mathcal{A}_2}(\gamma) \end{cases}$$

and

$$\begin{cases} \mathtt{T}^-_{\mathcal{A}_2}(\alpha) \leq \mathtt{T}^-_{\mathcal{A}_1}(\alpha) \text{ and } \mathtt{T}^+_{\mathcal{A}_2}(\alpha) \leq \mathtt{T}^+_{\mathcal{A}_1}(\alpha) \\ \mathtt{I}^-_{\mathcal{A}_2}(\beta) \geq \mathtt{I}^-_{\mathcal{A}_1}(\beta) \text{ and } \mathtt{I}^+_{\mathcal{A}_2}(\beta) \geq \mathtt{I}^+_{\mathcal{A}_1}(\beta) \\ \mathtt{F}^-_{\mathcal{A}_2}(\gamma) \geq \mathtt{F}^-_{\mathcal{A}_1}(\gamma) \text{ and } \mathtt{F}^+_{\mathcal{A}_2}(\gamma) \geq \mathtt{F}^+_{\mathcal{A}_1}(\gamma) \end{cases}$$

Therefore, we directly obtain that

$$\begin{cases} \mathtt{T}^-_{\mathcal{A}_1}(\alpha) = \mathtt{T}^-_{\mathcal{A}_2}(\alpha) \text{ and } \mathtt{T}^+_{\mathcal{A}_1}(\alpha) = \mathtt{T}^+_{\mathcal{A}_2}(\alpha) \\ \mathtt{I}^-_{\mathcal{A}_1}(\beta) = \mathtt{I}^-_{\mathcal{A}_2}(\beta) \text{ and } \mathtt{I}^+_{\mathcal{A}_1}(\beta) = \mathtt{I}^+_{\mathcal{A}_2}(\beta) \\ \mathtt{F}^-_{\mathcal{A}_1}(\gamma) = \mathtt{F}^-_{\mathcal{A}_2}(\gamma) \text{ and } \mathtt{F}^+_{\mathcal{A}_1}(\gamma) = \mathtt{F}^+_{\mathcal{A}_2}(\gamma) \end{cases}$$

which means that $\mathcal{A}_1 = \mathcal{A}_2$.

(iii) *Transitivity: For* $\mathcal{A}_1, \mathcal{A}_2, \mathcal{A}_3 \in \mathcal{E}$, *let us assume that*

$$\mathcal{A}_1 \preceq \mathcal{A}_2 \text{ and } \mathcal{A}_2 \preceq \mathcal{A}_3$$

Then, by the definition of the relation $\preceq$, *we have that*

$$\begin{cases} \mathtt{T}^-_{\mathcal{A}_1}(\alpha) \leq \mathtt{T}^-_{\mathcal{A}_2}(\alpha) \text{ and } \mathtt{T}^+_{\mathcal{A}_1}(\alpha) \leq \mathtt{T}^+_{\mathcal{A}_2}(\alpha) \\ \mathtt{I}^-_{\mathcal{A}_1}(\beta) \geq \mathtt{I}^-_{\mathcal{A}_2}(\beta) \text{ and } \mathtt{I}^+_{\mathcal{A}_1}(\beta) \geq \mathtt{I}^+_{\mathcal{A}_2}(\beta) \\ \mathtt{F}^-_{\mathcal{A}_1}(\gamma) \geq \mathtt{F}^-_{\mathcal{A}_2}(\gamma) \text{ and } \mathtt{F}^+_{\mathcal{A}_1}(\gamma) \geq \mathtt{F}^+_{\mathcal{A}_2}(\gamma) \end{cases}$$

and

$$\begin{cases} \mathtt{T}^-_{\mathcal{A}_2}(\alpha) \leq \mathtt{T}^-_{\mathcal{A}_3}(\alpha) \text{ and } \mathtt{T}^+_{\mathcal{A}_2}(\alpha) \leq \mathtt{T}^+_{\mathcal{A}_3}(\alpha) \\ \mathtt{I}^-_{\mathcal{A}_2}(\beta) \geq \mathtt{I}^-_{\mathcal{A}_3}(\beta) \text{ and } \mathtt{I}^+_{\mathcal{A}_2}(\beta) \geq \mathtt{I}^+_{\mathcal{A}_3}(\beta) \\ \mathtt{F}^-_{\mathcal{A}_2}(\gamma) \geq \mathtt{F}^-_{\mathcal{A}_3}(\gamma) \text{ and } \mathtt{F}^+_{\mathcal{A}_2}(\gamma) \geq \mathtt{F}^+_{\mathcal{A}_3}(\gamma), \end{cases}$$

which follows that

$$\begin{cases} \mathtt{T}^-_{\mathcal{A}_1}(\alpha) \leq \mathtt{T}^-_{\mathcal{A}_3}(\alpha) \text{ and } \mathtt{T}^+_{\mathcal{A}_1}(\alpha) \leq \mathtt{T}^+_{\mathcal{A}_3}(\alpha) \\ \mathtt{I}^-_{\mathcal{A}_1}(\beta) \geq \mathtt{I}^-_{\mathcal{A}_3}(\beta) \text{ and } \mathtt{I}^+_{\mathcal{A}_1}(\beta) \geq \mathtt{I}^+_{\mathcal{A}_3}(\beta) \\ \mathtt{F}^-_{\mathcal{A}_1}(\gamma) \geq \mathtt{F}^-_{\mathcal{A}_3}(\gamma) \text{ and } \mathtt{F}^+_{\mathcal{A}_1}(\gamma) \geq \mathtt{F}^+_{\mathcal{A}_3}(\gamma), \end{cases}$$

or equivalently, the relation $\mathcal{A}_1 \preceq \mathcal{A}_3$ *holds.*

As a consequence, we have following results

Proposition 4.2. *Some properties of SVNNs with respect to the partial ordering* $\preceq$ *can be listed as below*

(i) *If* $\mathcal{A}_1 \preceq \mathcal{A}_2$ *then* $\mathcal{A}_1 + \mathcal{A}_3 \preceq \mathcal{A}_2 + \mathcal{A}_3$ *for all* $\mathcal{A}_1, \mathcal{A}_2, \mathcal{A}_3 \in \mathcal{E}$.

(ii) *For every non-decreasing sequence* $\{\mathcal{A}_n\}_{n\geq 1} \subset \mathcal{E}$, *if* $\mathcal{A}_n \to \mathcal{A}$ *then* $\mathcal{A}_n \leq \mathcal{A}$, *for all* $n \in \mathbb{N}$.

(iii) *For each* $\mathcal{A}_1, \mathcal{A}_2, \mathcal{A}_3 \in \mathcal{E}$, *if* $\mathcal{A}_1 \preceq \mathcal{A}_2$ *then* $\mathcal{A}_3 \ominus^{gr} \mathcal{A}_2 \preceq \mathcal{A}_3 \ominus^{gr} \mathcal{A}_1$.

Definition 4.6. *Let* $f, g : J \subset \mathbb{R} \to \mathcal{E}$ *be continuous neutrosophic functions. Then, we say that* $f \preceq g$ *if and only if* $f(t) \preceq g(t)$ *for all* $t \in J$, *that means*

$$\begin{cases} \mathtt{T}^-_{f(t)}(\alpha) \leq \mathtt{T}^-_{g(t)}(\alpha) \text{ and } \mathtt{T}^+_{f(t)}(\alpha) \leq \mathtt{T}^+_{g(t)}(\alpha) \\ \mathtt{I}^-_{f(t)}(\beta) \geq \mathtt{I}^-_{g(t)}(\beta) \text{ and } \mathtt{I}^+_{f(t)}(\beta) \geq \mathtt{I}^+_{g(t)}(\beta) \\ \mathtt{F}^-_{f(t)}(\gamma) \geq \mathtt{F}^-_{g(t)}(\gamma) \text{ and } \mathtt{F}^+_{f(t)}(\gamma) \geq \mathtt{F}^+_{g(t)}(\gamma) \end{cases}$$

for all $\alpha, \beta, \gamma \in [0, 1]$ *and* $t \in J$.

5. Metric Space of Neutrosophic Numbers and the Continuity of Neutrosophic Functions

In this section, the topological structures on the set of neutrosophic numbers based on Hamming metric is firstly introduced. Moreover, we obtain that the space $\mathcal{E}$ endowed with the metric D^{gr} is a complete metric space. In addition, based on the granular metric, we can give the concepts of limit and continuity of neutrosophic functions.

5.1. Metric Space of Neutrosophic Numbers

Definition 5.1. *A real-valued function $D^{gr}: \mathcal{E} \times \mathcal{E} \to \mathbb{R}^+ \cup \{0\}$, given by*

$$D^{gr}(\mathcal{A}, \tilde{\mathcal{A}}) = \sup_{\alpha,\beta,\gamma} \max_{\mu_i,\tilde{\mu}_i} \frac{1}{3}\Big\{\Big|\mathrm{T}^{gr}_{\mathcal{A}}(\alpha,\mu_1) - \mathrm{T}^{gr}_{\tilde{\mathcal{A}}}(\alpha,\tilde{\mu}_1)\Big| + \Big|\mathrm{I}^{gr}_{\mathcal{A}}(\beta,\mu_2) - \mathrm{I}^{gr}_{\tilde{\mathcal{A}}}(\beta,\tilde{\mu}_2)\Big| + \Big|\mathrm{F}^{gr}_{\mathcal{A}}(\gamma,\mu_3) - \mathrm{F}^{gr}_{\tilde{\mathcal{A}}}(\gamma,\tilde{\mu}_3)\Big|\Big\}, \tag{3}$$

is said to be a distance between neutrosophic numbers $\mathcal{A}$ and $\tilde{\mathcal{A}} \in \mathcal{E}$.

Theorem 5.1. *Such real-valued function D^{gr} is a metric on $\mathcal{E}$.*

Proof. The proof of this theorem will be divided into three steps:

(i) Let $\mathcal{A}$ and $\tilde{\mathcal{A}}$ be two arbitrary neutrosophic numbers with respective horizontal membership functions

$$\mathcal{A}^{gr}(\alpha,\beta,\gamma,\mu) = (\mathrm{T}^{gr}_{\mathcal{A}}(\alpha,\mu_1), \mathrm{I}^{gr}_{\mathcal{A}}(\beta,\mu_2), \mathrm{F}^{gr}_{\mathcal{A}}(\gamma,\mu_3)),$$
$$\tilde{\mathcal{A}}^{gr}(\alpha,\beta,\gamma,\tilde{\mu}) = (\mathrm{T}^{gr}_{\tilde{\mathcal{A}}}(\alpha,\tilde{\mu}_1), \mathrm{I}^{gr}_{\tilde{\mathcal{A}}}(\beta,\tilde{\mu}_2), \mathrm{F}^{gr}_{\tilde{\mathcal{A}}}(\gamma,\tilde{\mu}_3)).$$

Since the formula (3), it implies that $D^{gr}(\mathcal{A}, \tilde{\mathcal{A}}) \geq 0$ for all $\mathcal{A}$, $\tilde{\mathcal{A}} \in \mathcal{E}$. Moreover, if $D^{gr}(\mathcal{A}, \tilde{\mathcal{A}}) = 0$ then we immediately get that

$$\begin{cases} \max_{\mu_1,\tilde{\mu}_1} \Big|\mathrm{T}^{gr}_{\mathcal{A}}(\alpha,\mu_1) - \mathrm{T}^{gr}_{\tilde{\mathcal{A}}}(\alpha,\tilde{\mu}_1)\Big| = 0 \\ \max_{\mu_2,\tilde{\mu}_2} \Big|\mathrm{I}^{gr}_{\mathcal{A}}(\beta,\mu_2) - \mathrm{I}^{gr}_{\tilde{\mathcal{A}}}(\beta,\tilde{\mu}_2)\Big| = 0 \\ \max_{\mu_3,\tilde{\mu}_3} \Big|\mathrm{F}^{gr}_{\mathcal{A}}(\gamma,\mu_3) - \mathrm{F}^{gr}_{\tilde{\mathcal{A}}}(\gamma,\tilde{\mu}_3)\Big| = 0, \end{cases}$$

or equivalently,

$$\begin{cases} \mathrm{T}^{gr}_{\mathcal{A}}(\alpha,\mu_1) = \mathrm{T}^{gr}_{\tilde{\mathcal{A}}}(\alpha,\tilde{\mu}_1) \\ \mathrm{I}^{gr}_{\mathcal{A}}(\beta,\mu_2) = \mathrm{I}^{gr}_{\tilde{\mathcal{A}}}(\beta,\tilde{\mu}_2) \\ \mathrm{F}^{gr}_{\mathcal{A}}(\gamma,\mu_3) = \mathrm{F}^{gr}_{\tilde{\mathcal{A}}}(\gamma,\tilde{\mu}_3), \end{cases}$$

for all $\mu_i = \tilde{\mu}_i \in [0,1]$ $(i = 1,2,3)$ which means $\mathcal{A} = \tilde{\mathcal{A}}$.

(ii) From the definition of metric D^{gr}, the symmetry of D^{gr} is straightforward and hence omitted.

(iii) The rest of proof is to show that

$$D^{gr}(\mathcal{A}_1, \mathcal{A}_2) \leq D^{gr}(\mathcal{A}_1, \mathcal{A}_3) + D^{gr}(\mathcal{A}_3, \mathcal{A}_2) \quad \text{for all } \mathcal{A}_1, \mathcal{A}_2, \mathcal{A}_3 \in \mathcal{E}. \quad (4)$$

Indeed, since the following inequality

$$\begin{aligned} &|\mathrm{T}^{gr}_{\mathcal{A}_1}(\alpha, \mu_1) - \mathrm{T}^{gr}_{\mathcal{A}_2}(\alpha, \tilde{\mu}_1)| \\ &\qquad \leq \left|\mathrm{T}^{gr}_{\mathcal{A}_1}(\alpha, \mu_1) - \mathrm{T}^{gr}_{\mathcal{A}_3}(\alpha, \overline{\mu}_1)\right| + \left|\mathrm{T}^{gr}_{\mathcal{A}_3}(\alpha, \overline{\mu}_1) - \mathrm{T}^{gr}_{\mathcal{A}_2}(\alpha, \tilde{\mu}_1)\right| \end{aligned}$$

holds for each $\alpha \in [0, \alpha_{neu}]$ and $\mu_1, \tilde{\mu}_1, \overline{\mu}_1 \in [0, 1]$, we immediately get

$$\begin{aligned} \left|\mathrm{T}^{gr}_{\mathcal{A}_1}(\alpha, \mu_1) - \mathrm{T}^{gr}_{\mathcal{A}_2}(\alpha, \tilde{\mu}_1)\right| \leq \max_{\mu_1, \tilde{\mu}_1, \overline{\mu}_1} &\left\{\left|\mathrm{T}^{gr}_{\mathcal{A}_1}(\alpha, \mu_1) - \mathrm{T}^{gr}_{\mathcal{A}_3}(\alpha, \overline{\mu}_1)\right|\right. \\ &\left. + \left|\mathrm{T}^{gr}_{\mathcal{A}_3}(\alpha, \overline{\mu}_1) - \mathrm{T}^{gr}_{\mathcal{A}_2}(\alpha, \tilde{\mu}_1)\right|\right\}. \end{aligned}$$

Therefore,

$$\begin{aligned} \sup_{\alpha} \max_{\mu_1, \tilde{\mu}_1} \left|\mathrm{T}^{gr}_{\mathcal{A}_1}(\alpha, \mu_1) - \mathrm{T}^{gr}_{\mathcal{A}_2}(\alpha, \tilde{\mu}_1)\right| &\leq \sup_{\alpha} \max_{\mu_1, \overline{\mu}_1} \left|\mathrm{T}^{gr}_{\mathcal{A}_1}(\alpha, \mu_1) - \mathrm{T}^{gr}_{\mathcal{A}_3}(\alpha, \overline{\mu}_1)\right| \\ &+ \sup_{\alpha} \max_{\overline{\mu}_1, \tilde{\mu}_1} \left|\mathrm{T}^{gr}_{\mathcal{A}_3}(\alpha, \overline{\mu}_1) - \mathrm{T}^{gr}_{\mathcal{A}_2}(\alpha, \tilde{\mu}_1)\right|. \end{aligned}$$

Similarly, we also have

$$\begin{aligned} \sup_{\beta} \max_{\mu_2, \tilde{\mu}_2} \left|\mathrm{I}^{gr}_{\mathcal{A}_1}(\beta, \mu_2) - \mathrm{I}^{gr}_{\mathcal{A}_2}(\beta, \tilde{\mu}_2)\right| &\leq \sup_{\beta} \max_{\mu_2, \overline{\mu}_2} \left|\mathrm{I}^{gr}_{\mathcal{A}_1}(\beta, \mu_2) - \mathrm{I}^{gr}_{\mathcal{A}_3}(\beta, \overline{\mu}_2)\right| \\ &+ \sup_{\beta} \max_{\overline{\mu}_2, \tilde{\mu}_2} \left|\mathrm{I}^{gr}_{\mathcal{A}_3}(\beta, \overline{\mu}_2) - \mathrm{I}^{gr}_{\mathcal{A}_2}(\beta, \tilde{\mu}_2)\right|, \\ \sup_{\gamma} \max_{\mu_3, \tilde{\mu}_3} \left|\mathrm{F}^{gr}_{\mathcal{A}_1}(\gamma, \mu_3) - \mathrm{F}^{gr}_{\mathcal{A}_2}(\gamma, \tilde{\mu}_3)\right| &\leq \sup_{\gamma} \max_{\mu_3, \overline{\mu}_3} \left|\mathrm{F}^{gr}_{\mathcal{A}_1}(\gamma, \mu_3) - \mathrm{F}^{gr}_{\mathcal{A}_3}(\gamma, \overline{\mu}_3)\right| \\ &+ \sup_{\gamma} \max_{\overline{\mu}_3, \tilde{\mu}_3} \left|\mathrm{F}^{gr}_{\mathcal{A}_3}(\gamma, \overline{\mu}_3) - \mathrm{F}^{gr}_{\mathcal{A}_2}(\gamma, \tilde{\mu}_3)\right|. \end{aligned}$$

Finally, by adding both sides of three above inequalities, we obtain the inequality (4). Hence, the proof is complete. □

Remark 5.1. *Such metric D^{gr} is said to be the granular metric on the space $\mathcal{E}$ of all single valued triangular neutrosophic numbers. Hence, the space $\mathcal{E}$ equipped with the metric D^{gr} is a metric space.*

The next theorem is to prove that the space $\mathcal{E}$ endowed with metric D^{gr} is a complete space, that ensures the convergent of every Cauchy sequence in $\mathcal{E}$. Hence, as a consequence, we can further study the quantitative nature to some dynamical systems and processes arising in science and engineering.

Theorem 5.2. *The metric space* $(\mathcal{E}, D^{gr})$ *is complete space.*

Proof. Let $\{\mathcal{A}_n\}_{n\geq 1} \subset \mathcal{E}$ be a Cauchy sequence in $\mathcal{E}$, that means

$$\forall \epsilon > 0,\ \exists N \in \mathbb{N}^* \text{ such that } \forall n \geq N,\ p \geq 1,\ D^{gr}(\mathcal{A}_{n+p}, \mathcal{A}_n) < \epsilon,$$

or equivalently,

$$\sup_{\alpha,\beta,\gamma} \max_{\mu_1,\tilde{\mu}_1} \left| \mathrm{T}^{gr}_{\mathcal{A}_{n+p}}(\alpha, \mu_1) - \mathrm{T}^{gr}_{\mathcal{A}_n}(\alpha, \tilde{\mu}_1) \right| < \epsilon,$$
$$\sup_{\alpha,\beta,\gamma} \max_{\mu_2,\tilde{\mu}_2} \left| \mathrm{I}^{gr}_{\mathcal{A}_{n+p}}(\beta, \mu_2) - \mathrm{I}^{gr}_{\mathcal{A}_n}(\beta, \tilde{\mu}_2) \right| < \epsilon,$$
$$\sup_{\alpha,\beta,\gamma} \max_{\mu_3,\tilde{\mu}_3} \left| \mathrm{F}^{gr}_{\mathcal{A}_{n+p}}(\gamma, \mu_3) - \mathrm{F}^{gr}_{\mathcal{A}_n}(\gamma, \tilde{\mu}_3) \right| < \epsilon.$$

As a result, we directly obtain that

$$\left| \mathrm{T}^{gr}_{\mathcal{A}_{n+p}}(\alpha, \mu_1) - \mathrm{T}^{gr}_{\mathcal{A}_n}(\alpha, \tilde{\mu}_1) \right| < \epsilon,$$
$$\left| \mathrm{I}^{gr}_{\mathcal{A}_{n+p}}(\beta, \mu_2) - \mathrm{I}^{gr}_{\mathcal{A}_n}(\beta, \tilde{\mu}_2) \right| < \epsilon,$$
$$\left| \mathrm{F}^{gr}_{\mathcal{A}_{n+p}}(\gamma, \mu_3) - \mathrm{F}^{gr}_{\mathcal{A}_n}(\gamma, \tilde{\mu}_3) \right| < \epsilon.$$

Thus, it follows that $\left\{\mathrm{T}^{gr}_{\mathcal{A}_n}(\alpha, \mu_1)\right\}_{n\geq 1}$, $\left\{\mathrm{I}^{gr}_{\mathcal{A}_n}(\beta, \mu_2)\right\}_{n\geq 1}$, $\left\{\mathrm{F}^{gr}_{\mathcal{A}_n}(\gamma, \mu_3)\right\}_{n\geq 1}$ are Cauchy sequences in the space of real numbers $\mathbb{R}$, and then, these sequences are convergent in $\mathbb{R}$. In particular, let us consider the sequence $\left\{\mathrm{T}^{gr}_{\mathcal{A}_n}(\alpha, \mu_1)\right\}_{n\geq 1}$. By Definition 4.1, we can rewrite as

$$\mathrm{T}^{gr}_{\mathcal{A}_n}(\alpha, \mu_1) = \mathrm{T}^{-}_{\mathcal{A}_n}(\alpha) + \left(\mathrm{T}^{+}_{\mathcal{A}_n}(\alpha) - \mathrm{T}^{-}_{\mathcal{A}_n}(\alpha)\right) \mu_1$$

Since the convergence $\left\{\mathrm{T}^{gr}_{\mathcal{A}_n}(\alpha, \mu_1)\right\}_{n\geq 1}$ and $\mu_1 \in [0,1]$, it follows that the sequences $\left\{\mathrm{T}^{-}_{\mathcal{A}_n}(\alpha)\right\}$ and $\left\{\mathrm{T}^{+}_{\mathcal{A}_n}(\alpha)\right\}$ are also convergent. No loss generality, we assume that $\lim_{n\to\infty} \mathrm{T}^{-}_{\mathcal{A}_n}(\alpha) = \mathrm{T}^{-}_{\mathcal{A}}(\alpha)$, $\lim_{n\to\infty} \mathrm{T}^{+}_{\mathcal{A}_n}(\alpha) = \mathrm{T}^{+}_{\mathcal{A}}(\alpha)$ and due to the fact that $\mathrm{T}^{-}_{\mathcal{A}_n}(\alpha) \leq \mathrm{T}^{+}_{\mathcal{A}_n}(\alpha),\ \forall n \geq 1$, we obtain that $\mathrm{T}^{-}_{\mathcal{A}}(\alpha) \leq \mathrm{T}^{+}_{\mathcal{A}}(\alpha)$. At last, by employing analogous method as in proof of Theorem 8.5 in [15], we deduce that the interval $[\mathrm{T}^{-}_{\mathcal{A}}(\alpha), \mathrm{T}^{+}_{\mathcal{A}}(\alpha)]$ is the $\alpha-$ cuts of a fuzzy number. As a consequence, the similar results are also obtained for the sequences $\left\{\mathrm{I}^{gr}_{\mathcal{A}_n}(\beta, \mu_2)\right\}_{n\geq 1}$ and $\left\{\mathrm{F}^{gr}_{\mathcal{A}_n}(\gamma, \mu_3)\right\}_{n\geq 1}$. Therefore, we can see that if $\{\mathcal{A}_n\}_{n\geq 1}$ is a Cauchy sequence in $\mathcal{E}$, and then, $\{\mathcal{A}_n\}$ converges to an element $\mathcal{A} \in \mathcal{E}$. Hence, this achieves the proof. □

5.2. The Continuity

First of all, we introduce the concept of limit of a neutrosophic function at a point $t_0 \in [a, b] \subset \mathbb{R}$.

Definition 5.2. *[19] Let $f : [a, b] \subset \mathbb{R} \to \mathcal{E}$ be a neutrosophic function and a point $t_0 \in [a, b]$. Then, an element $\mathrm{L} \in \mathcal{E}$ is said to be the limit of the function $f(t)$ as t tends to t_0 and written by $\lim_{t\to t_0} f(t) = \mathrm{L}$ if and only if*

$$\forall \epsilon > 0,\ \exists \delta > 0 \text{ such that } \forall t \in [a, b] : |t - t_0| < \delta,\ D^{gr}(f(t), \mathrm{L}) < \epsilon.$$

In special case,

- *If $t_0 = a$ then $\lim_{t\to a^+} f(t) = \mathrm{L}$ means that*

$$\forall \epsilon > 0,\ \exists \delta > 0 \text{ s.t } \forall t \in [a, b] : 0 < t - a < \delta,\ D^{gr}(f(t), \mathrm{L}) < \epsilon.$$

- *If $t_0 = b$ then $\lim_{t\to b^-} f(t) = \mathrm{L}$ means that*

$$\forall \epsilon > 0,\ \exists \delta > 0 \text{ s.t } \forall t \in [a, b] : 0 < b - t < \delta,\ D^{gr}(f(t), \mathrm{L}) < \epsilon.$$

Definition 5.3. *[19] A function $f : (a, b) \to \mathcal{E}$ is said to be continuous on (a, b) if and only if f is continuous at every point $t_0 \in (a, b)$, i.e.,*

$$\forall t_0 \in (a, b),\ \forall \epsilon > 0,\ \exists \delta > 0 \text{ s.t } \forall t \in [a, b] : |t - t_0| < \delta,\ D^{gr}(f(t), f(t_0) < \epsilon.$$

6. The Neutrosophic Derivatives

Definition 6.1. *Let $f : (a,b) \subset \mathbb{R} \to \mathcal{E}$ be a function defined on (a,b). Then, the function f is said to be granular differentiable (gr-differentiable for short) at a point $t_0 \in (a,b)$ if there exists an element $f'_{gr}(t_0) \in \mathcal{E}$ such that the following limit*

$$\lim_{h \to 0} \frac{f(t_0+h) \ominus^{gr} f(t_0)}{h} = f'_{gr}(t_0),$$

exists for h sufficiently near 0 and then, the value $f'_{gr}(t_0)$ is called the granular derivative (gr-derivative) of neutrosophic function f at the point t_0.

Especially,

- *A function f is said to be granular differentiable (gr-differentiable) on (a,b) if the gr-derivatives of f exist for all points $t_0 \in (a,b)$.*
- *The mapping $t \in (a,b) \mapsto f'_{gr}(t)$ is called the granular derivative of function f on (a,b) and denoted by f'_{gr} or $\frac{d_{gr} f(t)}{dt}$ or $df_{gr}(t)$.*

Proposition 6.1. *[19] A necessary and sufficient condition for the granular differentiability of a neutrosophic function $f(t)$ at a point $t_0 \in (a,b)$ is the differentiability of its horizontal membership function at that point. Moreover, we have*

$$\mathcal{H}\left(f'_{gr}(t_0)\right) = \frac{\partial f^{gr}(t_0, \alpha, \beta, \gamma, \mu_f)}{\partial t}.$$

Proof. By the assumption that the neutrosophic function f is gr-differentiable at the point $t_0 \in (a,b)$, it follows that

$$\forall \epsilon > 0,\ \exists \delta > 0 \text{ such that } \forall h : 0 < h < \delta,$$
$$D^{gr}\left(\frac{f(t_0+h) \ominus^{gr} f(t_0)}{h}, f'_{gr}(t_0)\right) < \epsilon.$$

Here, for simplicity in representation, we denote $\frac{f(t_0+h)\ominus^{gr} f(t_0)}{h}$ and $f'_{gr}(t_0)$ by $\mathcal{A}$ and $\mathcal{A}'$, respectively. Then, by employing the concept of granular metric, the

above statement can be rewritten as follows
$\forall \epsilon > 0,\ \exists \delta > 0$ such that $\forall h : 0 < h < \delta,$ we have

$$\sup_{\alpha,\beta,\gamma} \max_{\mu_i,\tilde{\mu}_i} \frac{1}{3}\Big\{ \left|\mathrm{T}^{gr}_{\mathcal{A}}(\alpha,\mu_1) - \mathrm{T}^{gr}_{\mathcal{A}'}(\alpha,\tilde{\mu}_1)\right| + \left|\mathrm{I}^{gr}_{\mathcal{A}}(\beta,\mu_2) - \mathrm{I}^{gr}_{\mathcal{A}'}(\beta,\tilde{\mu}_2)\right| + \left|\mathrm{F}^{gr}_{\mathcal{A}}(\gamma,\mu_3) - \mathrm{F}^{gr}_{\mathcal{A}'}(\gamma,\tilde{\mu}_3)\right| \Big\} < \epsilon,$$

that is equivalent to

$$\left\| \frac{f^{gr}(t_0+h,\alpha,\beta,\gamma,\mu_f) - f^{gr}(t_0,\alpha,\beta,\gamma,\mu_f)}{h} - \left(f'_{gr}\right)^{gr}(t_0,\alpha,\beta,\gamma,\mu_{f'}) \right\|$$

is getting as small as h tends to 0. Thus, it follows that the gr-differentiability of $f(t)$ implies the differentiability of its horizontal membership function. By analogous arguments, we also obtain the converse statement. Hence, the proof is complete. □

Proposition 6.2. *Let $f,\, g : [a,b] \to \mathcal{E}$ be gr-differentiable functions. Then, based on the horizontal membership function approach, the following statements are fulfilled:*

(i) $(\mathcal{A})'_{gr} = \tilde{0}$, *where $\mathcal{A} \in \mathcal{E}$ and $\tilde{0}$ is zero neutrosophic number.*

(ii) $(\alpha f(t) \pm \beta g(t))'_{gr} = \alpha f'_{gr}(t) + \beta g'_{gr}(t)$ *for all $t \in [a,b]$ and $\alpha,\ \beta \in \mathbb{R}$.*

(iii) $(fg)'_{gr}(t) = f'_{gr}(t)g(t) + f(t)g'_{gr}(t)$ *for all $t \in [a,b]$.*

Example 6.1. *Let two triangular neutrosophic numbers of type 1*

$$\mathcal{A} = (5,10,15;3,6,9;10,16,22)$$
$$\tilde{\mathcal{A}} = (4,6,8;1,3,5;9,11,13).$$

Let us recall from Example 4.2 the function $f(t) = \mathcal{A}\sin t + \tilde{\mathcal{A}}\sin 2t$, defined on the closed interval $[0,2\pi]$. Here, we can see that the horizontal membership function $f^{gr}(t,\alpha,\beta,\gamma,\mu,\tilde{\mu})$ of the function $f(t)$ is differentiable on

$0, 2\pi]$. *Moreover, by Proposition 4.1, we have*

$$\begin{aligned}
&\frac{\partial f^{gr}(t,\alpha,\beta,\gamma,\mu,\tilde{\mu})}{\partial t}\\
&= ([5+5\alpha+(10-10\alpha)\mu_1]\cos t+[8+4\alpha+(8-8\alpha)\tilde{\mu}_1]\cos 2t;\\
&\quad [6-3\beta+6\beta\mu_2]\cos t+[6-4\beta+8\beta\tilde{\mu}_2]\cos 2t;\\
&\quad [16-6\gamma+12\gamma\mu_3]\cos t+[22-4\gamma+8\gamma\tilde{\mu}_3]\cos 2t),
\end{aligned}$$

for all $\alpha,\beta,\gamma\in[0,1]$ *and* $\mu=(\mu_1,\mu_2,\mu_3)$, $\tilde{\mu}=(\tilde{\mu}_1,\tilde{\mu}_2,\tilde{\mu}_3)\in[0,1]^3$. *As a consequence, we deduce that the neutrosophic function* f *is gr-differentiable on the interval* $[0,2\pi]$.

In addition, the (α,β,γ) *- cuts of its derivative can be given as below*

$$\begin{aligned}
[f'_{gr}(t)]_{(\alpha,\beta,\gamma)} &= \mathcal{H}^{-1}\left(\frac{\partial f^{gr}(t,\alpha,\alpha_u,\alpha_v)}{\partial t}\right)\\
&= ([5+5\alpha,15-5\alpha]\cos t+[8+4\alpha,16-4\alpha]\cos 2t;\\
&\quad [6-3\beta+6+3\beta]\cos t+[6-4\beta,6+4\beta]\cos 2t;\\
&\quad [16-6\gamma,16+6\gamma]\cos t+[22-4\gamma,22+4\gamma]\cos 2t).
\end{aligned}$$

By using $(\alpha,\beta,\gamma)-$*cuts representation theorem, we obtain*

$$\begin{aligned}
f'_{gr}(t) = \Bigg(&\bigcup_{\alpha}\{[5+5\alpha,15-5\alpha]\cos t+[8+4\alpha,16-4\alpha]\cos 2t\};\\
&\bigcup_{\beta}\{[6-3\beta,6+3\beta]\cos t+[6-4\beta,6+4\beta]\cos 2t\};\\
&\bigcup_{\gamma}\{[16-6\gamma,16+6\gamma]\cos t+[22-4\gamma,22+4\gamma]\cos 2t\}\Bigg)\\
= (5,10,15;3,6,9;&10,16,22)\cos t+(8,12,16;2,6,10;18,22,26)\cos 2t.
\end{aligned}$$

Therefore, we have the gr-derivative $f'_{gr}(t)$ *is* $\mathcal{A}\cos t+2\tilde{\mathcal{A}}\cos 2t$, *whose graphical representation can be shown in Figure 7.*

7. The Neutrosophic Integral

This section presents the concept of granular integrability of $\mathcal{E}-$ valued functions and related properties. Especially, we introduce the fundamental the-

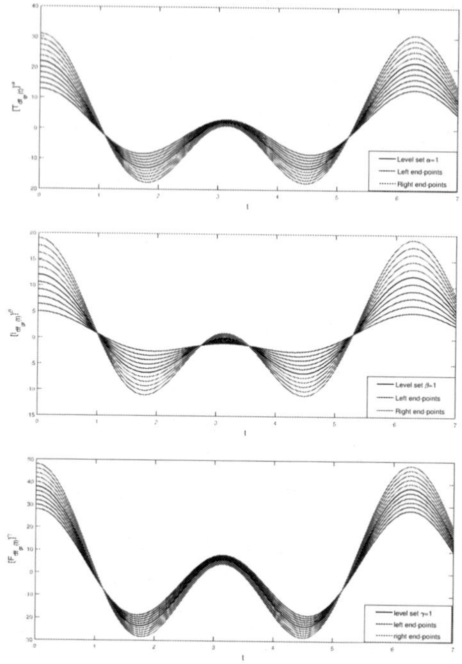

Figure 7. The $(\alpha, \beta, \gamma)-$ cuts of the granular derivative of neutrosophic function $f(t)$ on the interval $[0, 2\pi]$, where the black curve corresponds to the certain values, while the blue and red curves represent for the left and right end-points

orem of integral that play an important role in analysis theory.

Definition 7.1. *Let $f : [a, b] \to \mathcal{E}$ be a continuous function whose horizontal membership function $\mathcal{H}(f(t))$ is integrable on $[a, b]$. Then, if there exists a neutrosophic number $\mathcal{A} \in \mathcal{E}$ such that*

$$\mathcal{H}(\mathcal{A}) = \int_a^b \mathcal{H}(f(\tau))\, d\tau$$

then such number $\mathcal{A}$ is said to be the granular integral (gr-integral for short) of function f on $[a, b]$ and denoted by $\mathcal{A} = \int_a^b f(\tau) d\tau$. Moreover, the function f is then said to be gr-integrable on $[a, b]$.

Remark 7.1. *Similar to Proposition 6.1, we can easily prove that the integrability of the function f is equivalent to the integrability of its horizontal membership function.*

Next, we will present some characteristic properties of the gr-integral of a neutrosophic function f on an interval $[a, b]$

Proposition 7.1. *Let $f, g : [a, b] \to \mathcal{E}$ be gr-integrable functions on $[a, b]$. Then, the following statements hold*

(i) For all $\alpha, \beta \in \mathbb{R}$, function $\alpha f + \beta g$ is also gr-integrable on $[a, b]$ and

$$\int_a^b [\alpha f(\tau) + \beta g(\tau)]\, d\tau = \alpha \int_a^b f(\tau) d\tau + \beta \int_a^b g(\tau) d\tau.$$

(ii) If the function f is also gr-integrable on $[b, c]$ then f is gr-integrable on $[a, c]$ and

$$\int_a^c f(\tau) d\tau = \int_a^b f(\tau) d\tau + \int_b^c f(\tau) d\tau.$$

(iii) Due to the function f is gr-integrable on $[a, b]$, f is also gr-integrable on each sub-interval $[a, x] \subset [a, b]$ for all $x \in [a, b]$.

The following theorem is a fundamental theorem of analysis theory, that determines the form of the anti-derivative of a neutrosophic function f on an interval $[a, b]$.

Theorem 7.1. *Let $f : [a, b] \to \mathcal{E}$ be a continuous function on $[a, b]$. For each $t \in [a, b]$, denote*

$$\Phi(t) = \int_a^t f(\tau) d\tau.$$

Then, the function Ψ is a well-defined and gr-differentiable function on $[a, b]$. Such function is called an anti-derivative of the function f on $[a, b]$.

Proof. Indeed, let an arbitrary point $t_0 \in [a, b]$. Since the assumption that the function f is continuous at t_0, it implies that

$$\forall \epsilon > 0,\ \exists \delta > 0 \text{ such that } \forall\, t \in [a, b] : |t - t_0| < \delta,\ D^{gr}(f(t), f(t_0)) < \epsilon,$$

or equivalently, $\lim_{t \to t_0} f(t) = f(t_0)$.

Next, for h sufficiently near 0, we consider the following quotient

$$\begin{aligned}\frac{\Delta\Phi}{\Delta t_0} &= \frac{1}{h}\left[\Phi(t_0 + h) \ominus^{gr} \Phi(t_0)\right] \\ &= \frac{1}{h}\left[\int_a^{t_0+h} f(\tau)d\tau \ominus^{gr} \int_a^{t_0} f(\tau)d\tau\right],\end{aligned}$$

whose granular representation can be given as follows

$$\begin{aligned}\mathcal{H}\left(\frac{\Delta\Phi}{\Delta t_0}\right) &= \frac{1}{h}\left[\int_a^{t_0+h} \mathcal{H}(f(\tau))d\tau - \int_a^{t_0} \mathcal{H}(f(\tau))d\tau\right] \\ &= \frac{1}{h}\int_{t_0}^{t_0+h} \mathcal{H}(f(\tau))d\tau \\ &= \frac{1}{h}\int_{t_0}^{t_0+h} f^{gr}(\tau, \alpha, \beta, \gamma, \mu_f)d\tau.\end{aligned}$$

Then, using Mean Value Theorem for integrals, we immediately get

$$\begin{aligned}\mathcal{H}\left(\frac{\Delta\Phi}{\Delta t_0}\right) &= \frac{1}{h}\int_{t_0}^{t_0+h} f^{gr}(t_0 + \xi h, \alpha, \beta, \gamma, \mu_f)d\tau, \quad \xi \in (0, 1) \\ &= f^{gr}(t_0 + \xi h, \alpha, \beta, \gamma, \mu_f),\end{aligned}$$

Thus, due to the fact that $t_0 + \xi h$ tends to t_0 as $h \to 0$, it follows that

$$\begin{aligned}\lim_{h \to 0} \mathcal{H}\left(\frac{\Delta\Phi}{\Delta t_0}\right) &= \lim_{h \to 0} f^{gr}(t_0 + \theta h, \alpha, \beta, \gamma, \mu_f) \\ &= f^{gr}(t_0, \alpha, \beta, \gamma, \mu_f),\end{aligned}$$

that means $\mathcal{H}\left(\Phi'_{gr}(t_0)\right) = f^{gr}(t_0, \alpha, \beta, \gamma, \mu_f)$. Finally, by applying inverse transformation introduced in Proposition 4.1, we obtain that

$$[\Phi'_{gr}(t_0)]_{(\alpha,\beta,\gamma)} = \mathcal{H}^{-1}(f^{gr}(t_0, \alpha, \beta, \gamma, \mu_f)) = [f(t_0)]_{(\alpha,\beta,\gamma)}$$

for all α, β, $\gamma \in [0, 1]$. Hence, the proof is completed. □

Corollary 7.1. *For a neutrosophic function $f : [a, b] \to \mathcal{E}$ and $x \in [a, b]$, we have that*

$$\frac{d_{gr}}{dx}\left(\int_a^x f(\tau)d\tau\right) = f(x).$$

Next, we present an other theorem of integral, namely Newton-Leibniz formula, that give a calculation of gr-integral of a neutrosophic function f on a finite interval $[a, b]$.

Theorem 7.2 (Newton-Leibniz Formula). *Assume that $\Phi : [a, b] \to \mathcal{E}$ is a gr-differentiable function and the $\mathcal{E}-$ function $f(t) := \Phi'_{gr}(t)$ is continuous on $[a, b]$. Then, the function f is gr-integrable and*

$$\int_a^b f(\tau)d\tau = \Phi(b) \ominus^{gr} \Phi(a).$$

Proof. Firstly, we denote

$$\omega(t) = \int_a^t f(\tau)d\tau, \quad t \in [a, b].$$

As a consequence of Theorem 7.1, we deduce that the function $\omega(t)$ is an anti-derivative of function f on $[a, b]$ whose granular representation can be given as follows

$$\omega^{gr}(t, \alpha, \beta, \gamma, \mu_\omega) = \int_a^t f^{gr}(\tau, \alpha, \beta, \gamma, \mu_f)d\tau, \text{ for each } \alpha, \beta, \gamma \in [0, 1],$$

which means $\omega^{gr}(t, \alpha, \beta, \gamma, \mu_\omega)$ is an anti-derivative of $f^{gr}(t, \alpha, \beta, \gamma, \mu_f)$ on the interval $[a, b]$.

On the other hand, by the assumption that $f(t) := \Phi'_{gr}(t)$, we can see that $\Phi(t)$ is an anti-derivative of the neutrosophic function $f(t)$ on $[a, b]$. Thus, we deduce that $\Phi^{gr}(t, \alpha, \beta, \gamma, \mu_\Phi)$ is another anti-derivative of the function $f^{gr}(t, \alpha, \beta, \gamma, \mu_f)$ on $[a, b]$ and hence,

$$\begin{aligned}\Phi^{gr}(t, \alpha, \beta, \gamma, \mu_\Phi) &= \omega^{gr}(t, \alpha, \beta, \gamma, \mu_\omega) + C \\ &= \int_a^t f^{gr}(\tau, \alpha, \beta, \gamma, \mu_f)d\tau + C, \end{aligned} \tag{5}$$

where C is a constant vector. In addition, if we substitute $t = a$ then

$$\Phi^{gr}(a, \alpha, \beta, \gamma, \mu_\Phi) = \int_a^a f^{gr}(\tau, \alpha, \beta, \gamma, \mu_f) d\tau + C,$$

or equivalently, $\Phi^{gr}(a, \alpha, \beta, \gamma, \mu_\Phi) = C$. Thus, the equality (5) becomes

$$\Phi^{gr}(t, \alpha, \beta, \gamma, \mu_\Phi) = \int_a^t f^{gr}(\tau, \alpha, \beta, \gamma, \mu_f) d\tau + \Phi^{gr}(a, \alpha, \beta, \gamma, \mu_\Phi).$$

Next, by letting $t = b$, we obtain that

$$\int_a^b f^{gr}(\tau, \alpha, \beta, \gamma, \mu_f) d\tau = \Phi^{gr}(b, \alpha, \beta, \gamma, \mu_\Phi) - \Phi^{gr}(a, \alpha, \beta, \gamma, \mu_\Phi).$$

Then, by applying the inverse transformation in Proposition 4.1, we have the (α, β, γ) – cuts can be given as follows

$$\int_a^b \mathcal{H}^{-1}\left(f^{gr}(\tau, \alpha, \beta, \gamma, \mu_f)\right) d\tau = \mathcal{H}^{-1}\left(\Phi^{gr}(b, \alpha, \beta, \gamma, \mu_\Phi) - \Phi^{gr}(a, \alpha, \beta, \gamma, \mu_\Phi)\right).$$

This follows that the following integral equality is fulfilled

$$\int_a^b f(\tau) d\tau = \Phi(b) \ominus^{gr} \Phi(a).$$

□

Example 7.1. *For two SVNNs*

$$\begin{aligned} \mathcal{A}_1 &= (4, 7, 10; 0, 1, 2; 3, 5, 7) \\ \mathcal{A}_2 &= (2, 4, 6; 1, 2, 3; 1, 3, 5) \end{aligned}$$

whose horizontal membership functions are given by

$$\begin{aligned} A_1^{gr}(\alpha, \beta, \gamma, \mu) &= (4 + 3\alpha + (6 - 6\alpha)\mu_1; 1 - \beta + 2\beta\mu_2; 5 - 2\gamma + 4\gamma\mu_3) \\ A_2^{gr}(\alpha, \beta, \gamma, \tilde{\mu}) &= (2 + 2\alpha + (4 - 4\alpha)\tilde{\mu}_1; 2 - \beta + 2\beta\tilde{\mu}_2; 3 - 2\gamma + 4\gamma\tilde{\mu}_3), \end{aligned}$$

we consider a neutrosophic function

$$\begin{aligned} \Phi : [0, 5] &\to \mathcal{E} \\ t &\mapsto \Phi(t) = \mathcal{A}_1 e^{-t} + \mathcal{A}_2 t^2, \end{aligned}$$

Then, the horizontal membership function of $\Phi(t)$ can be given by

$$\begin{aligned}\mathcal{H}(\Phi(t)) &= \Phi^{gr}(t,\alpha,\beta,\gamma,\mu,\tilde{\mu})\\ &= A_1^{gr}(\alpha,\beta,\gamma,\mu)e^{-t} + A_2^{gr}(\alpha,\beta,\gamma,\tilde{\mu})t^2\\ &= \big([4+3\alpha+(6-6\alpha)\mu_1]\,e^{-t} + [2+2\alpha+(4-4\alpha)\tilde{\mu}_1]\,t^2;\\ &\quad [1-\beta+2\beta\mu_2]\,e^{-t} + [2-\beta+2\beta\tilde{\mu}_2]\,t^2;\\ &\quad [5-2\gamma+4\gamma\mu_3]\,e^{-t} + [3-2\gamma+4\gamma\tilde{\mu}_3]\,t^2\big).\end{aligned}$$

By using similar arguments as in Example 4.2, we can prove that the neutrosophic function $\Phi(t)$ is granular differentiable on the closed interval $[0,5]$ and moreover, its granular derivative $f(t) := \Phi'_{gr}(t)$ has the granular representation as follows

$$\begin{aligned}f^{gr}(t,\alpha,\beta,\gamma,\mu,\tilde{\mu}) &= \frac{\partial\Phi^{gr}(t,\alpha,\beta,\gamma,\mu,\tilde{\mu})}{\partial t}\\ &= -A_1^{gr}(\alpha,\beta,\gamma,\mu)e^{-t} + 2A_2^{gr}(\alpha,\beta,\gamma,\tilde{\mu})t\\ &= \big([-4-3\alpha-(6-6\alpha)\mu_1]\,e^{-t} + [4+4\alpha+(8-8\alpha)\tilde{\mu}_1]\,t;\\ &\quad [-1+\beta-2\beta\mu_2]\,e^{-t} + [4-2\beta+4\beta\tilde{\mu}_2]\,t;\\ &\quad [-5+2\gamma-4\gamma\mu_3]\,e^{-t} + [6-4\gamma+8\gamma\tilde{\mu}_3]\,t\big),\end{aligned}$$

where $\alpha,\beta,\gamma \in [0,1]$ and $\mu = (\mu_1,\mu_2,\mu_3)$, $\tilde{\mu} = (\tilde{\mu}_1,\tilde{\mu}_2,\tilde{\mu}_3) \in [0,1]^3$. Finally, by employing $(\alpha,\beta,\gamma)-$ level sets representation theorem, we obtain that

$$f(t) = \mathcal{A}_3 e^{-t} + \mathcal{A}_4 t,$$

where $\mathcal{A}_3$ and $\mathcal{A}_4$ are SVNNs defined by

$$\begin{aligned}\mathcal{A}_3 &= (-10,-7,-4;-2,-1,0;-7,-5,-3)\\ \mathcal{A}_4 &= (4,8,12;2,4,6;2,6,10).\end{aligned}$$

In addition, it is easy to see that the function $f(t)$ is continuous on the interval $[0,5]$. Thus, since the fact that all assumptions of Theorem 7.2 are fulfilled, we immediately obtain

$$\begin{aligned}\int_0^5 f(\tau)d\tau &= \Phi(5) \ominus^{gr} \Phi(0)\\ &= \left(\mathcal{A}_3 e^{-5} + 5\mathcal{A}_4\right) \ominus^{gr} \mathcal{A}_3\\ &= \left(1-e^{-5}\right)\mathcal{A}_3 + 5\mathcal{A}_4.\end{aligned}$$

As a consequence of Definition 4.6, we have following result

Proposition 7.2. *Let $f, g : [a, b] \to \mathcal{E}$ be granular integrable functions defined on $[a, b]$. Then,*

(i) If $f \preceq g$ on $[a, b]$ then

$$\int_a^b f(\tau)d\tau \preceq \int_a^b g(\tau)d\tau.$$

(ii) If $\tilde{0} \preceq f$ on $[a, b]$ then

$$\tilde{0} \preceq \int_a^b f(\tau)d\tau,$$

where $\tilde{0}$ is zero neutrosophic number whose membership functions are all equal to 0.

Conclusion

In conclusion, this chapter is devoted to presenting some basic concepts on the space of single valued triangular neutrosophic numbers and constructing a foundation on neutrosophic calculus. Based on horizontal membership functions approach, a new representation of triangular neutrosophic number is introduced and additionally, the metric on space of single valued triangular neutrosophic numbers and the continuity of neutrosophic valued functions are also presented. Especially, a new type of derivative of neutrosophic function, namely granular derivative, is firstly defined based on granular difference beside the foundation of granular integral.

References

[1] Smarandache, F. (1998), *Neutrosophy: Neutrosophic Probability, Set, and Logic*, American Research Press, Rehoboth.

[2] Smarandache, F. (2015), *Neutrosophic Precalculus and Neutrosophic Calculus*, Europa-Nova, Brussels.

[3] Çevik, A., Topal, S., Smarandache, F. (2018), Neutrosophic Logic Based Quantum Computing, *Symmetry*, 10 No.11, 656-667.

[4] Çevik, A., Topal, S., Smarandache, F. (2018), Neutrosophic Computability and Enumeration, *Symmetry*, 10, No.11, 643-656.

[5] Smarandache, F. (2013), *Introduction to Neutrosophic Measure, Neutrosophic Integral, and Neutrosophic Probability*, Sitech & Education Publisher, Craiova.

[6] Smarandache, F. (2014), *Introduction to Neutrosophic Statistics*, Sitech & Education Publishing.

[7] Taç, F., Topal, S., Smarandache, F. (2018), Clustering Neutrosophic Data Sets and Neutrosophic Valued Metric Spaces, *Symmetry*, 10, No.10, 430-442.

[8] Ye, J. (2013), Multicriteria decision-making method using the correlation coefficient under single-valued neutrosophic environment, *International Journal of General Systems* 42, 386-394.

[9] Ye, J. (2014), Improved correlation coefficients of single valued neutrosophic sets and interval neutrosophic sets for multiple attribute decision making, *Journal of Intelligent and Fuzzy Systems*, 27, 2453-2462.

[10] Ye, J. (2014), Clustering methods using distance-based similarity measures of single-valued neutrosophic sets, *Journal of Intelligent Systems* 23, 379-389.

[11] Ye, J. (2014), Single valued neutrosophic cross-entropy for multicriteria decision making problems, *Applied Mathematics Modelling* 38, 1170-1175.

[12] Sahin, R., Liu, P. (2017), Possibility-induced simplified neutrosophic aggregation operators and their application to multi-criteria group decision-making. *Journal of Experimental & Theoretical Artificial Intelligence*, 29, No. 4, 769-785.

[13] Ye, J. (2017), Projection and bidirectional projection measures of single-valued neutrosophic sets and their decision-making method for mechanical design schemes. *Journal of Experimental & Theoretical Artificial Intelligence*, 29, No.4, 731-740.

[14] Mazandarani, M., Pariz, N., Kamyad, A.V. (2018), Granular differentiability of fuzzy-number-valued functions, *IEEE Tran. Fuzzy Syst.*, 26, No.1, 310-323.

[15] Bede, B. (2013), *Mathematics of Fuzzy Sets and Fuzzy Logic*, Springer.

[16] Chakraborty, A., Mondal, S.P., Ahmadian, A., Senu, N., Alam, S., Salahshour, S. (2018), Different forms of triangular neutrosophic numbers, De-Neutrosophication techniques, and their applications, *Symmetry*, 10, No.8, 327.

[17] Piegat, A., Landowski, M. (2016), Aggregation of inconsistent expert opinions with use of horizontal intuitionistic membership functions, *Novel Developments in Uncertainty Representation and Processing*, Springer, 215-223.

[18] Piegat, A., Landowski, M. (2017), Fuzzy arithmetic type-1 with HMFs, *Uncertainty Modeling*, Springer, 233-250.

[19] Son, N.T.K., Dong, N.P., Long, H.V., Son, L.H. (2019), Towards granular calculus of single-valued neutrosophic functions under granular computing, *Multimedia Tools and Applications*, 1-37. DOI.10.1007/s11042-019-7388-8.

In: An Essential Guide to Fuzzy Systems
Editor: Michael Gr. Voskoglou
ISBN: 978-1-53616-128-1

Chapter 10

APPLICATION OF HYBRID FUZZY POTENTIAL FIELD FOR NAVIGATION OF SUMO ROBOTS

Carlos Erlan Olival Lima *
Department of Mechanical Engineering,
Toyohashi University of Technology,
Toyohashi, Aichi, Japan

ABSTRACT

An improvement of the conventional potential field method on an autonomous sumo robot class is presented in this chapter. The potential field method can be applied in dynamic environments where the target and the obstacles are in movement. However, in this specific application it is considered that just the target is in movement and the obstacles are static. Using this method it is possible to determine the velocity and the path taken by the robot in order to reach the desired target in the shortest possible time. This improvement on the conventional potential field is used to determine the velocity and position of an autonomous sumo robot class, which is intended to push the opponent out of the arena. This hybrid method is composed of potential field approach with fuzzy logic,

* Corresponding Author's Email: erlanolival@gmail.com.

and the Mamdani model is used to obtain the necessary variables to determine the velocity and direction to be followed by the robot. Two fuzzy systems are used, both with three inputs and one output. The difference between the two fuzzy systems is the number of linguistics terms. The simulations have been done by the Software MATLAB to verify the enhancement of the hybrid method regards to the conventional potential field. The simulation in this work represents the arena used in sumo robot competitions and it is possible to determine the trajectory of the robot until finding the target. Accordingly to the simulation done, the hybrid method using Potential Field and Fuzzy system was 14% more effective than the simulation using just conventional Potential Field. This simulation is presented in a Cartesian coordinate system and from it the results are obtained regarding the relative distance from the robot to the target and velocity of the robot in every single moment. An analysis is also made of the movements of the robot along the trajectory. The movements of the robot are described in three parts. The first part of the movement corresponds to the initial acceleration motion. In this case the acceleration is quite high due to the high torque applied. This high torque provides the slip of the wheels during a certain time where the robot is accelerating but it is not changing its position in the space. When the wheels of the robot stop slipping, the robot starts to change its position in space. In this moment the robot accelerates to reach its maximum speed, which depends of the engine that is used. This is the second part of the movement. Finally, when the robot has reached its maximum speed, it begins to move through the values established by the method used.

Keywords: potential field, velocity planning, direction planning, mobile robot, fuzzy system

INTRODUCTION

Robots have been used for specialized tasks, which are mainly too dangerous for human beings. The robots can be remote-controlled, semiautonomous or autonomous. Autonomous robots act completely on their own in performing tasks, using microcontrollers or computers to control the movements of the robot and many different sensors to get the environment data [1]. The design of an autonomous robot is a complex task and the criteria of success are evaluated in terms of its capabilities to make decisions and to act by itself in a reliable and satisfactory manner [2].

When autonomous robots have to deal with unknown environments and imprecise information, the navigational problem becomes extremely complex; therefore different approaches for navigation have been studied. The main approaches are: visibility graph or roadmap, probabilistic road map, path planning and velocity planning methods and potential field approaches [3-5].

The potential field approach is the only method that deals with dynamic environment where the obstacles and the targets are in movement [6]. The potential field has been widely used in robotics for generating paths and for mobile robots navigation. This method creates a field that directs the robot to the goal position. This approach treats the robot as a point under the influence of an artificial potential field. The robot moves by following the field, the goal acts as an attractive force on the robot, and the obstacle acts as repulsive force [7]. Several drawbacks are reported for this approach such as: convergence to local minimum, heavy calculation requirement, and goal non reachable with obstacles nearby [8].

Therefore a modified model of the potential field, which was proposed in [9] has been used in this chapter. This model overcomes the convergence to local minimum and the goal non reachable with obstacles nearby problems. The model proposed by [9] has the disadvantage with respect to the navigation properly when the target is very close and the velocity of the robot is much greater than the velocity of the target. In this case the robot maintains the higher speed, which is not desirable, therefore the robot has difficulty to find the target in this situation.

The disadvantages of the model proposed by [9] have been overcome by using a fuzzy system. This new approach is developed using a fuzzy system to establish guidelines to be followed by the robot, regardless of the distance and speed of the robot relative to the target. Therefore, the hybrid method showed in this paper does not have the disadvantage to find the target when it is very close and the robot velocity is greater than the target velocity.

Another disadvantage of the potential field is the heavy calculation. This problem has been overcome with fuzzy systems, which reduce the high dimensionality of the system from eight inputs (relative position between robot and target, target velocity, target angle related to X-axis, relative position between robot and obstacle, obstacle velocity, obstacle angle related to X-axis, relative position angle between robot and obstacle, relative position angle between robot abd target) to only four inputs (relative position between robot and target, target velocity, relative position angle between robot and target, relative position angle between robot and obstacle). Furthermore, the computing power required with this reduction will be less than that used in mathematical modeling [5].

The approach used in this work has the objective of planning the velocity and direction of an autonomous sumo robot class. In this work the opponent is the target and the edge of the arena is the obstacle. The arena is the environment where the robot is moving. This arena is a ring with black surface and an edge painted with white color. The edge represents the obstacle and the black surface is the region of the arena where the robot and the target are moving. As the edge is considered the obstacle, it is defined as a stationary obstacle. Although this article applies the hybrid fuzzy-potential field method for autonomous sumo robots class, its application can be extended for other types of robots, such as the follower's line and soccer class. It can also be deployed in industry in unstructured or semi structured environments with limited information regarding the provision of static or dynamic objects.

In this chapter a simulation of a sumo robot contest is performed in order to show the advantages of the method applied. The simulation tracks the trajectory of the robot and the target until the position where the robot finds the target. From this simulation it was generated the graph of the robot velocity related to the target and the graph of the distance between robot and target. Those simulations have been done by the software MATLAB.

Modeling and Control Architecture

Modeling System

Figure 1 shows the trajectory of the robot, as well as the vectors of the velocity and the relative positions between robot and target and between robot and obstacle. To simplify the modeling only one target and one obstacle are considered in the calculations. The following notations are used to describe the system:

XoY: world coordinates in the workspace;
p_{tar} ε R_2: target position;
v_{tar} ε R_2: target velocity;
p ε R_2: robot position;
v ε R_2: robot velocity;
p_{obs} ε R_2: obstacle position;
v_{obs} ε R_2: obstacle velocity;
$p_{rt} = p_{tar} - p$: relative position between robot and target;
$p_{ro} = p_{obs} - p$: relative position between robot and obstacle;
$p_{ot} = p_{tar} - p_{obs}$: relative position between obstacle and target;
ψ: angle of p_{rt};
θ_{tar}: angle of v_{tar};
θ_{ro}: angle of p_{ro};
θ_{ot}: angle of p_{ot};
θ_{obs}: angle of v_{obs};
θ: angle of v;

In this paper it was considered three assumptions concerning the application of the potential field. The assumptions are:

Assumption 1. The positions p, p_{tar} and p_{obs} and velocities v, v_{tar} and v_{obs} are all known.

Assumption 2. The obstacle is covered by a circle or a convex polygon. In this paper is considered that the obstacle is covered by a circle.

Assumption 3. The robot, the target and the obstacle are treated as point mass.

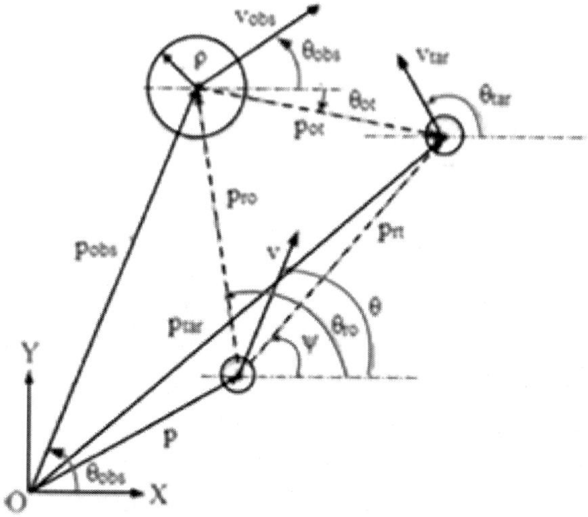

Figure 1. Representation of Robot Positioning with respect to Target and Obstacle [9].

The following assumptions on the limits of the operation of the robot are also made:

Assumption 4. The maximum linear and angular speeds of the robot are v_{max} and ω_{max} respectively.

Assumption 5. The speed of the target is limited by $||v_{tar}|| <= v_{max}$.

In the simulation presented in this paper, the position and velocity of the target and obstacle are obtained by ultrasound sensors. The v_{max} and ω_{max} are determined according to the engine used.

Robot Velocity Planning

The robot velocity varies according to the distance of influence of the obstacle. Attractive potential and repulsive potential act in the robot when it is in the edge of the arena. However, only attractive potential acts in the robot when it is moving on the black surface of the arena, because the obstacle has no influence in that moment. Therefore, there is no repulsive potential when the robot has no contact with the edge of the arena.

Figure 1 shows the angles formed between the robot and the horizontal axis and the angle between the horizontal axis and the path projected by the target. Figure 1 also shows the angle between the horizontal axis and velocity vectors of each element.

According with L. Huang in [9] the velocity and the direction of the robot once there is attractive potential and repulsive potential, they are represented by the equations (1) and (2), respectively.

$$||v|| = \left(||v_{tar}||\cos(\theta_{tar}-\psi)-\beta||v_{obs}||\cos(\theta_{obs}-\theta_{ro}\right) + \varepsilon_1||p_{rt}||)^2 + [||v_{tar}||^2(\sin(\theta_{tar}\bar{\psi}))^2]^{1/2} \quad (1)$$

$$\theta = \bar{\psi} + \sin^{-1}((v_{tar}sen(\theta_{tar} - \bar{\psi}))/(||v||)) \quad (2)$$

Where

$$\bar{\psi} \triangleq \tan^{-1} (\sin\psi - \beta\sin\theta_{ro}) \Big/ (\cos\psi - \beta\cos\theta_{ro}) \quad (3)$$

$$\beta = (\eta||p_{ro}||) \Big/ (\varepsilon_1||p_{rt}||) \quad (4)$$

$$\eta = \begin{cases} \varepsilon_2\rho^{-2}||\rho_{ro}||^{-1}(\rho^{-1} - \rho_o^{-1}), if\ \rho \le \rho_o \\ 0 \qquad , \quad else \end{cases}$$

$$\rho = ||\rho_{ro}|| - r_{obs}$$

The velocity and the direction of the robot as there is only attractive potential are represented by the equations (5) and (6), respectively.

$$||v|| = \left(||v_{tar}||^2 + 2\varepsilon_1||p_{rt}||||v_{tar}||\cos(\theta_{tar} - \psi\right) + \varepsilon_1{}^2||p_{rt}||^2)^{1/2} \quad (5)$$

$$\theta = \psi + \sin^{-1}((v_{tar} sen(\theta_{tar} - \psi))/(||v||)) \tag{6}$$

In equations (1) and (5) ε_l is an empirical constant that corresponds to the scaling factor for attractive potential. The constant ε_l is directly proportional to the speed of the robot, and then its value is defined mainly based in the maximum speed of the robot. In the equation (4) the variable η is showed, which depends of ε_2, ρ, ρ_o and ρ_{ro} to be determined. In the equation (4) ε_2 correspond to the scaling factor for repulsive potential, ρ denotes the minimum distance between the robot and the obstacle and ρ_o is the distance of influence of the obstacle.

Applying the Hybrid Fuzzy Potential Field

The potential field method has a large number of inputs needed to determine the velocity and the direction that the robot must follow. Because of this, a fuzzy system is used to reduce the high dimensionality of the system [10, 11]. Originally the method has eight inputs, but using the fuzzy system only four inputs are required by the potential field, and among them only three are inputs fuzzy.

The inputs required by the potential field method are: relative position between robot and target (p_{rt}), target velocity (v_{tar}), target angle related to X-axis (θ_{tar}), relative position between robot and obstacle (p_{obs}), obstacle velocity (v_{obs}), obstacle angle related to X-axis (θ_{obs}), angle of the relative position between robot and obstacle (θ_{ro}) and the angle related to p_{rt} (ψ).

By using the fuzzy system the only inputs needed are p_{rt}, v_{tar}, and ψ. The output of this fuzzy system is the difference between θ_{tar} and ψ. The value of the output fuzzy is used in the equation 1 to determine the robot velocity. The output of the fuzzy system used has a range of [- $\pi/2$, $\pi/2$]. By using the difference between θ_{tar} and ψ as output of the fuzzy system, the efficiency of the direction planning is better than the one obtained using the conventional potential field method. In the potential field method the variables θ_{tar} and ψ are determined separately. Therefore, the error is greater in the conventional potential field because there are two variables to determine. In the simulations showed in this paper is verified that using the fuzzy system the robot reaches the target faster than using only conventional potential field. Therefore, the fuzzy system reduces the number of inputs required and improves efficiency.

For determining the direction of the robot it is also necessary to establish θ_{ro}. This variable is obtained through the position of the infrared sensors. It is only necessary to set this variable when the robot detects the edge, because in this case there is repulsive potential. While the robot does not detect the edge, there is only attractive potential, in this case η is zero and $\overline{\psi}$ is equal to ψ.

The fuzzy system is applied for a robot with two active wheels and one support wheel. The configuration of the robot is shown in Figure 2.

In this research two fuzzy systems have been used. The two fuzzy systems have three input variables and one output variable. In the first system each variable has three linguistics terms with two triangular membership functions for the two first terms and one trapezoidal membership function for the last term. In the second system two fuzzy input variables have three linguistics terms and one has five linguistics terms with a triangular membership function for each term.

In the first fuzzy system each variable of the system has three linguistic terms. The inputs fuzzy and related linguistic terms are: Distance (p_{rt}) (Small, Medium, High), Target Velocity (v_{tar}) (Low, Medium, High), Tilt Straight Line (ψ) (Low, Medium, High). The output fuzzy and related linguistics terms are represented as Difference (Low, Medium, High), which corresponds to the difference between the variables ψ and θ_{tar}. The two first membership functions of each linguistic term are triangular and the last membership function is trapezoidal.

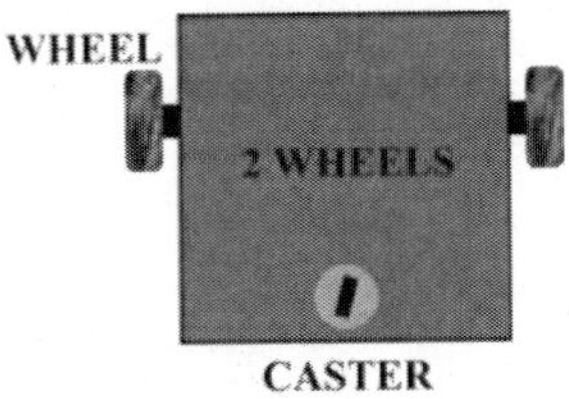

Figure 2. Robot Locomotion System [12].

The inputs of the second fuzzy system are equivalent the inputs of the first fuzzy system. However, in the second fuzzy system all membership functions are triangular and the input (ψ) has five fuzzy linguistic terms. The fuzzy linguistic terms of (ψ) are: Very Low, Low, Medium, High and Very High.

A total of 27 rules are required to implement the first fuzzy system while the second fuzzy system requires 45 rules. The model is based on fuzzy IF-THEN rules. A sample of the rules is described below:

R1: IF Distance is Small AND Target Velocity is Low AND Tilt Straight Line is Low THEN Difference is Low

R2: IF Distance is Small AND Target Velocity is Low AND Tilt Straight Line is Medium THEN Difference is Low

R3: IF Distance is Small AND Target Velocity is Low AND Tilt Straight Lime is High THEN Difference is Medium

The inference output from those rules is computed by the Mamdani (max-min) operator for composition, minimum operator for implication and centroid for defuzzification [5]. The fuzzy systems used are shown in Figure 3 (a) and Figure 3 (b), which presents a little scheme with the inputs and output of the fuzzy system.

To validate this system, a simulation is performed with respect to the direction the robot must follow to find its target. This simulation is presented in a Cartesian coordinate system and from this simulation the following results are obtained: relative distance between robot and target, velocity of the robot at each moment.

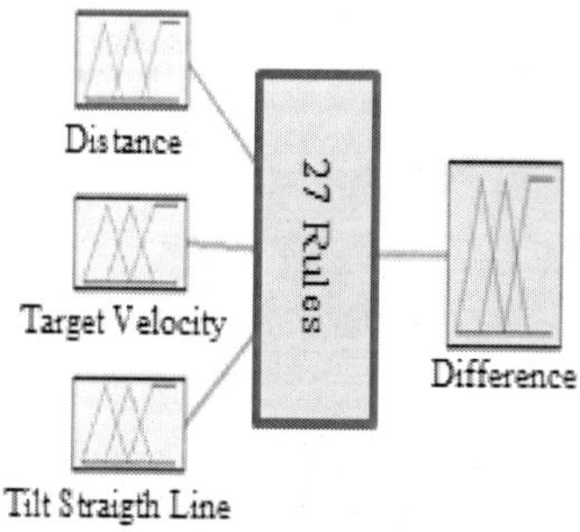

Figure 3 (a). Fuzzy Inference System, with three inputs and one output (First Fuzzy System).

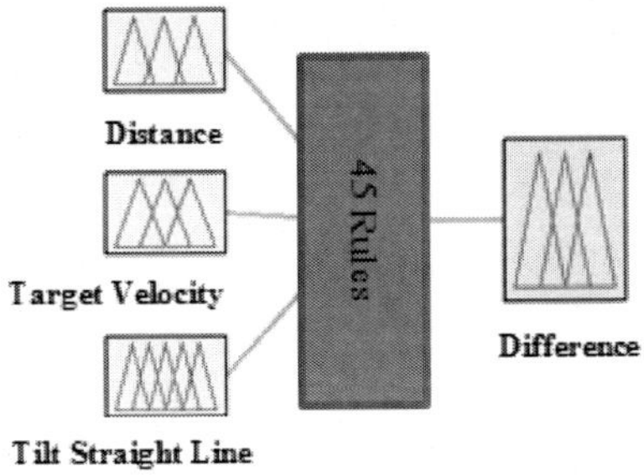

Figure 3 (b). Fuzzy Inference System, with three inputs and one output (Second Fuzzy System).

In this simulation, the movement of the robot can be described in three parts. The first part of the movement corresponds to the initial acceleration motion. In this moment the acceleration is higher due to high torque applied. This provokes the slippage of the wheels during a certain amount of time. In this first moment the robot is accelerating but it is not changing its position.

When the wheels of the robot stop slipping, the robot starts to change its position. In this moment the robot accelerates until it reaches its maximum speed, and this speed depends of the engine used. This is the second part of the movement. Finally, after the robot has reached its maximum speed, it begins to move through the values established by the path planning method used.

SIMULATION OF THE TRAJECTORY OF A SUMO ROBOT CLASS

Two simulations have been done by the software MATLAB to validate the hybrid method used. The first simulation tracks the target movement in a straight path parallel to X-axis. This simulation is used to determine the most effective scaling factor. Then, the value determined for the scaling factor is used in the second simulation. The first simulation is shown in the Figure 4. In this simulation the target describes a linear motion with a constant velocity and the target is initially set in the position [0 70] while the robot is set in the position [0 0].

The maximum speed defined to the target is 20 cm/s, which is the average speed achieved by the majority of robots used in the competition. [13].

Two graphs were generated from this environment, the first one describes the relative distance between robot and target, and the latter one describes the relative velocity of the robot. Those graphs are generated in function of the time spent to reach the target. They can be observed in the Figure 5 and Figure 6, respectively.

Three values for the constant ε_1 were fixed to determine the shortest time required for the robot to find the target. This constant has influence in determining the velocity of the robot and the trajectory that it will follow. The constant ε_1 is directly proportional to the velocity of the robot. Because of that, if a high value is set for the constant ε_1, the speed of the robot at each moment will increase. The value for this constant is set accordingly the maximum speed of the robot, which depends of the motor, battery and reduction ratio gears.

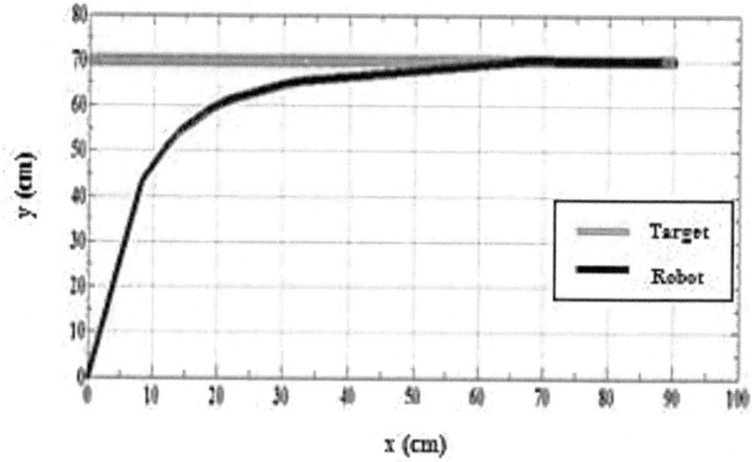

Figure 4. Trajectory described by the robot to find the target.

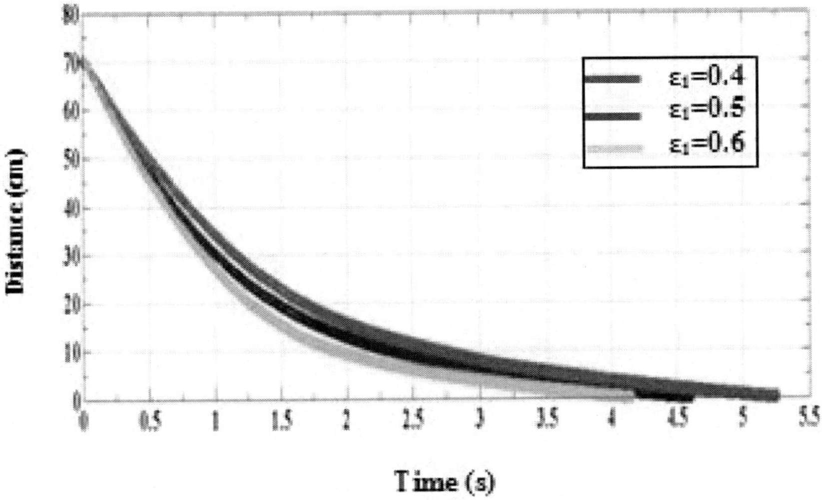

Figure 5. Distance between Robot and Target – Red Line ($\varepsilon_1 = 0.4$), Blue Line ($\varepsilon_1 = 0.5$), Cyan Line ($\varepsilon_1 = 0.6$).

In Figure 5 the robot can quickly find the target when the value for the constant ε_1 is higher. Therefore, the defined velocity for the robot is higher when the value of ε_1 is higher. Nevertheless the largest possible value of ε_1 is 0.5. If a higher value than that is applied, the robot speed would be greater than the maximum speed allowed by the engine used. The engine used for the calculus was the Pittman [14] with reduction ratio of 29.5:1. Using this motor, the highest speed defined for the robot of this simulation is 51 cm/s. The time required to reach the target is 4.616 seconds when the scaling factor is 0.5. For other values shown in the graph, the time is 4.161 and 5.266 seconds and the scaling factors are equivalent to 0.6 and 0.4, respectively.

The torque of the robot is 4.52 N.m when it is using the highest speed, which enables the robot to carry a weight five times its own. Therefore, this robot can carry up to 15 kg since its mass is 3 kg.

Figure 6 illustrates how the velocity of the robot changes in function of time and it should be also noted that the movement of the robot is divided into three parts. Figure 7 illustrates the first part of the movement, in which the wheels are slipping. In this moment the robot is not changing its position.

In Figure 6 the second and third part of the movement are shown. The second part of the movement is correlated to the moment in which the robot starts to accelerate until it reaches its maximum speed. During this part of the movement the hybrid fuzzy-potential method acts only to define the direction that the robot should follow. The third part of the movement occurs when the robot reaches its maximum speed. From this moment, the path planning method used will define the velocity and the trajectory of the robot.

In Figure 7 the robot is slipping during a time interval of 0.015 seconds, and its speed at the end of this first step is 0.1766 cm/s. From this point, the robot starts to move with a speed of 0.1766 cm/s, which is represented by the red square in the graph.

The movement that the robot performs is uniformly accelerated, and the initial acceleration is given by the equation presented by [12]. It represents the acceleration in which the robot is slipping, and its value is 11.772 m/s^2.

About the second part of the movement, it occurs during a time interval of 0.036 seconds, and it does not depend of the scaling factor. During this time, the robot is accelerated accordingly the equation presented by Meggiolaro in [12]. From this equation is possible to obtain the acceleration without slipping, and the obtained value was 12.5 m/s^2.

The second simulation has been done to validate the hybrid method used. In this simulation is created an environment that represents the arena of a sumo robot contest. A circle with diameter of 152 centimeters representing the sumo arena is shown in Figure 8. In this representation the target firstly moves in a straight line and its acceleration is 8 m/s^2. This occurs until it reaches its maximum speed.

When the target reaches its maximum speed, the motion becomes linear and from this point the target is moving at a constant speed until it reaches a certain point inside the circle. After that, the target attempts to escape from the robot instead of trying to find it. In this moment the target is tracking a sinusoidal path and its acceleration is 8 m/s^2.

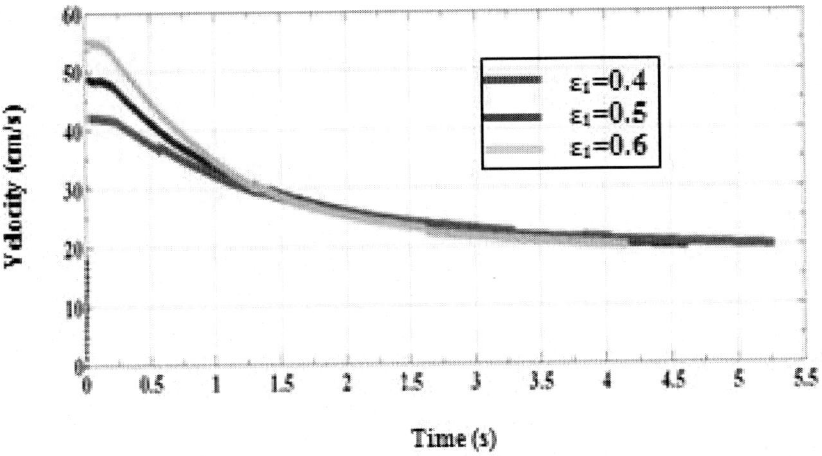

Figure 6. Velocity of the Robot - Red Line ($\varepsilon_1 = 0.4$), Blue Line ($\varepsilon_1 = 0.5$), Cyan Line ($\varepsilon_1 = 0.6$)

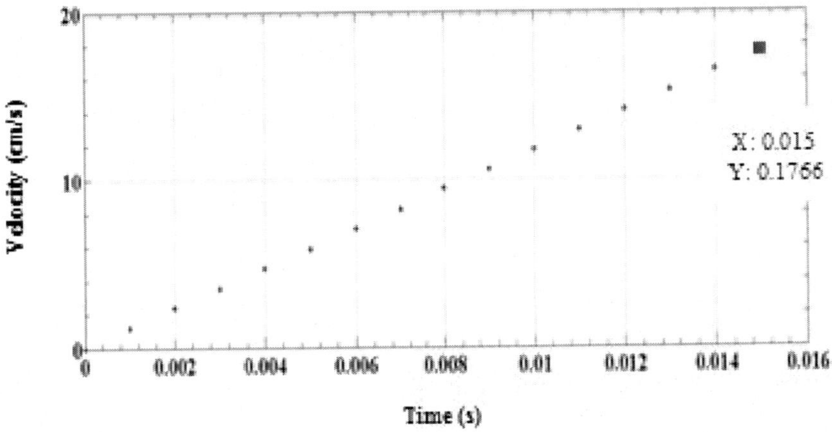

Figure 7. Representation of the time interval in which the wheels of the robot are slipping.

In this simulation the target is initially set in the position [50 114] while the robot is in the position [50 38]. The positions determined for the robot and the target were chosen in a way that the minimum distance between robot and target in the beginning of the simulation is 24 centimeters. Therefore, the simulation was initialized with the robot and target placed in opposite sides of the arena.

The maximum speed of the target is 20 cm/s, which is the average speed achieved by the majority of robots used in the competition. The obstacle is represented for the circle and its velocity is zero [13].

The movement of the target was designed in order to replicate some of the possible movements of a robot during the match. In the first moment, the target tries to find the robot. This movement occurs while the target is moving in straight line. About the moment where the target is tracking a sinusoidal path, this corresponds to the time in which the target is trying to escape from the robot. In this time interval, the target describes its movement in the form of a parabola in order to make it difficult for the robot to reach the target.

The motion that the robot describes to find the opponent is shown in Figure 8. To better illustrate the movement that the target describes, its movement is shown in more details in Figure 9.

In Figures 10 and 11 it is shown the following graphs: relative position between robot and target, velocity of the robot. Those graphs were generated in function of time, and they show the difference between the time obtained using only potential field and the time obtained using hybrid fuzzy potential field method.

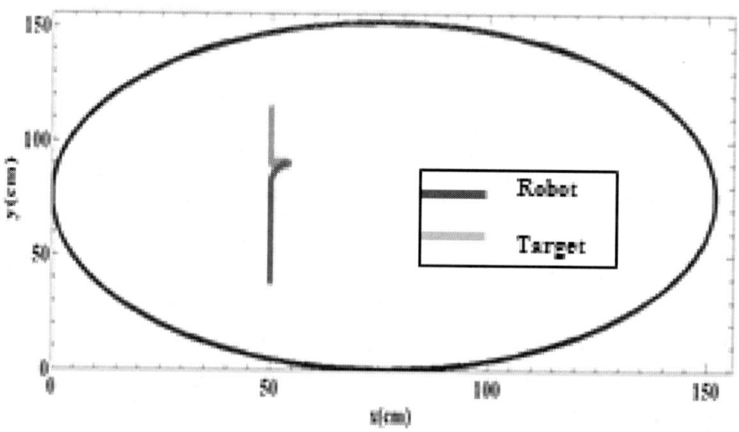

Figure 8. Trajectory described by the robot to find the target.

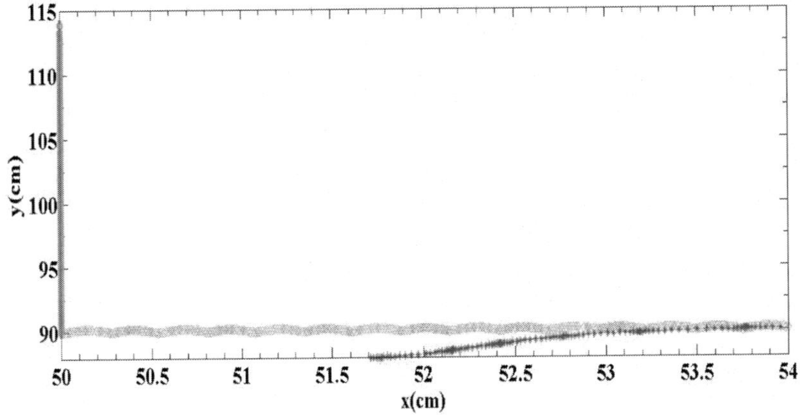

Figure 9. Trajectory Described by Target.

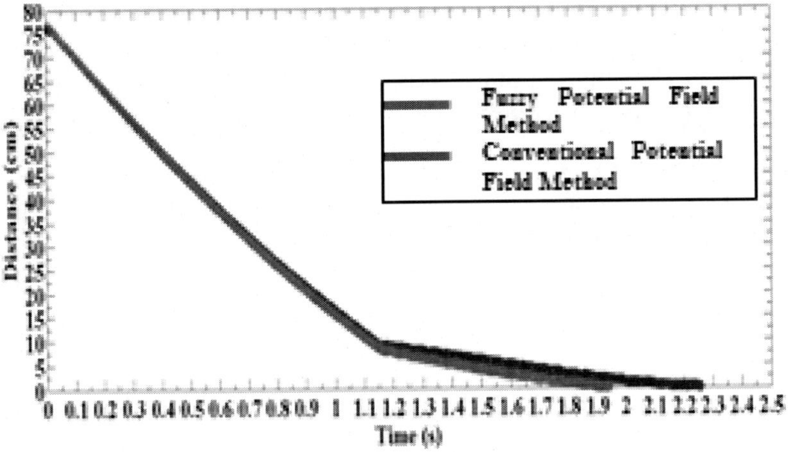

Figure 10. Distance between robot and target.

In the analysis of the graphs of Figure 10 and 11 it can be verified that the robot finds the target quickly when it is using the hybrid fuzzy potential field method. This occurs because the fuzzy system establishes specific conditions to determine the velocity and direction of the robot in order to decrease the measurement errors.

The measurement error occurs mainly due to the difficulty to establish the value of θ_{tar}. When just the conventional potential field is applied, this angle is obtained by deriving the movement equation. On the contrary, if the hybrid fuzzy potential field method is applied, it is not necessary to obtain the value of θ_{tar}. In this hybrid method it is necessary to obtain the difference between the angles θ_{tar} and ψ, and this difference is the output of the fuzzy system.

The output difference is obtained using equations (1), (2), (5) and (6). According to these equations, the higher the velocity of the robot the lower is the difference between the angles; therefore this value depends onn the cosine. The opposite occurs with the direction, which depends on the sine of the difference between the angles.

Apart from the value of the difference between the angles θ_{tar} and ψ, another factor that has influence in the navigation is the constant ε_1. As it was mentioned previously, the largest possible value for ε_1 is 0.5. This happens because, if a value greater than that is set for this constant, the robot speed would be higher than the maximum speed allowed by the engine used. The engine used in this simulation was the Pittman's [14], whose gear reduction ratio is 29.5:1. The highest speed that the robot can reach in this simulation is 51 cm/s. The time required to find the target by applying the conventional potential field method with this scaling factor is 2.261 seconds. Applying the hybrid fuzzy potential field method, the time required is 1.944 seconds. In conclusion, the hybrid fuzzy potential field method was 14.2% more efficient than the potential field method.

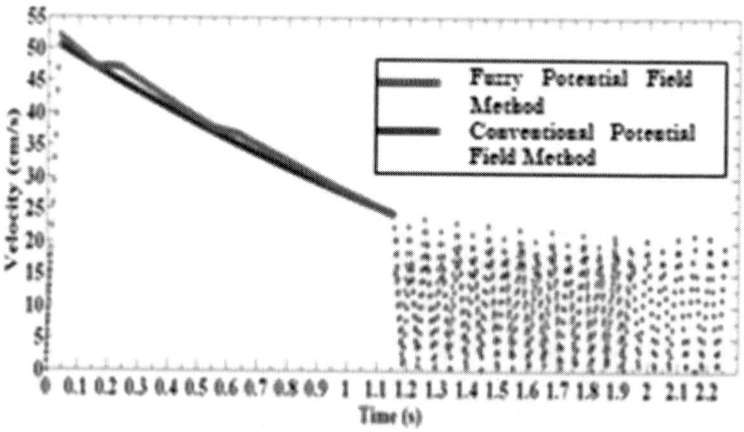

Figure 11. Velocity of the robot.

The performance of the robot using the hybrid fuzzy potential field and the conventional potential field method is shown in Table 1. Table 1 shows the total distance travelled by the robot until it reaches the target and the amount of time required to reach the target.

As it was mentioned previously, the movement that the robot performs is divided in three parts. Figure 12 represents the third part of the movement exactly in the moment in which the target velocity is not constant. The velocity of the robot depends on the velocity of the target as is shown in (1). As a result, if the target accelerates or decelerates, the robot does the same. This happens because those variables are directly proportional. As a consequence, it avoids unnecessary waste of energy supplied by the battery, which becomes a big advantage for the robot.

Table 1. Performance of the Robot

	Total Distance that the Robot Moves until it reaches the target (cm)	Total Time Required to Reach the Target (s)
Hybrid Fuzzy Potential Field Method	54.093	1.944
Conventional Potential Field	54.962	2.261

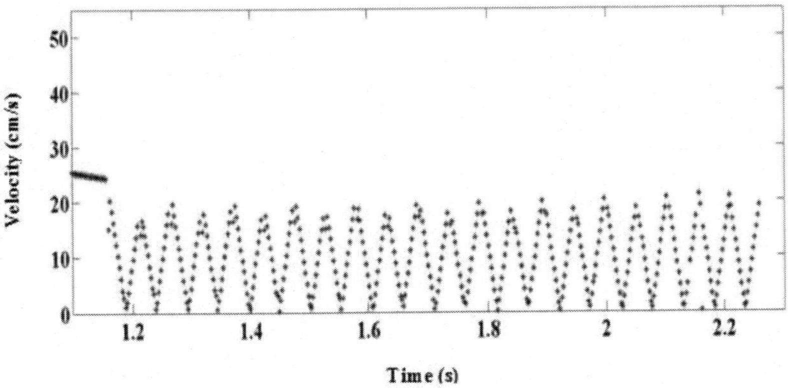

Figure 12. Velocity of the Robot When the Target Velocity is not constant.

APPLICATION

As it has already been mentioned, this method is applied for an autonomous sumo robot class aiming to improve the perception of the robot with respect to an opponent. The robot possess thirteen ultrasound sensors arranged side by side with a tilt angle of 15°. This configuration is set due to the opening angle of the signal emitted by the sensor, which is 15°. This configuration is used to cover up the front view of the robot. This sensor was chosen because of the small opening angle. As a result, the grain size is small, which allows more accurate perception of the environment. The location of the sensor that detects an object is used as input for the fuzzy system. This input is represented by the linguistic variable Tilt Straight Line. In Figure 13 an illustration is shown of how the sensors are supposed to be arranged in the robot. The ultrasonic sensor used is the HC - SR04, which has a range of 2 cm up to 4 meters [15].

CONCLUSION

An enhancement of the conventional potential field method by using fuzzy system was presented in this chapter. The hybrid fuzzy potential field method was applied to make the velocity and direction planning of an autonomous sumo robot class. Two fuzzy system models have been used to verify the efficiency of this hybrid method. The hybrid method was investigated based on simulations done by the software MATLAB.

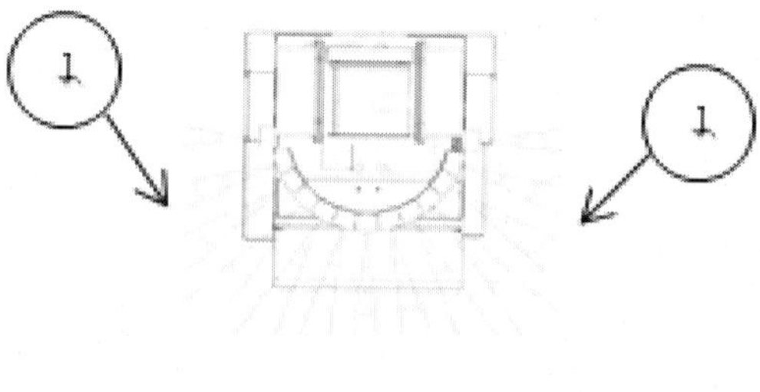

Figure 13. Arrangement of the Sensors; 1 – Opening Angle of the Sensor.

The simulations presented in this chapter verify that the hybrid fuzzy potential field method is more efficient than the conventional potential field method. By using the hybrid method the robot can find the target in only 1.944 seconds while by using the conventional potential field the time is 2.261 seconds. Therefore, the hybrid fuzzy potential field method is 14% more effective than the conventional potential field method.

The hybrid method presented in this chapter takes into account the coefficient of friction between the tire and the contact surface as well as the torque required to drive the robot. As a result, this hybrid method can be used in various types of robots, independently of the environment. In future it is expected to combine the hybrid fuzzy potential field method with neural networks aiming to improve the efficiency of the approach.

REFERENCES

[1] Carrol, T; Miles, P. *Build Your Own Combat Robot*, Ed. McGraw-Hill, 2002.

[2] Marc Boumedine, M; Alejandro Ramirez, S. "Fuzzy knowledge-based controller design for autonomous robot navigation," *Expert Systems with Applications*, 14, (1998), 179-186.

[3] Filliat, J. Meyer. "Map-based navigation in mobile robots: I. A review of localization strategies," *Cogn Syst Res*, (2003), 4, 243–282.

[4] Meyer, J; Filliat, D. "Map-based navigation in mobile robots: II. A review of map-learning and path-planning strategies," *Cogn Syst Res*, (2003), 4, 283–317.

[5] Jaradat, MAK; Garibeh, MH; Feilat, EA. "Autonomous Mobile Robot Dynamic Motion Planning Using Hybrid Fuzzy Potential Field," *Soft Comput*, (2012), 16, 153-164.

[6] Borenstein, J; Koren, Y. Real-time obstacle avoidance for fast mobile robots, *IEEE Transactions on Systems, Man, and Cybernetics*, 20 (4), (1989), 1179–1187.

[7] Siegwart, R; Nourbakhsh, IR; Scaramuzza, D. *Introduction to Autonomus Mobile Robots*, Ed. Massachusetts Institute of Technology, 2004.

[8] Pozna, C; Troester, F; Radu-Emil, P; Tar. József, K; Preitl, S. "On the Design of an Obstacle Avoiding Trajectory: Method and Simulation," *Mathematics and Computers in Simulation*, (2009), 2211 – 2226. Ed. Elsevier.

[9] Huang, L. "Velocity planning for a mobile robot to track a moving target — a potential field approach," *Robotics and Autonomous Systems*, 57, (2009), 55–63.

[10] Pradhan, SK; Parhi, DR; Pandar, AK. "Fuzzy Logic Techniques for Navigation of Several Robots," *Applied Soft Computing*, (2009), 290 – 304.

[11] Parhi, R; Mohanta, JC. "Navigational control of several mobile robotic agents using Petri-potential-fuzzy hybrid controller," *Applied Soft Computing*, 11, (2011), 3546-3557.

[12] Meggiolaro, MA. "RioBotz Combot Tutorial," *UFRJ*, 2009.

[13] *ROBOCORE*. (2019, March, 24) Availible: <http://www.robocore.net/modules.php?name=Forums&file=viewtopic&t=3979>.

[14] *PITTMANEXPRES*S. (2013, July) Availible: <http://www.ametektip.com/Ametek/PublicPDF/PittmanExpress.pdf>.

[15] *Ultrasonic Ranging Module* HC – SR04. (2019, March, 24) Availible: < https://www.mouser.com/ds/2/813/HCSR04-1022824.pdf >.

BIOGRAPHICAL SKETCH

Carlos Erlan Olival Lima

Toyohashi University of Technology, Toyohashi, Japan

Education:

- Oct. 2017 – Sep. 2019 – Master's Degree in Robotics and Automation and Researcher at Toyohashi University of Technology, in Toyohashi, Japan (Monbukagakusho Scholarship).

- Sept. 2014 – July. 2015 –Graduated in Mechatronics Engineering at Beijing Institute of Technology in Beijing, China (CSC Scholarship).
- Sept. 2013 – July. 2014 – Program to study Chinese Language at Tianjin University in Tianjin, China (CSC Scholarship).
- Mar. 2009 – Jan. 2017 – Graduated in Mechanical Engineering at Federal Institute of Education, Science and Technology of Piauí – IFPI in Teresina, PI, Brazil.

Research and Professional Experience:

Current Research:

Currently, I am taking my Master's degree in Robotics at Toyohashi University of Technology, in Japan. The research that I am developing is a power wheelchair using omnidirectional wheels and articulated suspension, similar with the suspension used in a rover robot.

The main objective of the project is the development of a power wheelchair able of moving in narrow spaces. The omnidirectional wheels are being used due to its high maneuverability. Moreover, the new design is been developed in a way that allows this wheelchair robot to transfer objects and perform dexterous tasks while moving around freely. In order to make the wheelchair to have a certain autonomy, Potential Field and Fuzzy Systems are applied to perform the navigation system. In this way, the user can perform other activities even if the wheelchair is moving around.

Professional Experience:

Mechanical Engineer with experience in areas such as: 3D Design, Maintenance, Robotics and Renewable Energies. I published a book about an application of Solar Energy in the Industry and I got two awards, one related to Robotics and the other one to Sustainability. Co-founder of a Start-up company considered one of the 15 best Start-up ideas in Brazil in the area of Social Business. Able to speak four languages (English, Japanese, Chinese and Portuguese) with degrees obtained in China and Japan.

Honors:

- Co-founder and business development of the Start-up *Leadservice* related to maintenance and development of new technologies for disabled people;
- Third Position in the XII Century Cup at Beijing Institute of Technology in China (Project: *Navigation of Robots Using Artificial Intelligence*);
- XXV Young Scientist Award, 2011 edition (Project: *Air-Conditioning System by Solar Energy Absorption*);

Publications of the Last 3 Years:

- *Design and Analysis of a new Type of Mecanum Wheel* (DOI: IASTEM.13032019.11781);
- *Vibration and Strength Analysis of a new Type of Mecanum Wheel Based on FEM* (DOI: IASTEM.20032019.11854);
- *Megabots Team: A Team for the Category Sumo Autonomous 3 kg* (DOI: 10.17265/2159-5275/2016.05.005);
- *Air-conditioning system by adsorption using solar energy* (*Sistema de Climatização por Adsorção Utilizando Energia Solar*) ISBN-13: 978-3-639-75358-5;
- *An enhancement in conventional potential field using a fuzzy system for navigation of a sumo robot* (DOI: 10.1109/ICAR.2013.6766506);
- *Velocity and direction planning in a sumo robot type using the method of potential field with fuzzy systems* (DOI: 10.1109/ICAR.2013.6766490);

In: An Essential Guide to Fuzzy Systems ISBN: 978-1-53616-128-1
Editor: Michael Gr. Voskoglou

Commentary

GENERALIZATIONS OF FUZZY SETS AND RELATIVE THEORIES

Michael Gr. Voskoglou*
Graduate Technological Educational Institute of Western Greece,
Koukouli, Patras, Greece

From the time that Zadeh introduced the theory of *Fuzzy Set (FS)* in 1965 [20] many efforts have been made for improving its effectiveness to deal with uncertain, ambiguous and vague situations. As a result a series of extensions and generalizations of the ordinary FS followed and relevant theories have been proposed as alternatives to the FS theory. The target of the present commentary is to offer a framework for those wanting to study deeper those theories that have been found many and important applications during the last 60 years to almost all sectors of the human activity.

An introduction to the FS theory and the infinite-valued *Fuzzy Logic (FL)* that has been generated by it can be found in Chapter 1 of the present book. It is recalled here that a FS A on the crisp set of the discourse U is

* Corresponding Author's Email: mvoskoglou@gmail.com.

defined with the help of its *membership function m:* U ⟶ [0, 1] as the set of the ordered pairs A = {(x, m(x)): x ∈ U}. For reasons of simplicity many authors identify a FS with its membership function. FSs are mathematical models representing vagueness and imprecise information that have the capability of recognizing, representing, manipulating, interpreting, and utilizing data and information which are vague and lack certainty. However, the creditability of a FS in representing the corresponding real situation depends on the successful definition of its membership function. For general facts on FSs we refer to the book [8].

Zadeh introduced also the *Fuzzy Numbers (FNs)* [21] as a special form of FSs on the set of the real numbers. He defined the basic arithmetic operations on them in terms of his *extension principle,* which provides the means for any function mapping the crisp set X to the crisp set Y to be generalized so that to map fuzzy subsets of X to fuzzy subsets of Y. FNs play an important role in fuzzy mathematics analogous to the role of the ordinary numbers in the traditional mathematics. For general facts on FNs we refer to the book [7]. In Chapter 1 of the present book Triangular Fuzzy Numbers (TFNs), the simplest form of FNs, are used for the assessment of computational thinking skills.

Forms of multi-valued logic have been studied since the 1920s, notably by the Polish philosopher Jan Lukasiewicz (1878-1956) and by the Polish-Jew logician and mathematician Alfred Tarski (1901-1983). FL, in contrast to the traditional bi-valued logic of Aristotle, is an infinite-valued logic in which the truth values of variables may be any real number between 0 and 1. It is employed to handle the concept of partial truth, where the truth value may range between completely true (1) and completely false (0). FL is based on the observation that people make frequently decisions in terms of imprecise and non-numerical information.

As it was expected, the far-reaching theory of FL aroused some objections to the scientific community. While there have been generic complaints about the fuzziness of assigning values to linguistic terms, the most cogent criticisms come from Haack [6] in 1979, who argued that there are only two areas – the nature of Truth and Falsity and the fuzzy systems' utility – in which FL could be possibly needed, and then maintained that in both cases it can be shown that FL is unnecessary. Fox [5] responded against to her objections, his most powerful argument being that traditional and FL need not be seen as competitive, but as complementary and that FL, despite the objections of classical logicians, has found its way into practical applications to almost all fields of human activity, from control theory to artificial intelligence and has been proved very successful there.

In 1975 Zadeh generalized the ordinary FS, otherwise called *type-1 FS*), to the *type-2 FS* [21], so that more uncertainty can be handled connected to the membership function. The membership function of a type-2 FS is three - dimensional, its third dimension being the value of the membership function at each point of its two – dimensional domain, which is called *Footprint of Uncertainty (FOU)*. The FOU is completely determined by its two bounding functions, a *lower* membership function and an *upper* membership function, both of which are type-1 FSs. When no uncertainty exists about the membership function, then a type-2 FS reduces to a type-1 FS, in a way analogous to probability reducing to determinism when unpredictability vanishes. In order to distinguish between a type-1 and a type-2 FS, a tilde symbol is put over the FS, so that A denotes the type-1 FS and $\tilde{A}$denotes the comparable type-2 FS. Nevertheless, Zadeh didn't stop there, but in the same paper [21] generalized the type-2 FS to the *type-n FS* n = 1, 2, 3, … However, when Zadeh proposed type-2 FS in 1975, the time was not right for researchers to drop what they were doing with type-1 FS and focus on type-2 FS. This changed in the late 1990s as a result of Prof. Jerry Mendel and his student's works on type-2 FS [10]. Since then, more and more researchers around the world are writing articles about type-2 FS and systems.

Another application for FS that has also been inspired by Prof. Zadeh is the *Computing with Words (CWW),* a methodology in which the objects of computation are words and propositions drawn from a natural language [22]. The idea was that computers would be activated by words, which would be converted into a mathematical representation using FSs and that these FSs would be mapped by a CWW engine into some other FS, after which the latter would be converted back into a word. Much research is under way about CWW. As Mendel has argued [11] a type-2 fuzzy set should be used as a model for a word.

The concept of the *interval-valued FS (IVFS)* was introduced in 1975 independently by Zadeh, Sambuc, Jahn and Grattan Guiness [4]. An IVFS is defined by a mapping F from the universe U to the set of closed intervals in [0, 1]. The idea behind IVFs is that the membership degrees of the traditional FSs can hardly be precise.

In 1982 Julong Deng, professor of the Huazhong University of Science and Technology, Wuhan, China, introduced the theory of *Grey System (GS)* [2] for handling the approximate data that are frequently appear in the study of large and complex systems, like the socio-economic, the biological ones, etc. The systems which lack information, such as structure message, operation mechanism and behaviour document, are referred to as GSs. Usually, on the grounds of existing grey relations and elements one can identify where "grey" means poor, incomplete, uncertain, etc. The GS theory was mainly developed in China and it has found many applications in agriculture, economy, management, industry, ecology and in many other fields of the human activity [3].

An effective tool of the GS theory is the use of *Grey Numbers (GNs)* that are indeterminate numbers defined in terms of the closed real intervals. More explicitly, a GN, say A, is of the form A $\in$ [a, b], where a and b are real numbers with $a \leq b$. In other words, the range in which A lies is known, but not its exact value. A GN may enrich its uncertainty representation with respect to the interval [a, b] by a function g: [a, b] $\rightarrow$ [0, 1], which defines a *degree of greyness* g(x) for each x in [a, b]. The well-known arithmetic of the real intervals introduced by Moore et al.[13] has been used to define the basic arithmetic operations among the GNs. The real number with the greatest probability to be the representative real value of the GN A [a, b] is denoted by W(A). The technique of determining the value of w(A) is called *whitening* of A. When the distribution of A is unknown (i.e., no function g has been defined for it) one usually takes $W(A) = \frac{a+b}{2}$. For general facts on GNs we refer to the book [9]. In Chapter 1 of the present book GNs are used for the assessment of computational thinking skills.

Kassimir Atanassov, professor of mathematics at the Bulgarian Academy of Sciences, introduced in 1986 as a complement of Zadeh's membership degree m(x), x $\in$ U, the *degree of non-membership* n(x) and proposed the notion of *intuitionistic FS (IFS)* for a more accurate quantification of the uncertainty [1]. An IFS A is formally defined as the set of the ordered triples A = {(x, m(x), n(x)): x $\in$ U, 0 $\leq$ m(x) + n(x) $\leq$ 1}.

One can write m(x) + n(x) + h(x) = 1, where h(x) is called the *hesitation* or *uncertainty degree* of x. If h(x) = 0 for all x in U, then m(x) = 1 - n(x) and A becomes an ordinary FS.

Similar to the concept of IFS is the *hesitant FS (HFS)* introduced by Torra and Narukawa in 2009 [18]. The difference in the definition of a HFS with respect to an ordinary FS is that the *hesitant degree* *h*(x) of an element x of U is not a single value like its membership degree, but a set of some values in [0, 1]. For example, if U = {a, b, c}, then we could have *h*(a) = {0.2, 0.3}, *h* (b) = {0.75, 0.8, 0.82} and *h* (c) = {0.9}.

A *Pythagorean FS (PFS)*, introduced by Yager in 2013 [19], considers the membership degree *m*(x) and non-membership degree *n*(x) satisfying the condition $m^2(x) + n^2(x) \leq 1$. HFS and PFS have stronger ability than IFS to manage the uncertainty in real-world decision-making problems [23].

A *rough set*, first described by the Polish computer scientist Zdzislaw Pawlak in 1991 [15] is a formal approximation of a crisp set in terms of a pair of sets which give the *lower* and the *upper* approximation of the original set. In the standard version of rough set theory the lower and upper-approximation sets are crisp sets, but in other variations, the approximating sets may be FSs. The theory of rough sets has found important applications in many scientific fields and in particular in Informatics.

The Romanian – American writer and mathematician Florentin Smarandache, professor at the branch of Gallup of the New Mexico University, introduced in 1995 the *degree of indeterminancy / neutrality (i)* and defined the *neutrosophic set* in three components (M, N, I), where M = {m(x): x $\in$ U}, N = {n(x): x $\in$ U} and I = {i(x): x $\in$ U} are subsets of the interval [0, 1] [17]. In other words, if A is a neutrosophic set on U, then each element x of U is expressed with respect to A in the form (m(x), n(x), i(x)). A neutrosophic set generalizes the notions of FS and of IFS. When the components m, n and i are independent, they are leaving room for incomplete information when their sum is <1, for paraconsistent information when their sum is >1 and for complete information when their sum is equal to 1.

In 1999 Dmtri Molodstov, professor of the Computing Center of the Russian Academy of Sciences in Moscow, in order to overcome the existing in each particular case difficulty of defining the proper membership function of a FS, proposed the *Soft Sets* as a new mathematical tool for dealing with the uncertainties [12]. Let E be a set of parameters, then a pair (F, E) is called a soft set on the universal set U, if, and only if, F is a mapping of E into the set of all subsets of U. In other words, the soft set is a paramametrized family of subsets of U. Every set F(ε) of this family, ε $\in$ E, may be considered as the set of the ε-elements of the soft set (F, E). As an example, let U be the set of the girls of a high school and let E be the set of the characterizations {pretty, ugly, tall, short, clever} assigned to each of them. It becomes evident that for an ε in E the corresponding set F(ε) could be arbitrary depending on the observer's personal criteria, or empty, whereas some of them could have non empty intersection. A FS on U with membership function y = m(x) is a soft set on U of the form (F, [0. 1]), where F(α) = {x $\in$ U: m(x) $\geq$ α} is the corresponding α – cut of the FS, for each α in [0. 1].

Ramot et al. [16] introduced in 2002 the notion of *Complex FS (CFS)* characterized by a complex-valued MF, whose range is extended from the traditional fuzzy range of [0, 1] to the unit circle in the complex plane. More explicitly, the membership function of a CFS is of the form $m(x) = r(x)e^{i\theta(x)} = r(x)[\cos[\theta(x)]+i\sin[\theta(x)]$. In the above formula $r(x)$ is the *amplitude term* and $\theta(x)$ is the *phase term* of the membership function. The terms $r(x)$ and $\theta(x)$ are both real-valued and $r(x)$ is in [0, 1] for all x in the universal set U. Since $m(x)$ is a periodic function, one may only consider $\theta(x)$ in [0, 2π]. When $\theta(x) = 0$ for all x in U, then $m(x)$ reduces to the membership function of an ordinary FS.

The catalogue of the extensions of Zadeh's ordinary FS does not end here. Several other generalizations have been introduced, some of them being hybrid approaches of the above mentioned concepts. For example, if in the definition of the soft set the set of all subsets of U is replaced by the set of all fuzzy subsets of U, one gets the notion of the fuzzy soft set. Also, the notion of neutrosophic set has been combined with that of the IVFS to form a new hybrid set called *interval valued neutrosophic set*, etc.

Fuzzy systems are considered to be a part of the wider area of *Soft Computing*, which also includes probabilistic reasoning and *neural networks* [14]. We may say in general that fuzzy systems and neural networks try to emulate the operation of human brain. Neural networks, which are based on biological neural networks, have the ability to learn and also have a parallel structure which can rapidly process the information. In other words they concentrate on the structure of the human brain, i.e., on the "hardware", emulating its basic functions. On the other hand, the fuzzy logic systems concentrate on the "software", emulating fuzzy and symbolic reasoning. Fuzzy logic makes decision based on the raw and ambiguous data given to it, whereas neural networks try to learn from the data, incorporating the same way involved in the biological neural networks.

A *neuro-fuzzy system* is a fuzzy system that uses a learning algorithm derived from or inspired by neural network theory to determine its parameters (FSs and fuzzy rules) by processing data samples. Characteristic examples of such kind of systems are the Adaptive Neuro-Fuzzy Inference Systems (ANFIS).

REFERENCES

Atanassov, K.T., Intuitionistic Fuzzy Sets, *Fuzzy Sets and Systems*, 20(1), 87-96, 1986.

Deng, J., Control Problems of Grey Systems, *Systems and Control Letters,* 288-294, 1982.

Deng, J., Introduction to Grey System Theory, *The Journal of Grey System*, 1, 1- 24, 1989.

Dubois, D., Prade, H., *Interval-Valued Fuzzy Sets, Possibility Theory and Imprecise Probability,* retrieved from: www.eusflat.org/proceedings/EUSFLAT-LFA_2005/papers/S116-01.pdf, September 2018.

Fox, J., Towards a reconciliation of fuzzy logic and standard logic, *Int. J. of Man - Machine Studies,* 15, 213-220, 1981.

Haack, S., Do we need fuzzy logic? *Int. J. of Man-Machine Studies,*11, 437-445, 1979.

Kaufmann, A. & Gupta, M., *Introduction to Fuzzy Arithmetic*, Van Nostrand Reinhold Company, New York, 1991.

Klir, G. J. & Folger, T. A., *Fuzzy Sets, Uncertainty and Information*, Prentice-Hall, London, 1988.

Liu, S. F. & Lin, Y. (Eds.), *Advances in Grey System Research*, Berlin – Heidelberg: Springer, 2010.

Mendel, J. M., *Uncertain Rule-Based Fuzzy Logic Systems: Introduction and New Directions,* Prentice-Hall, Upper-Saddle River, NJ, 2001.

Mendel, J. M., Fuzzy Sets for Words: a New Beginning, *Proc. IEEE FUZZ Conference*, St. Louis, MO, May 26–28, pp. 37–42, 2003.

Molodtsov, D., Soft Set Theory – First Results, *Computers and Mathematics with Applications,* 37(4-5), 19-31, 1999.

Moore, R.A., Kearfort, R.B. & Clood, M.J., *Introduction to Interval Analysis,* 2nd Printing, Philadelphia, SIAM, 1995.

Paplinski, A. P., *Neuro-Fuzzy Computing,* Lecture Notes, Monash University, Australia, 2005.

Pawlak, Z., *Rough Sets: Aspects of Reasoning about Data,* Kluer Academic Publishers, Dordrecht, 1991

Ramot, D.; Milo, R.; Friedman, M.; Kandel, A. Complex fuzzy sets. *IEEE Transactions on Fuzzy Systems.*, *10*, 171–186, 2002.

Smarandache, F., *Neutrosophy / Neutrosophic probability, set, and logic,* Proquest, Michigan, USA, 1998

Torra V., Narukawa Y. On hesitant fuzzy sets and decision, *18th IEEE Int. Conference on Fuzzy Systems*, Jeju island, Korea, vol. 544; pp.1378–82, 2009.

Yager R.R., Pythagorean fuzzy subsets, in *Proceedings of Joint IFSA World Congress and NAFIPS Annual Meeting,* Edmonton, Canada, pp. 57–61, 2013.

Zadeh, L.A., Fuzzy Sets, *Information and Control*, 8, 338-353, 1965.

Zadeh, L.A., The Concept of a Linguistic Variable and its Application to Approximate Reasoning, Parts 1-3, *Information Science*, 8, 199-249, 301-357, and 9, 43-80, 1975.

Zadeh, L. A., Fuzzy logic = computing with words, *IEEE Trans. on Fuzzy Systems*, 4, 103–111, 1996.

Zhang, Z., Hu, Z., Extension of TOPSIS to Multiple Criteria Decision Making with Pythagorean Fuzzy Sets, *Int. J. of Information Systems,* 29(12), 1061-1078, 2014.

About the Editor

Michael Gr. Voskoglou, PhD

Professor Emeritus of Mathematical Sciences

Graduate Technological Educational Institute (T. E. I.) of Western Greece, School of Technological Applications, Patras, Greece

Education:

- BSc, Department of Mathematics, Aristotle University of Thessaloniki, Greece, 1972
- MSc, Department of Pure Mathematics, University of Leeds, UK, 1978
- MPhil, Department of Pure Mathematics, University of Leeds, UK, 1980

- PhD, Department of Mathematics, University of Patras, Greece, 1982

Research and Professional Experience:

- 3 Dissertations (MSc, MPhil, PhD)
- 11 authored Books and one edited Book (1984-2017)
- 266 Papers in English published in refereed academic journals and in Proceedings of refereed International Conferences of about 30 countries in the five continents (1983-2019).
- 33 Announcements in International Mathematical Conferences (publication of the Abstracts only, 1983-2019)
- 41 Reviews of the American Mathematical Society (2008-2019)
- 98 Articles in Greek published in Greek mathematical journals and in proceedings of Mathematical Conferences (1984-2018)
- 12 Announcements in Greek Mathematical Conferences (publication of the Abstracts only, 1983-2018)
- 5 Technical Reports for Programs of Applied Research (T. E. I. of Western Greece, Messolonghi (1990-1997)
- 14 Lecture Notes (1982-2010)
- His published research work has very many citations by other researchers around the Globe (Scholar Google H-index 2018 = 24, 2401 citations).
- Supervisor of more than 15 student dissertations (T. E. I. of Western Greece, 1990-2009)
- External Examiner of 6 Ph.D. dissertations at Universities of India, Saudi Arabia and Egypt.
- Editor in Chief of the "International Journal of Applications of Fuzzy Sets and Artificial Intelligence" (IJAFSAI: ISSN 2241-1240, http://eclass.teipat.gr/eclass/courses/523103) Impact Factor: SJIF 3.806, Universal 28.45.
- Reviewer of the AMS, Member of the Editorial Board or Referee in about 30 academic mathematical journals around the Globe.

- 1990-1994: Vice- President of the Researches' Committee and Institutional Coordinator of the European program "Erasmus" for student and Faculty exchanges, T. E. I. of Messolonghi, Greece
- 1995-1997: Dean of the Faculty of Management and Economics, T. E. I. of Messolonghi, Greece
- 2001-2005: Head of the Department and Scientific Director of the European Program "Expansion of Tertiary Education," Department of Renovation and Restoration of Buildings, T. E. I. of Patras, Greece.
- His research interests include Algebra, Finite Markov Chains, Fuzzy Sets, Artificial Intelligence and Mathematics Education.

Professional Appointments:

- 2010 - Today: Professor Emeritus of Mathematical Sciences of the School of Technological Applications of the Graduate T. E. I. of Western Greece, Patras, Greece.
- 2000 – 2010: Full Professor of Mathematics for Engineers/ Operations' Research, Department of Renovation and Restoration of Buildings, Graduate T. E. I. of Patras (and now of Western Greece)
- November 2009: Visiting Professor, MSc Course in International Business of the School of Management at the University of Warsaw, Poland.
- May 2010: Visiting Professor, MSc Course in Finance and Risk Management of the Department of Operational Mathematics at the University of Applied Sciences (HTW) of Berlin, Germany.
- October 2016: Visiting Professor at the Mathematics Department of the National Institute of Technology of Durgapur under a grand of the GIAN program (Course No. 161021K03) of the Indian Government.
- 2006 – 2008: Instructor in the Program of Business Administration of the Hellenic Open University (Quantitative Methods: Mathematics for Economists, Statistics, Operations' Research).

- 1989 – 2000: Full Professor of Mathematics for Economists/ Operations' Research, Department of Management of Cooperative Organizations, Graduate T. E. I. of Messolonghi (and now of Western Greece).
- 1997 – 2000 (under sabbatical): Visiting Researcher, Institute of Mathematics and Informatics, National Bulgarian Academy of Sciences, Sofia, Bulgaria.
- 1987 – 1989: Lecturer of Mathematics, Department of Management of Cooperative Organizations, Graduate T. E. I. of Messolonghi (and now of Western Greece).
- 1972 – 1987: Teacher of Mathematics in Public Secondary Education, Patras, Akhaia, Greece.
- 1977 – 1980 (under sabbatical): Postgraduate studies and research, University of Leeds, UK.
- 1980 – 1982: Tutor in Abstract Algebra, Department of Mathematics, University of Patras, Greece.
- 1983 – 1989: Instructor at the Schools of Primary and Secondary In-Service Teachers Training (of Elementary Mathematics from a Modern Optical Corner and of Linear Algebra respectively).

Honors:

- 1969-1972: Scholarship from the Greek National Scholarships' Institute during his undergraduate studies,
- 1977-1980: NATO's Fellowship for postgraduate studies in England.
- 1996: WHO IS WHO IN THE WORLD (Marquis: New Providence, NJ, USA)
- 1996: MEN OF ACHIEVEMENT (The International Biographical Centre: Cambridge, England).
- 1997: WHO IS WHO IN AMERICA (Marquis: New Providence, NJ, USA)
- 1999: RESEARCH BOARD OF ADVISORS (American Biographical Institute)

- 2009: THE INTERNATIONAL EINSTEIN AWARD (The International Biographical Centre: Cambridge, England).
- 2015: SINCEREST APPRECIATION PLAQUE (Ain Shams University, Faculty of Computer and Information Sciences, Cairo, Egypt).
- 2018: MEDAL OF THE EGYPTIAN COMPUTER SCIENCE SOCIETY (Academy of Sciences and Technology, Alexandria, Egypt)

Memberships:

- Greek Mathematical Society (Vice-President/General Secretary for a series of years at the Sector of Patras).
- American Mathematical Society
- International Study Group of Mathematical Modelling and Applications (ICTMA)
- Senior Member of the International Engineering and Technology Institute (IETI)
- Hellenic Bridge Federation

Links:

Home page: http://eclass.teipat.gr/eclass/courses/523102
Related pages: http://arxiv.org/a/voskoglou_m_1
https://mindreaderpublications.academia.edu/MichaelVoskoglou
https://researchgate.net/profile/Michael_Voskoglou
http://www.researcherid.com/rid/C-4504-2014;
http://orcid.org/0000-0002-4727-0089
https://www.amazon.com/author/michaelvoskoglou

Recent Publications (2016 – 2019)

1. *Finite Markov Chains and Fuzzy Models in Management and Education*, GIAN Program (Course No. 161021K03), National Institute of Technology, Durgapur, India, 2016.

2. Fuzzy Numbers: A Tool for the Assessment Processes (with I. Subbotin), *International Journal of Applications of Fuzzy Sets and Artificial Intelligence*, 6, 17-32, 2016.
3. New Technologies, Open Access and Author Ethics: Some Remarks on a Case of Plagiarism (with I. Subbotin), Special Editorial, *International Journal of Applications of Fuzzy Sets and Artificial Intelligence,* 6, 67-71, 2016.
4. Utilizing Fuzzy Numbers and the Cog Technique for Student Assessment, *Journal of Global Research in Education and Social Science,* 7(1), 43-54, 2016.
5. Triangular and Trapezoidal Fuzzy Assessment Models, in S. J. John (Ed.), *Handbook of Research on Generalized and Hybrid Set Structures and Applications for Soft Computing*, Chapter 18, 375-393, IGI-Global, Hershey, PA, USA, 2016.
6. Applications of Finite Markov Chain Models to Management, *American Journal of Computational and Applied Mathematics,* 6(1), 7-13, 2016.
7. An Application of Fuzzy Numbers to the Assessment of Mathematical Modelling Skills (with I. Subbotin), *International Journal of Mathematical Modelling and Computations,* 6(1), 83-103, 2016.
8. Fuzzy Numbers as an Assessment Tool in the APOS/ACE Instructional Treatment of Mathematics, *Journal of Physical and Mathematical Education,* Sumy State Pedagogical University, Ukraine, 1(7), 29-37, 2016.
9. An Application of the Generalized Rectangular Fuzzy Model to Critical Thinking Assessment (with I. Subbotin), *American Journal of Educational Research,* 4(5), 397- 403, 2016.
10. A Tool for Assessing the Ability of Understanding the Infinity Based on Triangular Fuzzy Numbers, *Egyptian Computer Science Journal,* 40(2), 11-23, 2016.
11. An Application of Fuzzy Numbers to Assessment of Bridge Players Performance, *American Journal of Business and Society,* 1(2), 32-41, 2016.

12. An Application of Fuzzy Logic for Learning Mathematics According to the Bloom's Taxonomy (with I. Subbotin), *HMS International Journal for Mathematics in Education,* 7, 157-174, 2015-16.
13. Application of Fuzzy Numbers to the Assessment of Cbr Systems, *Journal of Advances in Management Sciences and Information Systems,* 2, 53-62, 2016.
14. Computational Thinking and Fuzzy Logic in Problem Solving, in K. Newton (Ed.), *Problem-Solving: Challenges and Outcomes,* Nova Publishers, Chapter 1, 1-30, 2016.
15. Problem Solving in Our Knowledge Society and Future Perspectives, in K. Newton (Ed.), *Problem-Solving: Challenges and Outcomes,* Nova Publishers, Chapter 13, 243-258, 2016.
16. Problem Solving in the Forthcoming Era of the Third Industrial Revolution, *International Journal of Psychological Research,* 10(4), 2016.
17. An Application of Fuzzy Numbers To Assessment of Student Critical Thinking Skills, *International Journal of Mathematics, Game Theory and Algebra,* 25(1), 37-50, 2016.
18. On Student Understanding of the Concept of Infinity, Philosophy of Mathematics Education Journal, 30, 1-8, 2016. Available at: socialsciences.exeter.ac.uk/education/research/ centres/stem/publications/pmj.
19. Applying the Apos Theory To Study the Student Understanding of Polar Coordinates (with V. Borji), *American Journal of Educational Research,* 4(16), 1149-1156, 2016.
20. GIAN Course No. 161021ko3: Finite Markov Chain and Fuzzy Models In Management and Education, *International Journal of Applications of Fuzzy Sets and Artificial Intelligence,* 6, 183-189, 2016.
21. Comparison of the Cog Defuzzification Technique and Its Variations To the Gpa Index, *American Journal of Computational and Applied Mathematics,* 6(5), 187-193, 2016.

22. Fuzzy Numbers and Analogical Reasoning, SRI: *Journal of Mathematical Sciences and Applications,* 1(1), 20-32, 2016.
23. An Absorbing Markov Chain Model For Problem Solving, *American Journal of Applied Mathematics and Statistics,* 4(6), 173-177, 2016.
24. Use of Triangular Fuzzy Numbers For Assessing the Acquisition of the Van Hiele Levels in Geometry, *Journal of Physical and Mathematical Education, Sumy State Pedagogical University, Ukraine,* 4(10), 9-12, 2016.
25. Finite Markov Chain and Fuzzy Logic Assessment Models; Emerging Research and Opportunities, Createspace.com. – Amazon, Columbia, SC, USA, 2017 (ISBN-13:978-1548340070), https://tsw.createspace.com/title/7286560.
26. Application of Triangular Fuzzy Numbers For Assessing the Results of Iterative Learning (with I. Subbotin), *International Journal of Applications of Fuzzy Sets and Artificial Intelligence*, 7, 59-72, 2017.
27. Measuring the Uncertainty in the Van Hiele Levels of Geometric Reasoning, *American Journal of Educational Research,* 5(2), 109-113, 2017.
28. Application of Fuzzy Numbers For Assessing Student Learning Skills with the Bloom's Taxonomy, *Journal of Mathematical Sciences and Mathematics Education,* 12(1), 29-41, 2017.
29. An Application of the Generalized Rectangular Fuzzy Assessment Model To the Van Hiele Levels of Geometric Reasoning (with I. Subbotin), *Universal Journal of Applied Mathematics,* 5(1), 1-5, 2017.
30. Fuzzy Logic Computers and Mathematical Modelling, *Transactions on Mathematics,* 3(2), 14-26, 2017.
31. An Application of the Cog Technique To Assessment of Student Understanding of Polar Coordinates, *NAUN International Journal of Education and Information Technologies*, 11, 1-5, 2017.

32. Use of TFNS and TpFNS for Evaluating the Effectiveness of Cbr Systems, *NAUN International Journal of Fuzzy Systems and Advanced Applications,* 4, 1-7, 2017.
33. Managing the Uncertainty For Student Understanding of the Infinity, *NAUN International Journal of Fuzzy Systems and Advanced Applications,* 4, 8-13, 2017.
34. Assessing the Results of Exposure To Computers On Problem Solving Skills, *NAUN International Journal of Applied Mathematics and Informatics,* 11, 1-5, 2017.
35. Application of Finite Markov Chains To Decision Making, *NAUN International Journal of Economics and Statistics,* 5, 10-14, 2017.
36. Designing An Ace Approach For Teaching the Polar Coordinates (with V. Borji), *American Journal of Educational Research,* 5(3), 303-309, 2017.
37. Application of Fuzzy Numbers To Assessment of Human Skills, *International Journal of Fuzzy System Applications,* 6(3), 59-73, 2017.
38. Application of Fuzzy Numbers To Assessment Processes, In: M. Khosrow-Pour (Ed.), *Encyclopedia of Information Science and Technology, Fourth Edition,* Vol. IV, 3216-3225, IGI-Global, Hersey, PA, USA, 2017
39. Use of Fuzzy Numbers For Assessing Problem Solving Skills (with S. N. Mishra), *International Journal of Industrial Mathematics,* 92(3), 161-171, 2017.
40. A Fuzzy Logic Evaluation of Beliefs and Attitudes of Greek Secondary Teachers About Teaching the Euclidian Geometry (with A. Papadopoulou), *American Journal of Computational and Applied Mathematics,* 7(3), 80-85, 2017.
41. A Study On Smooth Varieties with Differentially Simple Coordinate Rings, *IARAS International Journal of Mathematical and Computational Methods,* 2, 53-59, 2017.
42. Variations of the Rectangular Fuzzy Assessment Model and Application To Human Activities, *IARAS International Journal of Education and Learning Systems,* 2, 115-124, 2017.

43. An Absorbing Markov Chain Model For Case-Based Reasoning, *IARAS International Journal of Computers,* 2, 99-105, 2017.
44. An Application of Fuzzy Numbers To Assessment of Human Activities, *IARAS International Journal of Control Systems and Robotics,* 2, 52-60, 2017.
45. Studying the Winger's "Enigma" About the Unreasonable Effectiveness of Mathematics in the Natural Sciences, *American Journal of Applied Mathematics and Statistics,* 5(3), 95-100, 2017.
46. A Note on the Graphical Representation of the Derivatives, *Journal of Physical and Mathematical Education, Sumy State Pedagogical University*, Ukraine, 2(12), 9-17, 2017.
47. Application of Grey Numbers To Assessment Processes (with Y. Theodorou), *International Journal of Applications of Fuzzy Sets and Artificial Intelligence,* 7, 59-72, 2017.
48. Fuzzy Logic in the Van Hiele Levels of Geometric Reasoning, *Transactions on Mathematics*, 3(4), 1-16, 2017.
49. Fuzzy Logic Models For Evaluating Student Understanding of Polar Coordinates, *Oriental Journal of Physical Sciences,* 2(2), 63-70, 2017.
50. Application of Fuzzy Numbers To Assessment of Computer Activities, *J. of Modern Technology & Engineering,* 2(3), 177-190, 2017.
51. Application of Grey Numbers To Assessment of Understanding the Derivative, *American Journal of Educational Research*, 5(11), 1167-1171, 2017.
52. A Study of the Problem Solving Process Using Fuzzy Relation Equations, *NAUN Int. Journal of Mathematical Models and Methods in Applied Sciences,* 11, 303-307, 2017.
53. Applications of Fuzzy Logic To Learning Contexts, Plenary Speech, *Int. Conference on Neural Networks and Fuzzy Systems, INASE,* Athens, Greece, April 2017, available in www.inase.org/conferences/2017/Athens/nn-fs.htm.
54. Use of Fuzzy Numbers For Assessing Human Skills, Plenary Speech, *Int. Conference on Pure Mathematics, Applied*

Mathematics and Computational Methods, INASE, Heraklion, Crete, Greece, July 2017, available in www.inase.org/conferences/2017/Crete/pmamcm.htm.

55. Is Mathematics Invented Or Discovered By Humans? *Philosophy of Mathematics Education Journal,* 33, 1-12, 2018. Available at: socialsciences.exeter.ac.uk/education/research/centres/stem/publications/pmj.
56. Fuzzy Assessment Methods for the Acquisition of the Van Hiele Levels In Geometry, *Journal of Uncertain Systems,* 12(1), 36-46, 2018.
57. Application of Fuzzy Numbers To Assessment of Human Skills, *Punjab University Journal of Mathematics,* 50(2), 91-102, 2018.
58. A Study of Student Learning Skills Using Fuzzy Relation Equations, *Egyptian Computer Science Journal,* 42(1), 80-67, 2018.
59. Application of Fuzzy Relation Equations To Assessment of Analogical Problem Solving Skills, *Journal of Physical and Mathematical Education, Sunny State Pedagogical University*, Ukraine, 1(15), 122-127, 2018.
60. Solving Systems of Equations with Grey Data, *International Journal of Applications of Fuzzy Sets and Artificial Intelligence*, 8, 103-111, 2018.
61. Application of Fuzzy Relation Equations To Student Assessment, *American Journal of Applied Mathematics and Statistics,* 6(2), 167-171, 2018.
62. Application of Triangular Fuzzy Numbers To Analogical Reasoning, *Int. J. of Quantitative Research in Education,* 4(3), 240-254, 2018.
63. Application of Grey Numbers To Computational Thinking, *Proceedings of Software Engineering Conference, ITHEA,* pp. 17-21, National Aviation University, Kyiv, Ukraine, 2018.
64. Fuzzy Logic: History, Methodology and Applications To Education, *Sumerianz Journal of Education, Linguistics and Literature,* 1(1), 10-18, 2018.

65. Solving Linear Programming Problems with Grey Data, *Oriental Journal of Physical Sciences*, 3(1), 17-23, 2018.
66. Use of Grey Numbers For Evaluating Mathematical Modelling Skills, *SF Computer Science and Applications,* 1(1), 16-22, 2018.
67. Evaluating the Acquisition of the Van Hiele Levels with the Cog Deruzzification Technique, *SF Journal of Artificial Intelligence*, 1(2), 1-8, 2018.
68. Decision Trees In Data Mining (with A.- B. M. Salem), *International Journal of Applications of Fuzzy Sets and Artificial Intelligence,* 8, 143-154, 2018.
69. Fuzzy Linear Programming, *Egyptian Computer Science Journal,* 42(4), 1-14, 2018.
70. Application of Grey Numbers To Student Assessment Under Fuzzy Conditions, *Journal of Mathematical Sciences and Mathematics Education,* 12(2), 33-42, 2018.
71. The Influence of the Schools of Mathematical Thought to the Development of Mathematics Education, *American Journal of Educational Research*, 6(9), 1283-1288, 2018.
72. The Effects of Formalism and Intuitionism on the Development of Mathematics Education, *Sumerianz Journal of Education, Linguistics and Literature,* 1(5), 121-127, 2018.
73. Solving Linear Programming Problems with Fuzzy Data, *Journal of Advances in Applied Mathematics,* 3(4), 127-138, 2018.
74. Application of Grey System Theory to Assessment with Approximate Data, *Proceedings of the 28th International Conference on Computer Theory and Applications*, 123-128, Alexandria, Egypt, 2018.
75. Use of Fuzzy Relation Equations For Evaluating Mathematical Modelling Abilities, *Oriental Journal of Physical Sciences,* 3(2), 2018
76. Current Problems and Future Perspectives of Mathematics Education, *Physical and Mathematical Education*, 3 (17), 13-19, 2018.

77. Fuzzy Sets, Fuzzy Logic and Their Applications, *Special Issue Information, Mathematics MDPI Journal,* available at www.mdpi.com/journal/mathematics/special_issue/Fuzzy_Sets.
78. Application of the Grey System Theory to the Assessment of Critical Thinking Skills, *American Journal of Applied Mathematics and Statistics,* 6(6), 253-261, 2018.
79. Teaching the Graphical Representation of the Derivative with the Apos/Ace Instructional Treatment For Mathematics, *First Congress of Greek Mathematicians (HMS)*, Abstracts, p. 82, Athens, Greece, June 2018.
80. Evaluating A System's Effectiveness with Approximate Data, Keynote speech, *28th International Conference of Computer Theory and Applications, Academy of Sciences, Technology and Maritime Transport*, Alexandria, Egypt, 30-10-2018, available at: https://iccta.aast.edu/Keynotespeakers.php.
81. Systems of Equations with Fuzzy Coefficients, *Journal of Physical Sciences,* 23, 77-88, 2018.
82. Application of Grey Numbers to Assessment of Mathematical Modelling Capacities, *International Journal of Trends in Mathematics Education Research*, 1(1), 9-12, 2018.
83. Multi-Valued Logics: A Review, *International Journal of Applications of Fuzzy Sets and Artificial Intelligence,* 9, 5-12, 2019.
84. Use of Fuzzy Numbers As Assessment Tools, in: M. Khosrow-Pour (Ed.), Advanced Methodologies and Technologies in *Artificial Intelligence, Computer Simulation and Human – Computer Interaction*, Chapter 30, pp.407-420, Information Resources Management Association, Hersey PA, USA, 2019.
85. Methods For Assessing Human-Machine Performance Under Fuzzy Conditions, *Mathematics,* 7(3), 230, 2019.
86. Derivations and Integrations on Rings, *American Journal of Applied Mathematics and Statistics*, 7(2), 75-78, 2019.

87. Application of Markov Chain to the Ace Teaching Style of Mathematics, *Sumerianz Journal of Education, Linguistics and Literature*, 2(2), 4-11, 2019.
88. Fuzzy Systems, Extensions and Relative Theories, *WSEAS Transactions on Advances in Engineering Education*, 16, 63-69, 2019.
89. Fuzzy Relation Equations on the Van Hiele Levels of Geometric Reasoning, *NAUN International Journal of Fuzzy Systems and Advanced Applications,* 6, 8-12, 2019.
90. On Properties of Differential Rings, *WSEAS Transactions on Mathematics,* 18, 112-117,l 2019.
91. Use of Fuzzy Relation Equations and the Bloom's Taxonomy For Learning For Evaluating Student Learning Skills, *WSEAS Transactions on Advances in Engineering Education,* 16, 1-6, 2019.
92. Fuzzy and Grey Assessment Methods, *NAUN International Journal of Fuzzy Systems and Advanced Applications*, 6, 1-7, 2019.
93. A Markov Chain Model for the Apos/Ace Instructional Treatment Of Mathematics, *IARAS International Journal of Education and Learning Systems,* 4, 1-6, 2019.
94. A Markov Chain Application on the Levels of the Bloom's Taxonomy of Learning Objectives, *American Journal of Educational Research*, 7(3), 294-298, 2019.
95. Data Evaluation In Fuzzy Systems, *WSEAS Transactions on Advances in Engineering Education,* 16, 70-74, 2019.
96. Management of Fuzzy Data In Education, *Scientific Journal of Physical and Mathematical Education,* 1(19), 13-17, 2019.
97. Probability, Fuzzy Sets, Generalizations and Relative Theories, Plenary Lecture, *21st Int. Conference on Mathematical and Computational Methods in Science and Engineering*, WSEAS, Venice, March 2019, available at: www.wseas.org/wseas/cms.action?id=19926.

INDEX

F

G

H

I

K

L

M

N

O

P

R

S

T

V